OXFORD MONOGRAPHS ON GEOLOGY AND GEOPHYSICS NO. 5

Series editors

P. Allen
H. Charnock
E. R. Oxburgh
B. J. Skinner

OXFORD MONOGRAPHS ON GEOLOGY AND GEOPHYSICS

Problematic Fossil Taxa

Edited by

ANTONI HOFFMAN
Lamont-Doherty Geological Observatory, Columbia University

MATTHEW H. NITECKI
Department of Geology, Field Museum of Natural History

OXFORD UNIVERSITY PRESS · New York
CLARENDON PRESS · Oxford · 1986

Oxford University Press

Oxford   New York   Toronto
Delhi   Bombay   Calcutta   Madras   Karachi
Petaling Jaya   Singapore   Hong Kong   Tokyo
Nairobi   Dar es Salaam   Cape Town
Melbourne   Auckland

and associated companies in
Beirut   Berlin   Ibadan   Nicosia

Library of Congress Cataloging-in-Publication Data
Main entry under title:
Problematic fossil taxa.
(Oxford monographs on geology and geophysics; no. 5)
Bibliography: p.   Includes indexes.
1. Paleontology.   2. Paleontology—Classification.
I. Hoffman, Antoni.   II. Nitecki, Matthew H.   III. Series.
QE899.P76   1986   560'.1'2   85-25857
ISBN 0-19-503992-0

9 8 7 6 5 4 3 2 1
Printed in the United States of America
on acid-free paper

FOREWORD: WHY AND HOW TO DO PROBLEMATICA

Reconstruction and explanation of the history of life on the Earth are the main goals of evolutionary biology, and we can approach these goals in two ways. One approach is to consider in statistical terms the large-scale patterns of change in global diversity and taxonomic composition of the biosphere through geological time, in the hope that the resulting macroevolutionary pattern will offer an insight into high-level processes that control the history. This approach may be called *uniformistic* because it regards individual taxa as particles that are indistinguishable, on some scale at least, from one another—just as individual electrons, protons, and neutrons are treated in physics and chemistry.

Perhaps the best examples of the uniformistic approach involve the analysis of taxonomic diversification of marine animals (and to a lesser degree terrestrial plants) in the Phanerozoic and the study of the temporal distribution of mass extinctions. The pattern of taxonomic diversity of marine animal families seems to agree with the logistic model developed on the assumption that the rates of family origination and extinction are conversely dependent on the number of families present. The fit becomes remarkable indeed when perturbations and stochastic noise are superimposed on the theoretical curve. Within the conceptual framework of the uniformistic approach, this result implies that intrinsic constraints on taxonomic diversification are the causal explanation for the pattern. On the other hand, the temporal pattern of mass extinctions in the marine biota suggests their periodicity in the Mesozoic and Cenozoic, which might imply an extrinsic control on the history of life on Earth. Such controls, for example, an extraterrestrial mechanism, may be—and indeed have been proposed to be—superimposed on the biotic processes of diversification and provide jointly with them the explanation for the empirical pattern.

The empirical data against which these theoretical models are tested, however, are severely biased not only by the taphonomic factors and the differential sampling intensity of particular geological time intervals, but also by taxonomic decisions. To give just one example, the logistic model of taxonomic diversification of marine animals has to be rejected if the Archaeocyatha are interpreted as metazoans and included in the analysis. Perhaps even more important, patterns strikingly similar to those derived statistically from paleontologic data can also be generated by various stochastic processes. This conclusion should not be surprising because random walks often tend to go somewhere and to give the impression of periodicity. And since we have only a very imperfect record of only one history of the biosphere to analyze, and only a very incomplete understanding of the processes of diversification and extinction to help us in choosing among various stochastic models, our statistical tests are not sufficiently powerful to reject the null hypothesis of randomness.

Thus, the apparent order in the history of the biosphere may merely be a by-product of the ecological and evolutionary behavior of individual organisms, populations, and species. This does not imply that the history of life is controlled by pure chance, in the metaphysical sense of the word, but only that evolutionary processes are so multifaceted that individual events are practically unpredictable, although deterministic in principle. By flipping a coin we obtain a pattern of heads and tails that can be described stochastically. However, each individual flip is causally determined by a number of physical forces, and if we knew all the boundary conditions we could, in theory, predict the result. This is the difference between macroscopic phenomena and the ontological indeterminacy at the subatomic level. In the macroscopic world, stochastic models can describe but not explain the pattern of coin flipping as well as the pattern of the history of life.

Consequently, however, analysis of the general characteristics of the historical pattern of life may be futile. The emphasis should, then, be on individual evolutionary events and chains of events. It is they that must be reconstructed and explained. This is the essence of the other—the *individualistic*—approach to the history of the biosphere.

But how to reconstruct particular evolutionary events and chains of events in the history of life? First, the phylogenetic relationships among taxa must be established. For as long as these relationships remain unknown, individual evolutionary events cannot be identified nor reconstructed. To give an extreme example, if humans were considered to be more closely related to New World monkeys than to apes, the inferred sequence of evolutionary events leading to the appearance of

Homo sapiens would differ radically from the picture given by modern paleoanthropology. Thus, the evolutionary scenario explaining our origins would also have to be different. The pattern of phylogeny is a necessary prerequisite to any evolutionary interpretation.

Comparative biology and cladistic analysis provide adequate means to recognize the phylogenetic relationships among major taxa. But they can also mislead easily. If living taxa are considered and extinct taxa are neglected, comparative biology and cladistics will allow only for description of the history of evolutionary successes. Yet an adequate reconstruction of the history of life on Earth must deal with its totality and not only with a biased sample chosen *a posteriori*. This is one of the reasons why paleontological data are indispensable for understanding the course of evolution. Problematic fossil taxa play a focal role because they may force us to reshape our views of the history of the biosphere as based on comparative biology of modern organisms. Problematic fossil taxa are those groups of extinct organisms that do not fit neatly into any extant systematic group and thus presumably represent a record of "experiments with life."

A second reason why paleontological data are indispensable for reconstruction of the history of the biosphere is that introduction of a taxon that was previously beyond the scope of cladistic analysis may considerably change the whole topology of interrelationships among taxa. Therefore, all related taxa must be considered during phylogenetic analysis or cladistics may fail to reconstruct adequately the phylogenetic pattern. In this context, again, the problematic fossil taxa assume a crucial role because it is their very oddity that is most likely to cause a radical transformation of the topology of phylogenetic relationships once they are taken into account.

The question is how to incorporate the problematic taxa into the conceptual framework of phylogenetic analysis? Or, how to know when to consider them as belonging to the analyzed group of taxa, and when to use them for comparison?

The answer to this question is obviously equivalent to a recognition of the proper systematic position of the problematic fossil taxa—that is, to the resolution of the enigma they represent. To address this fundamental problem, homologies between a given problematic group and other organisms must be established and distinguished from mere analogies, and the polarity of homologous character states must be determined. To this end, characters of the taxon in question must first be identified and analyzed. Identification of characters, however, is by no means a trivial matter. Characters that are accidental or arbitrarily separated out from an integrated functional complex are meaningless for phylogenetic analysis.

For instance, the number of ribs on a bivalve shell is very useful as a taxonomic trait, but the number of growth lines cannot be used in the same manner. This is why cladistic analysis of buttons, or nails and bolts, is very helpful as a training exercise in the logic of, but not an appropriate model for, phylogenetic systematics. After all, buttons, or nails and bolts do not represent a single nested hierarchy of relationships and, consequently, they can be arbitrarily dissected into character sets that result in absolutely incongruent webs of interrelationships. In contrast, the available biological evidence indicates that all taxa constitute a natural hierarchy which, in principle, should be unequivocally reconstructed by the cladistic analysis.

Adequate identification of taxonomic characters presupposes, then, a certain level of understanding of their functional meaning. We do not attribute any taxonomic or phylogenetic value to the number of growth lines on a bivalve shell precisely because we know that they are just growth lines and, hence, that their number depends on individual longevity of the organism. An insight into the functional meaning of characters is even more important for correct determination of their homology and polarity. Yet the functional meaning of traits and structures has to be seen always in a historical context. All characters arise at the intersection of immediate adaptive needs, phylogenetic inheritance, and constructional potential of the organism. Therefore, in order to achieve an understanding of the functional individuality of characters, at least some knowledge about phylogeny is prerequisite, which is ultimately impossible without a preconceived notion of characters, their homology, and polarity of character states.

This argument smacks of circular reasoning, and yet the impression of circularity is incorrect. For the corollary of this argument is simply that reconstruction of the history of life must follow the principles of Paul Ricoeur's hermeneutic circle, as all other historical analyses do. Each step of the analysis is potentially testable, although not in the sense of definite verification or falsification but merely of tentative corroboration or contradiction. The claim for testability assumes adaptive significance of characters, though not necessarily their origin by the process of adaptation for a given function. If the characters of a group of organisms were nonadaptive, any functional interpretations would be nonsensical and could neither corroborate nor contradict the contrasting phylogenetic inferences based on differential sets of characters, homologies, and polarity determinations. The assumption of adaptive significance of characters may be disputed, and is indeed rejected by the critics of functional interpretations in biology. But regardless of universal validity of this premise, its rejection is operationally meaningless in phylogenetic analysis, if only because it is impossible to prove that a trait or a structure is and was nonfunctional.

The contention that phylogenetic considerations must proceed in the hermeneutic circle implies that a rigorous phylogenetic analysis is possible only at a relatively high level of understanding of the investigated group of organisms. By definition, such a level is not achieved for problematic taxa, or else they would not be problematic. The present volume is devoted entirely to problematic fossil taxa. Therefore, it comes as no surprise that a reference to cladistic methodology is made only in the chapter on graptolites (Chapter 14, by Urbanek)—for this is the success story in the study of problematic taxa. We now know that the graptolites belong to the Hemichordata, and we can ask detailed questions concerning their development and functional morphology, their phylogenetic relationship to the pterobranchs, and the sequence of evolutionary events that led to their origin. And the history of their investigation clearly shows how an ingenious application of the principle of hermeneutics brings about a substantial improvement in understanding an enigmatic group.

Generally, however, the phylogenetic interpretation of problematic fossil taxa must rely on a more traditional approach. It must be based on an intuitive search for plausible sequences of anatomical and developmental transformations that could possibly bridge the gaps between problematica and other organisms. Presentation of such hypothetical scenarios may be labeled as pure storytelling but, in fact, this approach is both inevitable and testable, even though not rigorously, by compatibility with further discoveries. These discoveries may involve new anatomic characters that corroborate or contradict previous homologies and functional interpretations, or new fossils that fill in the morphological and temporal gaps in the record and thus corroborate or contradict the inferred sequences of evolutionary events.

Clearly, then, the first step in the study of problematic fossil taxa must involve a careful analysis and description in anatomical, functional, and developmental terms. As far as possible, this ought to be done without preconceived notions about the systematic position of the investigated group, even though a fully neutral descriptive language is unachievable because our ideas about how organisms operate are strongly shaped by the experience with modern biota. Adolf Seilacher's and Kenneth Towe's provocative hypotheses come to mind: What if the ancient heterotrophs, even those large-sized and apparently complex, were feeding entirely by diffusion? This warning is especially appropriate if perceived in the context of the old and deep-rooted principle that phyla are,

by definition, extant. The latter principle implies that an attempt should always be undertaken to fit the problematic fossil taxa into the systematic framework of extant phyla. However, as emphasized by Bengtson in Chapter 1 of this book, to assign a problematic group to the least dissimilar extant phylum is merely a cosmetic solution to a real biological problem. This is particularly true since the phyla themselves are problematic in the sense that their mutual phylogenetic relationships are unknown. And it is precisely their enigmatic systematic position that warrants their taxonomic status as phyla.

As a matter of fact, if there is any generalization that can be derived from the contributions to this volume, it is that a sweeping change in the paradigm of study of problematic fossil taxa is under way. The majority of the taxa considered in this book are now explicitly, or sometimes implicitly, regarded as new phyla. This is the case not only with many Precambrian metazoans (Chapter 6) and the incompletely known Early Cambrian organisms (Chapter 8), but also with the so-called alga *Tubiphytes* (Chapter 2), stromatoporoids (Chapter 12), conulariids (Chapter 11), the so-called tentaculitid *Styliolina* (Chapter 5), and a large number of excellently preserved problematica from the Burgess Shale (Chapter 13).

We must be very careful, however, and not allow the pendulum to swing too far to the extreme. To erect a new phylum may also be no more than a cosmetic solution to the problem because without an appropriate extant model for our fossils, the functional, and consequently phylogenetic, understanding of extinct organisms must indeed remain very limited. Or at least, our functional interpretations must rely entirely on the principle of optimization and the comparison to physical or engineering paradigms, with little potential for independent corroboration.

Conodonts may offer an instructive example. They provided the first indications of a new direction in the study of problematic taxa in that their status as a separate phylum was canonized in *The Treatise of Invertebrate Paleontology*. Yet a new consensus seems to emerge that they are in fact an extinct group of chordates, perhaps even more precisely—vertebrates (Chapters 15 and 16). Another example of this sort is represented by the Archaeocyatha. Recognized long ago as a distinct animal phylum but assigned even to plants or to the separate kingdom Archaeata, they now seem to fit quite comfortably in the Porifera (see comments and references in Chapters 3, 4, and 12). Even more striking a case is represented by chondrophorine hydrozoans. As suggested by Stanley (Chapter 7), their Precambrian representatives may be assigned to extant families! And the problem with the Machaeridia (Chapter 16) consists not so much in the fact that they do not fit any extant phylum, but that they can be reasonably well accommodated within more than one phylum.

Thus no single prescription can tell us how to deal with problematic fossil taxa. Ingenuity and openmindedness must certainly come at the top of the list of rules of thumb to follow, as suggested by Bengtson (Chapter 1) and Babcock (Chapter 2), but also the ability to admit humbly that sometimes we just do not know how to fit together the few pieces of evidence we have (see Chapter 9 on *Microdictyon* by Bengtson et al.). Therefore, riddles remain. This is well illustrated in this book by highly divergent interpretations of a number of problematica; compare, for instance, the views on *Wiwaxia* in Chapter 13 by Briggs and Conway Morris and Chapter 16 by Dzik. But the need for restraint is most clearly evident in the case of the algal wastebasket (Chapters 2 and 3), partly because the algae are such a diverse and poorly understood group of organisms but partly because the critical information may be lost forever as a result of taphonomic processes.

Our intent as the editors of this volume was to provide the readers with updated ideas on most of the major problematic fossil taxa, although, for obvious reasons, we could not even hope to cover them all. Among the problematic fossil taxa, Precambrian groups have certainly attracted the most attention, if only because of our natural curiosity for "the first": the first living organism, the first eukaryote, the first metazoan. They have been recently considered in great detail in *The Earth's Earliest Biosphere: Its Origin and Evolution,* edited by J. William Schopf (Princeton University Press, 1983), and in *The Dawn of Animal Life. A Biohistorical Study,* by Martin F. Glaess-

ner (Cambridge University Press, 1984). Therefore, the major focus of this book is on Paleozoic, especially Early Paleozoic organisms, although Precambrian biota are also discussed (see Chapters 6 and 7). We decided to omit the famous Carboniferous problematica from Mazon Creek because essentially no research has been done on those organisms since they were reviewed exhaustively in *The Mazon Creek Fossils,* edited by Matthew H. Nitecki (Academic Press, New York 1979). In contrast, an overview of the ongoing research on the problematic fossils from the Burgess Shale is included (Chapter 13).

Naturally, we wished to have each fossil group discussed by an expert. A variety of conditions must be met in order to permit a paleontologist to achieve expertise on a problematic fossil taxon. Perhaps the most fundamental prerequisite is that one has the opportunity to find and study the fossils. It is in the very nature of all fossils that political boundaries do not determine their distribution in rocks. This is why the contributors in this volume include paleontologists from the United States, Great Britain, the Soviet Union, Poland, Sweden, and Australia. We hope that this may also help to bridge the gap between different traditions of research, which sometimes present insurmountable difficulties. For science is indeed a human endeavor, not confined to just one language, culture, or political system.

We would like to thank all the colleagues who kindly reviewed the individual manuscripts and made valuable suggestions: Zbigniew Jastrzebski, who helped with the illustrative material, and Doris Nitecki, who prepared the index and patiently worked through the whole book as a master grammarian.

June 1986
New York, New York   A. H.
Chicago, Illinois   M. N.

CONTENTS

CONTRIBUTORS

RICHARD J. ALDRIDGE   Department of Geology, University of Nottingham, Nottingham NG7 2RD, England.

JACK A. BABCOCK   Research Center, Amoco Production Company, P.O. Box 591, Tulsa, Oklahoma 74102.

LOREN E. BABCOCK   Department of Geology, Kent State University, Kent, Ohio 44242.

STEFAN BENGTSON   Institute of Paleontology, Uppsala University, Box 558, 751 22 Uppsala, Sweden.

DEREK E. G. BRIGGS   Department of Geology, University of Bristol, Queen's Road, Bristol BS8 1RJ, England.

SIMON CONWAY MORRIS   Department of Earth Sciences, Cambridge University, Downing Street, Cambridge CB2 3EQ, England.

JERZY DZIK   Institute of Paleobiology, Polish Academy of Sciences, Al. Żwirki i Wigury 93, 02-089 Warszawa, Poland.

MIKHAIL A. FEDONKIN   Paleontological Institute, USSR Academy of Sciences, Profsoyuznaya ul. 113, Moscow 117868, USSR.

RODNEY M. FELDMANN   Department of Geology, Kent State University, Kent, Ohio 44242.

ANTONI HOFFMAN   Lamont-Doherty Geological Observatory, Columbia University, Palisades, New York 10964.

RICHARD H. LINDEMANN   Department of Geology, Skidmore College, Saratoga Springs, New York 12866.

S. CROSBIE MATTHEWS   (deceased) Department of Geology, University of Bristol, Bristol, England.

VLADIMIR V. MISSARZHEVSKY   Geological Institute, USSR Academy of Sciences, Pyzhevskiy per. 7, Moscow Zh-17, USSR.

MATTHEW H. NITECKI   Department of Geology, Field Museum of Natural History, Chicago, Illinois 60605.

A. YU. ROZANOV   Paleontological Institute, USSR Academy of Sciences, Profsoyuznaya ul. 113, Moscow 117868, USSR.

GEORGE D. STANLEY, JR.   Department of Geology, University of Montana, Missoula, Montana 59812.

ADAM URBANEK   Institute of Paleobiology, Polish Academy of Sciences, Al. Żwirki i Wigury 93, 02-089 Warszawa, Poland.

BARRY D. WEBBY   Department of Geology and Geophysics, University of Sydney, Sydney, N.S.W. 2006, Australia.

ELLIS L. YOCHELSON   U.S. Geological Survey, National Museum of Natural History, Washington, D. C. 20560.

ANDREY YU. ZHURAVLEV   Paleontological Institute, USSR Academy of Sciences, Profsoyuznaya ul. 113, Moscow 117868, USSR.

PROBLEMATIC FOSSIL TAXA

1. INTRODUCTION: THE PROBLEM OF THE PROBLEMATICA

Stefan Bengtson

Problematic fossils are fraught with problems—but what are the problems? It pays to consider this question carefully, for the most striking aspect about a large number of published studies on problematic fossils is that the problems appear to be viewed superficially. This has reduced the problematica from objects of great potential importance for our understanding of the evolutionary pattern to something little more than curiosities.

Although the problems may be many, there is one Problem that makes a problematic fossil: It does not fit our taxonomic concepts, in that it cannot be placed in any recognized phylum. This may be the fault of the investigator, in which case the Problem is trivial and does not deserve its capital P. It may be the fault of the fossil, in that it does not reveal enough of its character to the investigator. And it may be the fault of our taxonomic concepts, in which case it starts to become interesting.

For practical purposes this discussion will center on the phylum concept as applied to the kingdom Animalia and its bearing on the interpretation of problematic fossils of animal nature. The reasoning should be applicable to other kingdoms as well (see Chapter 2 by Babcock, this volume), including those that may not be recognized today.

PHYLA AS PROBLEMATIC TAXA

A commonly accepted definition of a problematic fossil is "a fossil that cannot be recognized as belonging to a known phylum." Although the concept of problematica is often stretched to take into account uncertainties at lower taxonomic levels ("problematic echinoderms," "problematic mollusks," etc.), this definition serves to delimit the core of the problematica and to highlight an often forgotten aspect of phyla: *A phylum is a group of organisms of uncertain taxonomic affinities, that is, a problematic taxon.*

This somewhat provocative description is adopted partly because it corresponds to actual practice in recognizing phyla, partly because it is suitable for the present discussion. If it appears to lack clarity and precision, it may be well to compare it with definitions given in modern dictionaries:

"One of the major kinds of group used in classifying animals" (Abercrombie et al. 1973).

"One of the large principle [*sic*] divisions of the animal kingdom" (Pennak 1964).

"A major unit in the taxonomy of animals, ranking above 'class' and below 'kingdom'" (Gary et al. 1972).

Many dictionaries will thus tell the reader nothing more than that a phylum is a phylum. A few are slightly more helpful:

"A major taxonomic unit comprising organisms sharing a fundamental pattern of organization and presumably a common descent" (*Webster's Third New International Dictionary of the English Language* 1964).

"Major division of animal or plant kingdom, containing species having same general form" (*The Concise Oxford Dictionary of Current English* 1976).

"A group of animals or plants constructed on a similar general plan, a primary division in classification" (Henderson et al. 1967)

Thus members of a phylum would seem to share some common pattern of organization acquired through common descent. This is, of course, what most biologists would state when asked to explain the concept. Simpson (1950, p. 23) provides an example:

The broadest of these themes are those formalized by zoologists as phyla, each of which represents a distinct plan or level of anatomical organization, with its attendant possibilities in the functioning of the animals. Put in another way, it may be said that at the base of each phylum lies the acquisition of some distinctive complex of characters which happened to be rich in evolutionary potentialities worked out, often in strikingly diverse ways, in the later members of the group.

A distinct "plan" or "complex of characters" may indeed characterize well-known phyla such as Echinodermata or Chordata. These groupings may seem so obvious that they can be recognized without any clear concept of what a phylum is or should be. But the power of a definition must be judged by its ability to set the boundaries—in other words, its ability to sort out the less clear cases. Here we are left completely in the dark by concepts such as "body plan" or "fundamental character complex." What is a "plan," and when does a character complex become fundamental? Like "phylum," they are easier to talk about than to define.

3

The phylogeny of phyla is a classic can of worms in zoology. The difficulty in assessing the relationships of the major groups of animals has led to such mutually exclusive schemes as those of Hadzi (1963), Jägersten (1968), and Salvini-Plawen (1978). The extreme case appears to be Nursall (1962), who derived nearly each metazoan phylum independently from a protozoan ancestor. On the other hand, if occasionally the mutual relationship of two or more phyla becomes established, the tendency has been to unite them. Thus phylum Pogonophora is now commonly regarded as belonging to the phylum Annelida (e.g., Southward 1980), and more than once the proposal has been made to unite the lophophorates into one phylum because of their presumed relationship (e.g., Emig 1977). Malakhov (1980) suggested a close relationship of the priapulids, kinorhynchs, and gordiaceans, and proposed to unite them in a new phylum Cephalorhyncha. Kristensen (1983) defined the new phylum Loricifera for a small group of newly discovered interstitial aschelminth-like animals with uncertain phylogenetic position. He concluded, however, that it may be possible to interpret the Loricifera as a stem group of the aschelminths, in which case "the Aschelminthes is a monophyletic taxon which should have the rank of phylum, and what I consider as the aschelminth phyla should be ranked as classes." These examples serve to emphasize the conceptual nature of phyla as major groups of uncertain systematic position.

Even though it is much easier to assign an organism, fossil or not, to phylum than to decide its more exact taxonomic status, "phylum" is a hazy concept that does not lend itself to clear definition. Nevertheless, the pretended clean cut division of all living organisms, or at least animals, into a small number of phyla has too often acquired the status of a natural law, with devastating effects on our understanding of the early fossil record: much has been said and written about the incompleteness of the fossil record for the understanding of evolutionary patterns, but almost nothing has been said about the equally serious incompleteness of the *living* record in this respect. It may be stated, again somewhat provocatively, that instead of using the present as the key to the past we often tend to use it as the *keyhole* to the past, letting it narrow our field of vision to a small fraction of what it could have been, had we instead used it as a key to open the door.

The essence of this argument is not that "phylum" is a useless concept. On the contrary, it is very useful in categorizing the major types of living animals that are known to us. However, the concept, as based on living organisms, is not a suitable tool to analyze the history of metazoan diversification *except* under the assumption that present-day phyla represent the full or almost full spectrum of body plans available to animals and that phyla never or seldom became extinct. Such an assumption could be regarded at least as a respectable working hypothesis, but the scientist applying it should be aware of doing so. Often, however, this hypothesis is applied uncritically simply because it is judged that the comparatively smaller amount of morphologic information that can be derived from fossils provides insufficient basis to challenge the established higher-level taxonomy. My contention, on the other hand, is that by forcing all fossils into the established phyla based on living organisms, we may largely obscure the information that may be gathered from the problematic fossils. We prune the phylogenetic bush into a phylogenetic tree.

In this paper, three questions will be considered: (1) What is the likelihood of now-extinct phyla existing in various environments at different times of Earth history? (2) How can we assess the probability of affinity of a problematic fossil to a given phylum? (3) How should data and hypotheses regarding the nature of problematic fossils be formulated so as to make them retrievable, understandable, and testable?

EXTINCT PHYLA

If a phylum has become extinct, its fossils will be regarded as problematic fossils. The question to what extent problematic fossils, conversely, represent extinct phyla is at the core of this discussion. Apparently personal preferences for or against the existence of extinct phyla often direct the interpretations of problematic fossils. An example is the Late Precambrian Ediacara fauna (see Chapter 6 by Fedonkin, this volume). The members of this unusual assemblage of early metazoans are with only a few exceptions placed by Glaessner (1984) in living phyla and living lower taxa, occasionally down to living families (see Chapter 7 by Stanley, this volume), whereas Seilacher (1984) regards almost none of them as representative of any extant phylum.

It is obvious that neither approach taken to the extreme (i.e., to see in a fossil assemblage only living or only extinct phyla) is valid research strategy, but who is more "right," the Glaessners or the Seilachers? This question cannot be answered with a simple verdict, but I would challenge Glaessner's philosophy, formulated as follows (Glaessner 1984, p. 133): "The basis of classifying Precambrian animals must be current zoological classification of living animals and their biohistorical predecessors despite the remoteness in time of the processes which produced them." This is the "keyhole approach" that I criticized above. Glaessner recognizes that a phylogenetic diagram for the Precambrian–Early Cambrian time is more like a bush than a tree, but concludes (p. 135) that in a historical perspective, "failures are not phyla." This is true (if somewhat circular), in so far as phyla are defined from living animals, but the conclusion does not logically follow that all "failures" belong to living phyla, even if the latter are also extended to cover closely related extinct groups. One could, theoretically, follow the lineages of each living phylum backward in time to their last common ancestors and use the resulting branching pattern to define an all-inclusive set of monophyletic or holophyletic (sensu Ashlock 1971) clades that could be called "phyla." However, if this was carried out as a cladistic analysis to define only holophyletic groups, the grouping of recent phyla could not be upheld. And if it was carried out according to principles of "evolutionary systematics," that is, also accepting paraphyletic groups as units of taxonomy, we would most certainly end up with early taxa that would be meaningless to include in any living phylum. (To presume otherwise would be, again, to make the teleologic assumption that metazoan diversity at a very early stage was channeled into exactly those and only those morphologic patterns that we see around us today.) In both cases there would be a number of taxa that did not belong to any now-recognized phylum. The point here is not to argue which type of classification is preferable, but to show that the established concept of phyla cannot be used to analyze patterns of early metazoan history even if the genealogy is completely known. And the latter not being the case, the approach of placing all problematic fossils into the least dissimilar extant phylum is likely to lead to serious distortion of the picture of metazoan diversification.

It is a striking aspect of the marine metazoan fossil record that higher taxa seem to have reached their maximum diversity at earlier times than lower ones (Valentine 1969). Thus, whereas the diversities of species, genera, and families are on a rising curve since the Permian extinction event and now surpass their maximum Paleozoic equivalents by factors of about two to four (Bambach 1977; Raup 1978; Sepkoski 1981; Sepkoski et al. 1981), higher taxa attained their maximum diversities earlier: orders reached a peak in the Late Ordovician, dropped slightly, and have since then remained at roughly the same level (Sepkoski 1978); 63 percent of classes and class-level *incertae sedis* cited by Sepkoski (1981) appeared already before the end of the Cambrian; and there is no indication that any phylum originated after the Cambrian or even the Precambrian.

This effect has been suggested to be due to the incomplete sampling of the fossil record: when samples are small it is easier to "hit" the larger than the smaller taxa (Raup 1972). Sepkoski (1978) simulated a stochastic pattern of diversification through time on three taxonomic levels (lineages, clades, and superclades) and found that, if all taxa are taken into account, the higher-level taxa tend to reach their diversity maxima later

than the lower-level ones, and also to suffer a slight decline after the maxima. When progressively incomplete sampling backward in time was introduced, however, the former reached their maxima earlier than the latter, and the effect of a subsequent decline was suppressed. Sepkoski thus suggested that the pattern of higher-taxon diversity seen in the fossil record may be interpreted as a simple reflection of the underlying diversity of species seen through a filter of stratigraphically biased sampling.

This result implies that the traditional picture of the appearance of phyla in the fossil record can be essentially accounted for by Sepkoski's kinetic model (1978) of Phanerozoic taxonomic diversity. This could be misleading, however, because the essence of established phyla, as argued above, is not that they are some kind of super-superclades, but that they are major units of uncertain systematic relationships. The existence of such units is only in an indirect sense related to the branching pattern of lineages.

Intuitively we would expect extinct phyla to be more frequent at earlier than later stages in the history of life, this being the normal "Lyellian" pattern shown by taxa at lower hierarchic levels (e.g., Kurtén 1972; Stanley 1980). In addition to the obvious geometric reasons for such a pattern, the following factors may have contributed to produce high frequencies of extinct phyla at early stages of life history, particularly in the Late Precambrian–Early Cambrian:

1. Conditions for life in the Late Precambrian were probably different from those of later periods, for example with regard to the level of atmospheric oxygen (e.g., Berkner and Marshall 1965; Towe 1970; Cloud 1976; Runnegar 1982b). This might have favored the evolution of body plans different from those represented by living phyla (e.g., Seilacher 1984).

2. The early radiations of metazoans probably took place in a relative "ecologic vacuum." This might have provided more room for quantum steps and morphologic "experiments" in evolution (e.g., Valentine 1975, 1980).

3. In a major radiation event, such as that producing the "Cambrian explosion," the number of major clades would be expected to reach a maximum at a fairly early stage followed by a decline because of stochastic effects or taxonomic "hindsight" (Raup et al. 1973; Sepkoski 1978), or because of increasing competition.

4. If one aspect of phyla, as argued here, is that they are major units of uncertain systematic position, we would expect that the diversification into phyla took place early and that it is not recorded in the traditionally known fossil record.

Conversely, one might argue that a new phylum is not likely to have arisen in the Phanerozoic because (1) conditions for life are more stable, (2) ecologic niches are closer to being saturated, (3) lacking taxonomic foresight we would be less likely to accept a more recently arisen clade as a new major type of animal, and (4) being able to recognize its phylogenetic origin, we would not be prone to erect a new phylum for it.

It deserves to be emphasized that the invasion of land by various types of animals did not give rise to a single new phylum. Interestingly, of the factors just listed, the first and second do not apply in this case, as the land certainly provided many new and unstable habitats with undersaturated ecologic niches. The third may be relevant but does not appear to carry particular weight, since the main radiation of land-living organisms took place as early as some 400 Ma ago. This leaves us essentially with the fourth explanation, again emphasizing the nature of phyla as problematic taxa.

J. J. Sepkoski points out (personal communication 1985) that there may be developmental constraints preventing any major reorganization of body plans after the initial stages of metazoan evolution. Thus Phanerozoic metazoans may have evolved a too-complex developmental pattern and simply be too "set in their ways" to give rise to anything that could be regarded as a new phylum. Although this idea, in the literature usually expressed in terms such as "the great ancestral potential of generalized early members" (Stanley 1976, p. 72), at first sight may seem dangerously close to idealistic concepts of evolution as a progressive march toward greater and greater complexity, it seems to be well worth considering whether there were even at the level of anatomic diversification of metazoans certain unique stages in the history of life that permitted the evolution of certain organizational patterns. Such unique stages appear to have been present in the molecular and early cellular evolution, or we would have all kinds of new kingdoms of microorganisms arising throughout the history of life.

Thus the picture is corroborated that new phyla most likely did not arise after the beginning of the Phanerozoic. This means that the highest diversity of phyla was probably reached already in the latest Proterozoic or earliest Phanerozoic.

How soon was the present-day pattern of phyla established; that is, when did the main extinction of "side phyla" take place? This is one of the questions that may be answered through a careful study of the problematic fossils. With regard to the Late Precambrian biota, it has already been mentioned that opinions differ widely regarding whether the Ediacara metazoans represent living phyla or not. Obviously a better understanding of their anatomy and biology is needed; analyses of the physiologic consequences of their morphologies, such as Runnegar's study of *Dickinsonia* (1982a), may help to show the way, but it is essential that such studies do not employ only models

derived from living phyla, but take the possibility of other structural and physiologic solutions into account as well.

Early Cambrian deposits are particularly rich in problematic fossils (see Chapter 8 by Rozanov, this volume). In most instances no convincing case has been made for the assignment to a living phylum of these fossils. Whereas many of the groups were extremely successful in the Cambrian, reaching great abundance and diversity as well as wide geographic distribution, few of them are known to have survived the Cambrian or even the Early Cambrian. Groups such as the Archaeocyatha (cf. Hill 1971), Anabaritida (cf. Abaimova 1978; Missarzhevsky 1982), Coeloscleritophora (cf. Bengtson and Missarzhevsky 1981; Bengtson and Conway Morris 1984), and Tommotiidae (cf. Rozanov et al. 1969; Bengtson 1977; Landing 1984) all had a time of dominance in the earliest Cambrian but disappeared before the end of the Cambrian. This is true also for a number of minor problematic groups. It should be pointed out that these groups do not form part of the "Cambrian evolutionary fauna" of Sepkoski (1981). The ones that do, have a somewhat later diversity peak and longer life. A few (Inarticulata, Monoplacophora) even survive as living fossils; the others (mainly Trilobita, Hyolitha, and Eocrinoidea) are to a certain measure "problematic," but on a level below that of phylum.

One of the earliest and most widespread groups of Cambrian fossils is the protoconodonts (*Protohertzina* and others). They lived at least through the Cambrian and may have given rise (via paraconodonts) to the extremely successful Paleozoic to Triassic euconodonts (Bengtson 1976, 1983b), classical Paleozoic problematica that plodded through the Cambrian at quite low diversities, then radiated rapidly toward a peak diversity in the Ordovician. Thus they do not quite conform to the pattern of the archaeocyathans and others. Lately Szaniawski (1982, 1983) has shown that the protoconodonts are probably closely related to the modern chaetognaths, and this conclusion may be extended to the whole taxon Conodonta as well (Briggs et al. 1983; Bengtson 1983a,b; but see Chapter 15 by Aldridge and Briggs, and Chapter 16 by Dzik, this volume). A similar picture is presented by the graptolites, which with a slow start in the Cambrian, possibly with roots among the Cambrian representatives of the extant and very conservative hemichordates (e.g., Kozłowski 1966; Rickards 1975, 1979; Bengtson and Urbanek, in press; Chapter 14 by Urbanek, this volume), had their high diversity in the Ordovician and then dwindled toward extinction in the Carboniferous.

A more complex question to analyze than the expected appearance and existence of extinct phyla in time is their possible relation to the physical and biologic environment. By virtue of their comparatively short life, extinct phyla are more likely to be found in the environment and geographic region where most of the phylum-level diversification originated. It is not clear, however, whether there was such a particular place. As in the case of the temporal origin of phyla, we may have a simple geometric relationship connected to the shape of the phylogenetic bush; areas where speciation and species turnover was rapid may have been more likely to produce also major ecologic innovations that could give rise to a diversity of phyla. This would draw the attention to offshore environments where rates of speciation are usually high (e.g., Bretsky and Lorenz 1970; Valentine 1975; Fortey 1980). However, the real picture may be quite different, as suggested by the findings of Jablonski et al. (1983) that during the Phanerozoic, major ecologic innovations might have first appeared in the nearshore environments and gradually spread offshore.

At present there is little agreement among zoologists with regard to where the large-scale pattern of metazoan evolution was mainly developed: in the infauna, meiofauna, epifauna, plankton, or all or none of these places. If this phylum cradle existed and produced short-lived phyla in at least the same numbers as long-lived ones, an unconditioned study of the problematic fossils in the Late Proterozoic and Early Phanerozoic may be able to indicate where the cradle stood. It may also be able to tell us which paths metazoan evolution took in addition to those represented in the living biota, and what the relative success of these lineages was.

ASSESSING AFFINITIES TO PHYLUM

A common confusion in systematic biology arises from failure to distinguish between criteria for definition and criteria for recognition of taxa. The appearance of a character or character state in a phylogenetic lineage may be used to define a holophyletic taxonomic unit (or as a necessary but not sufficient criterion to define a paraphyletic unit), but the *presence* of a character (or character state) cannot be used for definition of the unit, as the character may have been secondarily lost. The presence of the character (if it is unequivocal) may be used only as one criterion for *recognition* of the taxon, but there can be no guarantee that all members of the taxon have the character. *Definitions* of taxa based on the presence or absence of certain characters are form-classificatory, and the principles of form classification are fundamentally different from those of biologic taxonomy (Bengtson, 1985).

Thus a phylum (or any other unit in biologic taxonomy) cannot be defined by the presence (or, even less, absence) of a certain character in its members. A definition must be based on the inferred evolutionary history of the group, so as to unite forms on the basis of their genealogic relationship. This is true *even* if one does not follow a strictly cladistic approach: a paraphyletic group may well be unequivocally defined as containing "species A and all its descendants *except* species B and all its descendants." Such a concept would be based on a phenetic deviation of one clade from the rest of the parent clade (say, birds or mammals from reptiles) but would not employ presence or absence of characters (such as amniote eggs or feathers) in its *definition.*

The distinction between definition and recognition is important in the present context, because problematic fossils are often evaluated against so-called diagnostic characters, usually expressed in formal diagnoses, of known taxa. Such a comparison does not result in a perfect fit (or the fossils would not be problematic), but the partial agreement with existing diagnoses is considered significant. However, a diagnosis cannot define a taxon but can only attempt to describe it in comparison to the most closely related *known* taxa. Such a diagnosis is valuable as a practical tool for identification of already described taxa but is of little value when it comes to dealing with previously unknown taxa.

When attempting to classify an organism, fossil or Recent, one deals with a set of characters. The characters are used as guidance in the search for the nearest relatives, but in the end the taxonomic decision should not be the result of an assessment of similarities ("this taxon is most similar to taxon A") but of the formulation of a phylogenetic hypothesis ("this taxon is most closely related to taxon A" or "this taxon evolved from species B"). In practical work, the distinction is usually not crucial, because in most cases the character distribution of taxa is well enough known to allow the systematic placement of a new taxon on the basis of presence or absence of key characters. However, in the case of problematic fossils, the distinction does indeed become crucial, because *the existence of a problematic fossil indicates that current systematic concepts are at fault or, at least, incomplete.* This means that the problem of the systematic placement of the fossil cannot be solved within the framework of existing systematic concepts.

This statement may be somewhat circular but serves to highlight the importance of truly problematic fossils as indicators of the limits of our knowledge. It is important to recognize that to attempt to solve the problem as a purely taxonomic one, either by assigning the fossil to the least dissimilar higher-order taxon or by erecting a string of new monotypic higher-order taxa, is to apply a cosmetic solution to a scientific problem.

All biologic taxonomy, even at species level, continually modifies existing concepts as new taxa are added. Thus the principles just outlined are not different between problematic and non-problematic fossils. The difference is one of degree, in that the analysis of problematic fossils affects the established taxonomy at a much higher level, usually that of phylum. At this level the analysis of similarities and dissimilarities, often stated by practicing taxonomists to be the backbone of the science, can be particularly misleading. What we are really looking for is homologies rather than similarities. (I will not go into the very bulky literature that exists on the homology concept in systematics, only state that I believe that homologies are indeed real.) At low taxonomic levels, homologous characters are usually similar, and this is of course the reason why numerical taxonomy mostly works.

But homologous characters are sometimes not similar at all, and similar characters may not be homologous. This knowledge is more than a century old but needs to be reiterated. Neglect in this respect has led to many misinterpretations of problematic fossils. Thus Fedorov et al. (1979) interpreted phosphatic plates of the Tommotian genus *Tumulduria* Missarzhevsky 1969 (in Rozanov et al. 1969) as trilobite remains because of the similarity of the central fold to a trilobite rachis, despite the fact that the transverse banding is formed by growth lines rather than by articulation of the skeleton and thus cannot be homologous to trilobite pleura. [This occurrence of a Tommotian "trilobite" is now constantly invoked to discredit the concept of the Tommotian as a pre-trilobite stage, for example by Repina (1981), Qian (1984), and Yu (1984). The concept may well be wrong, but not on grounds of the presence of *Tumulduria*.] Bischoff (1976) identified the Early Cambrian phosphatic-plated *Dailyatia* as a lepadomorph cirripede because of the "striking similarity" in gross morphology between sclerites of the two groups, an identification not supported by any homologies (Bengtson 1977). Conversely, the proposal to unite the Cambrian Chancelloriidae and Halkieriidae (= "Sachitidae") on grounds of the structural uniqueness and proposed homology of their sclerites (Bengtson and Missarzhevsky 1981) has so far not won acceptance (and is criticized by Dzik in Chapter 10, this volume) because of the morphologic dissimilarities between sclerites of the two groups. [Ironically, the original description of the genus *Sachites* by Meshkova (1969) seems to have encompassed one chancelloriid, *S. proboscideus,* and one halkieriid, *S. sacciformis.* New well-preserved material from South Australia indicates that sclerites of *proboscideus*-type and rays of *Chancelloria* are almost indistinguishable in morphology and fine structure.]

In practice one is faced with a complex situation in which the most difficult problem may be to find a sufficient number of characters that are possible to interpret at all in terms of homology.

Obviously there can be no strict formula to follow in order to assess if a problematic fossil belongs to a certain phylum or not. It may be helpful to use the following set of questions as a checklist:

1. What are the observable characters of the fossil?
2. What were the original characters of the animal when preservational and diagenetic factors have been taken into account?
3. What constructional and functional significance can be attributed to the characters?
4. Can these characters be interpreted as homologous to characters formed by members of any known phylum?
5. Are there any of these possibly homologous characters that are unlikely to have arisen by convergence due to constructional or functional factors?
6. If any of the characters cannot be, or are not likely to be, homologous with characters in known phyla, can it be a derived character that evolved secondarily from a member of a known phylum?
7. Does the fossil show affinities with any other fossils from the same or any other period of time? Repeat questions 1–6 for this combined group.
8. If the fossil (group) on the basis of its characters can be interpreted as belonging to a known phylum, what are the consequences for our knowledge of the evolutionary history of the phylum?
9. If the fossil (group) cannot be interpreted as belonging to a known phylum, which testable hypotheses may be formulated regarding its biologic nature and phylogenetic origin?

These steps do not outline a proper phylogenetic analysis but may serve as preparatory measures for one. Their main purpose is to serve as a safeguard against casual misidentifications and to convey the message embodied in this book, that problematic fossils should be taken seriously.

PROVIDING FOR RETRIEVABLE AND TESTABLE RESULTS

Approaches to the analysis of problematic fossils may vary within a broad spectrum from pedestrian assignments to the least dissimilar living phylum to wild speculations about exotic animals that may never have lived on this earth. Nothing certain can be said about the most productive road to follow, except that it lies between those two extremes. The question then is how to organize the presentation of the information so that the results will be understandable and useful even to scientists of totally different temperament.

As in all other branches of science, the key is to present the data as clearly and correctly as possible and to formulate testable hypotheses for their interpretation. In taxonomic paleontology there is the particular concern that the structure of the nomenclatorial system more or less forces one to express an idea on affinities. Every taxonomic assignment implies a hypothesis regarding the pattern of evolution. This hypothesis must be clearly formulated in such a way as to make clear its underlying assumptions as well as its testable implications.

However, the rules of taxonomy do not force us to go all the way up to phylum. The discussion in this paper about problematic fossils as possible representatives of extinct phyla is not a recommendation to introduce a score of names for more or less hypothetical extinct phyla. In fact, only species and genus names are legally required, and even those may be expressed under open nomenclature if uncertainty exists. One should go only as far up in the taxonomic hierarchy as warranted by the analysis. The question mark is useful to indicate the existence of an insufficiently corroborated hypothesis, and the expression *incertae sedis* can be employed to mark the boundary between scientific hypothesis and the realm of speculations. To take any fossil of uncertain affinity as an excuse to erect a string of names of monotypic higher taxa (with the author's name shining after them forever in questionable glory) is no better than to tuck the fossil away in the least dissimilar known phylum. Both are cosmetic, nomenclatorial solutions to what should be a scientific problem, and both serve to disguise the Problem of the Problematica.

The affinities of a problematic fossil is only one of the questions we may ask of it, if perhaps the first one. Other questions concern such issues as the origin and early diversification of the various kingdoms of organisms, the nature and origin of body plans, the evolutionary sidetracks that are not represented in the living biota, and so on. Almost by definition, there can be no specialists on problematic fossils. Thus they more often end up in somebody's drawer than in scientific publications, and even if they are put to study, it is almost always as a sidetrack (unless they happen to be big and stratigraphically important groups). This has contributed to the lack of sophistication in the field. It is to be hoped that the present volume will help to inspire dedicated efforts into the nature of problematica. Then these fossils may start telling us something important about the living world rather than being curiosities to be swept under the systematic carpet.

Acknowledgments

My work has been supported by grants from the Swedish Natural Science Research Council. Jack Sepkoski read and commented on the manuscript.

REFERENCES

Abaimova, G. P. 1978. Anabaritidy—drevneishye iskopaemye s karbonatnym skeletom [Anabaritids—the oldest fossils with a calcareous skeleton]. *In* Novye materialy po stratigrafii i paleontologii Sibiri. *Trudy SNIIGGIMS* **260,** 77–83.

Abercrombie, M., Hickman, C. J., and Johnson, M. L. 1973. *A Dictionary of Biology.* 6th ed. Penguin Books, New York.

Ashlock, P. D. 1971. Monophyly and associated terms. *Syst. Zool.* **20,** 63–69.

Bambach, R. K. 1977. Species richness in marine benthic habitats through the Phanerozoic. *Paleobiology* **3,** 152–167.

Bengtson, S. 1976. The structure of some Middle Cambrian conodonts, and the early evolution of conodont structure and function. *Lethaia* **9,** 185–206.

————. 1977. Aspects of problematic fossils in the early Palaeozoic. *Acta Univ. Upsal., Abstr. Uppsala Diss. Fac. Sci.* **415,** 71 pp.

————. 1983a. A functional model for the conodont apparatus. *Lethaia* **16,** 38.

————. 1983b. The early history of the Conodonta. *Fossils and Strata* **15,** 5–19.

————. 1985. Taxonomy of disarticulated fossils. *J. Paleont.* **59,** 1350–1358.

———— and Conway Morris, S. 1984. A comparative study of Lower Cambrian *Halkieria* and Middle Cambrian *Wiwaxia. Lethaia* **17,** 307–329.

———— and Missarzhevsky, V. V. 1981. Coeloscleritophora, a major group of enigmatic Cambrian metazoans. *U.S. Geol. Surv. Open-File rep.* **81-743,** 19–21.

———— and Urbanek, A. In press. *Rhabdotubus,* a Middle Cambrian rhabdopleurid hemichordate. *Lethaia.*

Berkner, L. V. and Marshall, L. C. 1965. History of major atmospheric components. *Proc. Natl. Acad. Sci. U.S.A.* **53,** 1215–1225.

Bischoff, G. C. O. 1976. *Dailyatia,* a new genus of the Tommotiidae from Cambrian strata of SE. Australia (Crustacea, Cirripedia). *Senckenb. Leth.* **57,** 1–33.

Bretsky, P. W. and Lorenz, D. M. 1970. Adaptive response to environmental stability: a unifying concept in paleoecology. *Evolution of Communities. Proc. North Am. Paleontol. Conv.* **E,** 522–550.

Briggs, D. E. G., Clarkson, E. N. K., and Aldridge, R. J. 1983. The conodont animal. *Lethaia* **16,** 1–14.

Cloud, P. E. 1976. Beginnings of biospheric evolution and their biochemical consequences. *Paleobiology* **2,** 351–387.

The Concise Oxford Dictionary of Current English. 1976. 6th ed. Clarendon, Oxford.

Emig, C. C. 1977. Un nouvel embranchement: les Lophophorates. *Bull. Soc. Zool. Fr.* **102**, 341–344.

Fedorov, A. B., Egorova, L. I., and Savitsky, V. E. 1979. Pervaya nakhodka drevnejshikh trilobitov v nizhnej chasti stratotipa Tommotskogo Yarusa nizhnego kembriya (r. Aldan) [The first find of trilobites in the lower part of the stratotype of the Lower Cambrian Tommotian Stage (R. Aldan)]. *Doklady AN SSSR Geol.* **249**, 1188–1190.

Fortey, R. A. 1980. Generic longevity in Lower Ordovician trilobites: relation to environment. *Paleobiology* **6**, 24–31.

Gary, M., MacAfee, R., Jr., and Wolf, C. L. (eds.). 1972. *Glossary of Geology*. American Geological Institute, Washington, D.C.

Glaessner, M. F. 1984. *The Dawn of Animal Life. A Biohistorical Study*. Cambridge Univ. Press, Cambridge, 244 pp.

Hadzi, J. 1963. *The Evolution of the Metazoa*. Pergamon, Oxford, 499 pp.

Henderson, I. F., Henderson, W. D., and Kenneth, J. H. 1967. *A Dictionary of Biological Terms*. 8th ed. Oliver and Boyd, Edinburgh.

Hill, D. 1972. Archaeocyatha. *In* Moore, R. C. (ed.). *Treatise on Invertebrate Paleontology, Part E, Protista 1*. Geological Society of America and Univ. of Kansas Press, Lawrence, pp. E1–E158.

Jablonski, D., Sepkoski, J. J., Jr., Bottjer, D. J., and Sheehan, P. M. 1983. Onshore–offshore patterns in the evolution of Phanerozoic shelf communities. *Science* **222**, 1123–1125.

Jägersten, G. 1968. *Livscykelns evolution hos Metazoa*. Läromedelsförlagen, Stockholm (Eng. transl. 1972, Academic Press, London), 295 pp.

Kozłowski, R. 1966. On the structure and relationships of graptolites. *J. Paleont.* **40**, 489–501.

Kristensen, R. M. 1983. Loricifera, a new phylum with Aschelminthes characters from the meiobenthos. *Z. Zool. Syst. Evolutionsforsch.* **21**, 163–180.

Kurtén, B. 1972. The "half-life" concept in evolution illustrated from various mammalian groups. *In* Bishop, W. W. and Miller, J. A. (eds.). *Calibration of Hominoid Evolution*, Scottish Academic Press, Edinburgh. pp. 187–194.

Landing, E. 1984. Skeleton of lapworthellids and the suprageneric classification of tommotiids (Early and Middle Cambrian phosphatic problematica). *J. Paleont.* **58**, 1380–1398.

Malakhov, V. V. 1980. Cephalorhyncha—novyj tip zhivotnogo tsarstva, obyedinyayushchij Priapulida, Kinorhyncha, Gordiacea, i sistema pervichnopolostnykh chervej [Cephalorhyncha, a new phylum of the Animal Kingdom, uniting Priapulida, Kinorhyncha, Gordiacea,

and the classification of aschelminth worms]. *Zool. Zh.* **59**(4), 485–499.

Meshkova, N. P. 1969. K voprosu o paleontologicheskoj kharakteristike nizhnekembrijskikh otlozhenij Sibirskoj platformy [To the question of the palaeontological characteristics of the Lower Cambrian deposits of the Siberian Platform]. *In* Zhuravleva, I. T. (ed.). *Biostratigrafiya i paleontologiya nizhnego kembriya Sibiri i Dal'nego Vostoka*. Nauka, Moscow, pp. 158–174.

Missarzhevsky, V. V. 1982. Raschlenenie i korrelyatsiya pogranichnykh tolshch dokembriya i kembriya po nekotorym drevnejshim gruppam skeletnykh organizmov [Subdivision and correlation of the Precambrian–Cambrian boundary beds using some groups of the oldest skeletal organisms]. *Byulleten' Moskovskogo Obshchestva Ispytatelej Prirody, Otd. Geol.* **57**(5), 52–67.

Nursall, J. R. 1962. On the origins of the major groups of animals. *Evolution* **16**, 118–123.

Pennak, R. W. 1964. *Collegiate Dictionary of Zoology*. Ronald Press, New York.

Qian Yi. 1984. Early Cambrian–Late Precambrian small shelly faunal assemblage with a discussion on Cambrian–Precambrian boundary in China. *Developments in Geoscience*. Contribution to 27th Int. Geol. Cong., Moscow, Science Press, Beijing, pp. 9–20.

Raup, D. M. 1972. Taxonomic diversity during the Phanerozoic. *Science* **177**, 1065–1071.

———. 1978. Cohort analysis of generic survivorship. *Paleobiology* **4**, 1–15.

———, Gould, S. J., Schopf, T. J., and Simberloff, D. S. 1973. Stochastic models of phylogeny and the evolution of diversity. *J. Geol.* **81**, 525–542.

Repina, L. N. 1981. Trilobite biostratigraphy of the Lower Cambrian stages in Siberia. *U.S. Geol. Surv. Open-File Rep.* **81-743**, 173–180.

Rickards, R. B. 1975. Palaeoecology of the Graptolithina, an extinct class of the phylum Hemichordata. *Biol. Rev.* **50**, 397–436.

———. 1979. Early evolution of graptolites and related groups. *In* House, M. R. (ed.). *The Origin of Major Invertebrate Groups*. Academic Press, London, pp. 435–441.

Rozanov, A. Yu., Missarzhevsky, V. V., Volkova, N. A., Voronova, L. G., Krylov, I. N., Keller, B. M., Korolyuk, I. K., Lendzion, K., Michniak, R., Pykhova, N. G., and Sidorov, A. D. 1969. Tommotskij yarus i problema nizhnej granitsy kembriya [The Tommotian Stage and the problem of the lower boundary of the Cambrian]. *Trudy Geol. Inst. AN SSSR* **206**, 380 pp.

Runnegar, B. 1982a. Oxygen requirements, biology and phylogenetic significance of the late Precambrian worm *Dickinsonia*, and the evolution of the burrowing habit. *Alcheringa* **6**, 223–239.

———. 1982b. The Cambrian explosion: animals or fossils? *Geol. Soc. Aust. J.* **29**, 395–411.

Salvini-Plawen, L. V. 1978. On the origin and evolution of the lower Metazoa. *Z. Zool. Syst. Evolutionsforsch.* **16**, 40–88.

Seilacher, A. 1984. Late Precambrian and Early Cambrian Metazoa: preservational or real extinctions? *In* Holland, H. D. and Trendall, A. F. (eds.). *Patterns of Change in Earth Evolution.* Springer, Berlin, pp. 159–168.

Sepkoski, J. J., Jr. 1978. A kinetic model of Phanerozoic taxonomic diversity. I. Analysis of marine orders. *Paleobiology* **4**, 223–251.

———. 1981. A factor analytic description of the Phanerozoic marine fossil record. *Paleobiology* **7**, 36–53.

———. Bambach, R. K., Raup, D. M., and Valentine, J. W. 1981. Phanerozoic marine diversity and the fossil record. *Nature* **293**, 435–437.

Simpson, G. G. 1950. *The Meaning of Evolution. A Study of the History of Life and of the Significance for Man.* Oxford Univ. Press, London, 364 pp.

Southward, E. C. 1980. Regionation and metamerization in Pogonophora. *Zool. Jb. Anat.* **103**, 264–275.

Stanley, S. M. 1976. Fossil data and the Precam-brian–Cambrian evolutionary transition. *Am. J. Sci.* **276**, 56–76.

———. 1980 (date of imprint 1979). *Macroevolution. Pattern and Process.* Freeman, San Francisco, 332 pp.

Szaniawski, H. 1982. Chaetognath grasping spines recognized among Cambrian protoconodonts. *J. Paleont.* **56**, 806–810.

———. 1983. Structure of protoconodonts. *Fossils and Strata* **15**, 21–27.

Towe, K. M. 1970. Oxygen–collagen priority and the early metazoan fossil record. *Proc. Natl. Acad. Sci. U.S.A.* **65**, 781–788.

Valentine, J. W. 1969. Patterns of taxonomic and ecological structure of the shelf benthos during Phanerozoic time. *Palaeontology* **12**, 684–709.

———. 1975. Adaptive strategy and the origin of grades and ground-plans. *Am. Zool.* **15**, 391–404.

———. 1980. Determinants of diversity in higher taxonomic categories. *Paleobiology* **6**, 444–450.

Webster's Third New International Dictionary of the English Language. 1964. Merriam, Springfield, Mass.

Yu Wen. 1984. Early Cambrian molluscan faunas of Meischucun Stage with special reference to Precambrian–Cambrian boundary. *Developments in Geoscience. Contribution to 27th Int. Geol. Congr. Moscow.* Science Press, Beijing. pp. 21–35.

2. THE PUZZLE OF ALGA-LIKE PROBLEMATICA, OR RUMMAGING AROUND IN THE ALGAL WASTEBASKET

JACK A. BABCOCK

Like the other contributions to this volume, the subject of mine is fossils that do not fit. They don't fit our notions of the way the world is made, and they don't fit into our wonderfully complex schemes of classification and organization.

In preparing this essay, two themes seemed to form the framework. First, the problematica that are frequently assigned to the calcareous algae constitute one of the largest taxonomic wastebaskets in fossildom, and second, the proper identification and taxonomic assignment of these problematica will require a commitment to careful, thorough, and objective descriptions of both morphologic and nonmorphologic attributes, and comparisons with potential living analogs where possible.

THE USE OF THE ALGAE AS A TAXONOMIC WASTEBASKET

Although other taxonomic categories have been used as a dumping ground for fossils of unknown pedigree, the calcareous algae certainly are among the most abused this way. Wray (1977; p. viii) declared that "fossil calcareous algae frequently have been misinterpreted throughout the history of their study, and all too often, the group has served as a 'wastebasket' for various unidentifiable biotic and non-biotic constituents presumed to be algae."

Even if the multitude of stromatoform fossils are eliminated, the list of problematica that have been tentatively assigned to various algal taxa is long. The receptaculitids are one of the most frequently discussed groups (Chapter 3 by Nitecki, this volume). The spherical calcareous microfossils are locally abundant, enigmatic, and often have been interpreted to be sporangial structures from green algae, just as acritarchs (Tappan 1980) are typically characterized as algal encystment bodies.

There are many problematic forms that were significant to the formation of Paleozoic shallow marine carbonates, especially reefs and other buildups. Some of the genera of these "algal" problematica are *Keega* Wray, *Komia* Korde, *Dvinella* Khvorova, *Kamaena* Antropov, *Renalcis* Vologdin, and *Epiphyton* Korde. *Microcodium* Glück has long tantalized paleontologists and carbonate sedimentologists. In addition, there are

the Beresellidae, Stacheina, Ungdarellacea, Umbellinacea, Aougaliacea, and Moravamminidae, all of which are the subject of some controversy.

Why are the algae used as a taxonomic wastebasket? The obvious answer is that the algae are abundant, morphologically simple plants that have a very long geologic history. Therefore any morphologically simple, unidentified fossil is fair game. However, there are other reasons that small problematic fossils tend to be labeled "algal" more frequently than "foraminiferal" or some other well-known taxonomic category. Those reasons can be summarized under three headings: characteristics of the algae, characteristics of the samples, and the training and prejudices of the observer-interpreter.

Characteristics of the algae

Some features of the algae themselves have created problems, including the following:

1. The algae include an extremely diverse group of organisms in terms of their morphology, ecology, physiology, and potential for preservation in the rock record.

2. Algae are classified in two of the five kingdoms of organisms. Several different taxonomic divisions (equivalent to animal phyla) constitute the algae, from the morphologically simple prokaryotic blue-green algae (the Cyanophyta or, preferably, the Cyanobacteria) to the much more complex eukaryotic red algae (Rhodophyta).

3. Some of the morphologic features that are considered to be diagnostic in identification and classification of living algae are not preserved in fossil forms.

4. Calcareous algae are significant components of modern shallow carbonate-depositing seas. Apparently they have played a similar role throughout Phanerozoic time, and so when an odd fossil is observed in carbonate rocks the assumption is made frequently that it is probably algal.

5. Many Paleozoic fossils, including forms thought to have been algal, may represent extinct higher taxonomic groups that have no known modern analogs.

6. Many modern algae, and presumably their ancestors, exhibit morphologic plasticity under varying ecologic conditions (ecophenotypy).

Characteristics of the samples

Calcareous algal fossils are studied primarily through the use of petrographic thin sections of carbonate rocks, in which most of these forms are found. Therefore, a typical sample is limited to a thin, two-dimensional, random slice through the fossil of interest. The original morphology of the fossil must be interpreted by use of serial thin sections, statistical reconstructions, or more typically, by imaginative reconstructions done in the mind's eye. In association with the diagenetic adversities suffered by most calcareous fossils, this contributes to the taxonomic–systematic problems faced by those who deal with the calcareous algae.

Training and prejudices of the observer-interpreter

The training and prejudices of the observer-interpreter are perhaps the most important factors in the algal taxonomic wastebasket problem. Identifications and assignments of systematic affinities are creations of the interpreter, not of the fossil organism, and it is the interpreter who is responsible for creating the great algal wastebasket mess.

There are numerous aspects of paleontologic technique and psychology that have contributed to the algal wastebasket problem. However, among the most important factors are bias (a subjective mindset), the Fear of Problematica, and the failure to describe morphologic features objectively.

Unlike problems resulting from factors that are inherent to the algae, the bias factor is one that can be limited, if not removed entirely. However, bias is almost never apparent to the guilty paleontologist or geologist, and it is almost always a result of the best intentions, not the worst. In fact, some would argue that a biased approach to scientific inquiry is both inevitable and desirable. In the case of classification of Paleozoic problematica, however, I would argue for more objectivity and greater use of the scientific method for the following reasons:

1. *Bias influences descriptions.* To identify and classify problematica accurately requires clear, objective descriptions of all morphologic features. Preconceived ideas concerning biotic affinities of a problematic form may influence not only the descriptive terminology used, but also the selection of the morphologic traits to be described.

2. *Bias directs interpretation.* Although interpretation is a selection process, it can be based on objective observations, or it can be a deterministic exercise that inevitably arrives at a foregone conclusion. If interpretation follows the latter scenario, then arguments about biotic affinities of problematic unknowns tend to become circular.

The Fear of Problematica, or perhaps the fear of appearing indecisive, is a factor that may be at the root of much of the problem of using the algae as a taxonomic wastebasket. Even in cases where the assignment of proper taxonomic affinities to equivocal forms seems difficult or premature, many paleontologists apparently feel obliged to classify these entities.

Writing of this phenomenon in an essay devoted to problematica, Gould (1983) said, "Paleontologists are, in general, a conservative lot. Problematica of uncertain taxonomic affinity and few species are an embarrassment and an untidy bother; nothing makes an old-style paleontologist happier than the successful housing of problematical organisms within a well-known group." Gould (1984) further stated that "the traditional ploy of forcing old and problematical fossils into modern taxonomic categories often fails badly."

Gould's observation that paleontologists tend to abhor a systematic vacuum (the Fear of Problematica) and his hypothesis that because of experimentation, then standardization, problematica are much more frequently found among Paleozoic fossils than later ones, are important points to consider when evaluating the problem of the algal taxonomic wastebasket.

The failure to describe morphologic features clearly and objectively has contributed to the algal wastebasket puzzle in two ways. First, biased descriptions have misled those unfamiliar with a particular problematic form into believing that its affinities are established. Second, even unbiased but unclear, nonobjective descriptions have perpetuated problems with problematic taxa because of difficulties in reassessing these inadequately defined forms.

EXAMPLES OF ALGA-LIKE PROBLEMATICA FROM THE CAPITAN LIMESTONE

The following discussion focuses on concrete examples from the biota with which I am most familiar, that of the Permian Capitan Limestone.

Several unusual, controversial fossils are found in the Capitan Limestone (Permian, Guadalupian) of the Permian Reef Complex, West Texas and New Mexico. All of these forms have been classified previously as algae (e.g., Johnson 1942, 1951; Rigby 1957; Klement 1966). However, a reevaluation has suggested that some of these taxa may not be algal and that they all should be considered problematica for the time being (Babcock 1974, 1977, 1979). Among the most interesting forms in this biota are *Tubiphytes* Maslov and *Archaeolithoporella* Endo. These two forms are of greatest interest because they represent end-member problematic types and because they both are sedimentologically significant components of Capitan boundstones. *Tubiphytes* is found well

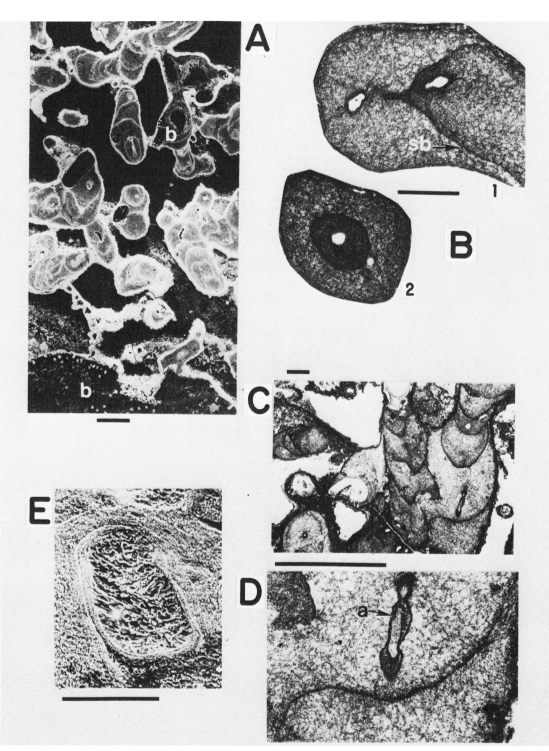

Fig. 1. A. Skeletal boundstone (framestone) formed by in-place encrustations of *Tubiphytes obscurus* colonies on and between an open meshwork of stony and fenestrate bryozoans (b). Notice that *Tubiphytes* colonies have grown freely in all directions within the constructional cavities. Scale bar = 2 mm. B. Morphologic characteristics of *Tubiphytes* Maslov. The irregularly ovoid to cylindrical gross morphology of *Tubiphytes* can be seen in B1 (oblique longitudinal cut) and B2 (cross section). In addition, the dark micritic boundaries of individual *Tubiphytes* subunits (sb), and the character and position of the axial tube can be observed. Scale bar = 1 mm. C. Multiple-unit encrustations of *Tubiphytes obscurus* demonstrating the typical superimposed and overlapping habit of the form. These encrustations are oriented downward in a cavity. Scale bar = 1 mm. D. Detail of lower-right segment of

14

preserved in the Capitan limestone but is a truly enigmatic fossil, while *Archaeolithoporella* is not as well preserved and is problematic principally because a critical diagnostic feature (cells) has not been found in specimens observed to date.

Tubiphytes Maslov 1956

Tubiphytes has been the subject of numerous papers, but thorough morphologic descriptions of this taxon are rare. Among the most useful descriptions are those in Maslov (1956; translated in Croneis and Toomey 1965), Flügel (1966), Rigby, (1958; a thorough description, but terminology specific to hydrozoan coelenterates was used throughout), Homan (1972), and Babcock (1974). A summary of the salient morphologic and nonmorphologic characteristics of *Tubiphytes* follows (see also Fig. 1):

Gross morphology: Cylindrical, elliptical, hemispherical, and irregularly ovoid masses that typically overlap and encrust each other as well as foreign hard substrates. Individual encrusting elements or units range from 0.3 to 1.0 mm in length and 0.5 to 1.0 mm in diameter. Encrusting aggregates are up to 7 mm in length.

Internal morphology and microstructure: In thin section two features stand out: (1) individual morphologic units of rounded, elongate, or irregular amoeboid shape delimited by dark micritic boundaries (walls), and (2) irregular cross sections of a central or axial smooth-walled tube. In addition, the general microfabric of the fossil consists of an irregular, discontinuous reticulate network of fine (1–2 μm) dark micritic sheets (lines in thin section).

Quality and mode of preservation: In reflected light it appears light (white); in transmitted light it appears dark. Preservation of Capitan specimens is excellent compared to associated fossils. Walls preserved as finely crystalline calcite.

Facies distribution and paleoenvironmental attributes: Occurs in abundance throughout the massive (reef) facies of the Capitan Limestone and irregularly in immediately adjacent facies. Typically found as benthic encrusting agents in skeletal boundstones; also occurs as encrusters in mixed skeletal–submarine cement boundstones. May be found free as skeletal grains, especially in outer-shelf grainstones adjacent to reef. Grew both surficially and in cryptic habitats.

Stratigraphic and geographic distribution: *Tubiphytes* ranges from the Carboniferous into the Jurassic (Flügel 1981a). It has been reported from western Europe, the Soviet Union, North Africa, the Middle East, Afghanistan, Burma, southern China, Central America, and the western United States.

As befits a truly enigmatic fossil, the systematic history of *Tubiphytes* has been marked by a wide variety of interpretations (Fig. 2). The most popular choices for the biotic affinities of *Tubiphytes* include blue-green alga and hydrozoan coelenterate, although it has also been assigned to the red algae and calcareous sponges. Recently, Flügel (1981a, b, 1983) has proposed that *Tubiphytes* may have been a consortium of green and blue-green algae.

Tubiphytes, though typically well preserved, has presented paleontologists with a difficult taxonomic–systematic problem for years. Why should this form be so controversial and so difficult to classify accurately? There are several possible explanations.

1. Critical morphologic features are missing, either because they were soft parts and were not preservable, or because diagenesis has severely modified them.

This explanation, however, is improbable because not only does *Tubiphytes* typically occur as well-preserved fossils, but *Tubiphytes* was also a benthic encruster that is found attached to original substrates and has filled some cavities in skeletal boundstones; there would have been little or no room for soft parts outside the preservable calcified portion of the fossil.

2. It is not analogous to the algae or hydrozoan coelenterates, the principal groups with which it has been compared. Therefore, comparisons have been unconvincing because they are inaccurate; or the comparisons have not been adequate evaluations based on the diagnostic features of living taxa.

This explanation may have some validity. Perhaps paleontologists have not compared *Tubiphytes* with the proper analog.

3. There is no other similar organism, either fossil or living (i.e., it may belong to an extinct phylum).

This explanation might be the best solution to the mystery of this well-preserved, widely distributed fossil of completely baffling pedigree.

C showing an oblique longitudinal section through a double-walled axial tube (a) and the irregular reticulate network of finely crystalline, dark micrite that forms the microstructure of *Tubiphytes*. Scale bar = 0.5 mm. E. Scanning electron micrograph showing double micrite-walled axial tube. Finely crystalline calcite of the *Tubiphytes* microstructure is visible surrounding the axial tube, while more coarsely crystalline calcite fills the cavity within the tube. Scale bar = 0.05 mm. All samples are from the Permian Capitan Limestone, Permian Reef Complex, Guadalupe Mountains, West Texas and New Mexico.

DATE	AUTHOR	INTERPRETATIONS						REMARKS
		ALGAE				HYDROZOAN	PROBLEMATICA	
		BLUE-GREEN	GREEN	RED	UNCERTAIN			
1951	Rauser–Chernoussova	(?)						Informal taxonomy. Proposed name *Shamovella*.
1953, 1955	Newell and others; Newell					o		First descriptions of form in Permian Reef Complex
1956	Maslov	o			x			Proposed name *Tubiphytes* with *T. obscurus* as type.
1958	Rigby					x		Proposed name *Nigriporella*.
1959	Konishi					o		Recognized synonymy of *Nigriporella* with *Tubiphytes*.
1965	Croneis & Toomey	(?)						
1966	Flügel			(?)	x			
1969	Toomey	o			x			
1972	Homann			o	x			
1974, 1977	Babcock						x	Problematica of zoologic affinities.
1980	Flügel & Flügel–Kahler	x			o			
1981, 1983	Flügel	o	o					Consortium hypothesis

Fig. 2. Highlights: Systematic history of *Tubiphytes*. **x,** Taxonomic opinion; **o,** formal taxonomic placement.

Apparently none of those who have discussed the affinities of *Tubiphytes* has examined the holotype of the type species *Tubiphytes obscurus,* including this writer. [However, the holotype seems to be missing from Maslov's collections at the Geological Institute in Moscow (P. L. Brenckle, personal communication, 1984).] In addition, apparently no one has compared critically the morphologic features of *Tubiphytes* with the diagnostic features of modern blue-green, green, or red algae, or of hydrozoan coelenterates. Therefore, the definitive work on *Tubiphytes* remains to be done. Among the questions that might be asked in that analysis are the following:

1. How does the morphology of *Tubiphytes* compare with that of the living algae or of the hydrozoan coelenterates? Are the basic body plan and fundamental morphologic features analogous with living algae? Is the body plan filamentous or cellular?

2. Are the sizes of body parts in *Tubiphytes* compatible with the sizes of analogous(?) parts in the living algae, or of hydrozoan coelenterates?

3. Are the microstructural elements and mineralogy of *Tubiphytes* analogous with living algae or hydrozoan coelenterates?

4. What are the outstanding morphologic attributes of *Tubiphytes?*

A number of features come to mind in answer to the last question. *Tubiphytes* has a very dark appearance in thin section and is very light (white) in reflected light; it has a distinctive rounded, amoeboid shape and is characterized by an overgrowing, overlapping habit that created 5- to 10-mm-long aggregate encrustations (Fig. 1); its fundamental morphologic units exhibit a consistent size range and internal features; its internal tissue has a dense, rather unstructured appearance in thin section; and it is typically found to be well preserved. Further, *Tubiphytes* apparently is not filamentous or cellular; it does not form simple irregular encrusting masses as are seen in *Girvanella* encrustations; it does not form nodular clumps of hollow filaments as in the *Ortonella-Cayeuxia* complex; and it is not chambered, coccoid, or tubular (except for its axial tube).

Taken altogether it seems that *Tubiphytes* richly deserves the taxonomic assignment of problematicum.

Archaeolithoporella Endo 1959

Archaeolithoporella has received much less attention than *Tubiphytes* for two reasons: first, it is a rather simple, small crustiform entity that frequently has been identified as a microstromatolite of uncertain type, if identified at all; and second, it is apparently restricted to Permian carbonates and therefore is more limited, stratigraphically, than *Tubiphytes*. Besides Endo's original description of *Archaeolithoporella* in 1959, additional ones are found in Flügel (1981a) and Babcock (1974, 1977). Discussions of *Archaeolithoporella's* affinities can also be found in Cronoble (1974), Mazzullo and Cys (1978), and Pisera and Zawidzka (1981). The critical features of *Archaeolithoporella* include the following (see also Fig. 3):

Gross morphology: Encrusting laminated habit. Laminae consist of alternating light and dark curving crusts. Doublets of light and dark laminae encrust one another, forming encrustations several millimeters thick. Dimensions of individual laminae: dark 15–20 μm, light 15–30 μm.

Internal morphology and microstructure: No apparent internal differentiation into cells or filaments. Two possible reproductive structures, without pores, were observed in Capitan specimens.

Quality and mode of preservation: Dark laminae preserved as microcrystalline calcite; light laminae as finely crystalline calcite. No internal structures, ornamentation, or other similar features have been observed. Quality of preservation is generally fair, but this form is quite susceptible to diagenetic alteration. It appears as light, wispy laminae in reflected light; as dark micritic laminae in transmitted light.

Facies distribution and environmental attributes: Like *Tubiphytes*, *Archaeolithoporella* is widely distributed in the massive (reef) facies of the Capitan Limestone and may extend into immediately adjacent shelf and foreslope facies. Also like *Tubiphytes*, *Archaeolithoporella* is a conspicuous encrusting agent in skeletal and mixed skeletal–submarine cement boundstones in the Capitan, and it also grew on the seafloor or in cavities. *Archaeolithoporella* also can be found lining submarine fissure fillings in the upper Capitan Limestone.

Stratigraphic and geographic distribution: *Archaeolithoporella* is apparently known only from Permian rocks, although it may be longer ranging. Endo's genus has been reported from Japan, the southwestern United States, the Philippines, central Europe, and the Northern Caucasus.

Endo (1959) erected the genus *Archaeolithoporella* for laminated structures found in the Shiroi Formation of Japan. Although Endo (1959) reported no tissue differentiation or cross partitions (which would indicate the presence of cells), he believed that this form was a fossil coralline red alga similar to the modern genus *Lithoporella*. Unlike *Tubiphytes,* there have been few formal descriptions or systematic discussions of *Archaeolithoporella* since 1959. This is probably because *Archaeolithoporella* typically occurs as a simple laminated structure that many observers have described as "stromatolitic." Simplicity seems to be

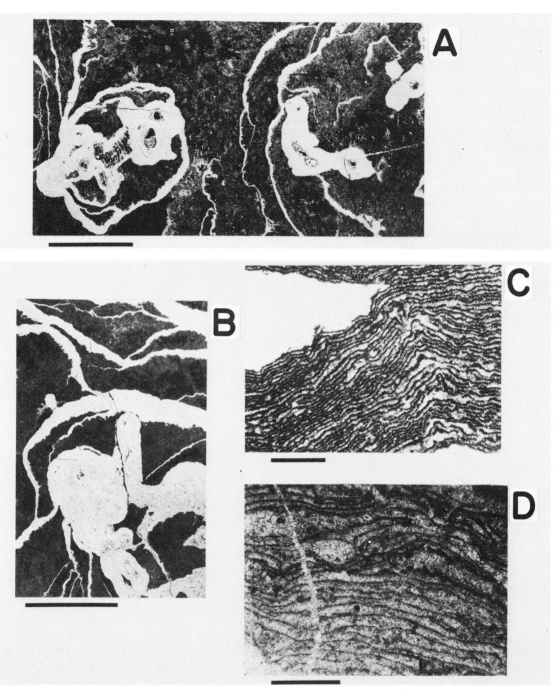

Fig. 3. A. Encrustations of multiple laminae of *Archaeolithoporella* Endo. Heaviest encrustations are on biotic nuclei, while thinner ones alternate with dark fibrous calcite cement crusts of submarine origin. *Archaeolithoporella* appears light because photograph is a thin-section negative print. Scale bar = 5 mm. B. Negative print as in A but at higher magnification. In this image, individual alternating dark and light laminae of the *Archaeolithoporella* entity can be seen. The characteristic multiple encrustations of *Archaeolithoporella* laminae are obvious, and the relationship between *Archaeolithoporella* and submarine cement crusts is clearly seen. Scale bar = 5 mm. C. Photomicrograph showing details of *Archaeolithoporella* including alternating light and dark laminae, the consistent size and shape of laminae, and the discontinuous, undulating nature of the form. Scale bar = 0.5 mm. D. Photomicrograph of holotype of *Archaeolithoporella hidensis* Endo 1959. Scale bar = 5 mm. All samples from the Permian Capitan Limestone, Permian Reef Complex, Guadalupe Mountains, West Texas and New Mexico (except D).

the crux of the *Archaeolithoporella* problem. It seems ironic that a form that is so simple—just alternating dark and light laminae, 10 or 15 μm thick—should possess such an imposing name. Because *Archaeolithoporella* is lacking diagnostic morphologic features, such as cells, and is so simple morphologically, it is problematic that it is an organism, let alone an alga (see Babcock 1974, 1977; Mazzullo and Cys 1978). However, if *Archaeolithoporella* is an organism, then the probability that it is a form of red alga, perhaps even a crustose coralline alga, is high. Why? Because of its gross morphology, the size and shape of its alternating laminations, its encrusting, overlapping habit, its restriction to reef facies, and its role in constructing boundstone (reef) frameworks.

The lack of cells in *Archaeolithoporella* is a critical problem, because the dark micritic laminae of this form are impossible to differentiate from diagenetic dust layers or micritic cements if only a single lamina is present (see Babcock et al. 1977; Schmidt 1977; Mazzullo and Cys 1978). However, it may be reassuring to know that early diagenetic destruction of cross partitions and other internal structures is apparently not unusual in calcareous red algae (Newell et al. 1953; Wolf 1965; Siesser 1972).

Archaeolithoporella was probably a red alga and may even be ancestral to modern monostromatic coralline red algae such as *Heteroderma, Melobesia, Fosliella,* or *Lithoporella*. However, until cells can be documented in this form, *Archaeolithoporella* should remain assigned to the problematica.

APPROACHES TO CLEANING UP THE ALGAL WASTEBASKET MESS

It should be clear that many odd little Paleozoic fossils and even some inorganic entities continue to be assigned to various algal taxa even though they probably should be classified as problematica. The controversies surrounding *Tubiphytes* and *Archaeolithoporella* illustrate some of the specific factors that contribute to this algal taxonomic wastebasket problem. However, the act of defining the problem and illustrating it with examples does not solve it. If the problem is constituted of factors that relate to the nature of the fossils themselves, to the types of samples that are used, and to the nature of our training and prejudices, then what can be done?

Recognition of the present situation and of the need to improve it must come before any progress can be made. Braithwaite's admonition (1973) to those involved in reef controversies seems appropriate here: "Unless geologists provide adequate descriptions, *followed* by interpretations, we can look forward only to further years of dispute." Recognition of the inherent characteristics and problems associated with calcereous algae, with thin-section samples, and with those who study the algae is the best way to begin a reexamination of these forms.

Recognize the nature of the algae

Calcareous algal fossils are not as well known to paleontologists or sedimentary geologists as are invertebrate animal fossils such as brachiopods, bryozoans, or echinoderms. Therefore, the identification and classification of calcareous algae are made typically without an adequate knowledge of algal morphology or systematics.

As was pointed out in an earlier section, the algae include a large and extremely diverse group of organisms that belong to several plant divisions and to more than one kingdom. In addition, many living algae are relatively simple morphologically, as compared with higher plants. However, their taxonomy is far from simple, and it is as well defined as that of higher plants or of invertebrate animals. So, dumping an odd-looking fossil into an algal classification is not acceptable from the standpoint of botanically educated algal specialists. Although there are major controversies concerning modern algal taxonomy, and there are some major gaps in our knowledge, the living algae still must be the benchmarks by which fossil unknowns thought to be algal are judged. Riding (1977b) made this point clearly when he stated that "criteria for recognition must derive from algal morphology in general and the morphology of Recent calcareous algae in particular. That such an obvious statement should have to be made reflects on the present state of systematic work on these fossils."

Unfortunately, the algologists have made the goal of a unified fossil-living algal taxonomy less and less attainable. Although the botanists have not emphasized nonpreservable criteria for algal identification and classification just to create problems for paleontologists, the result is the same. Fossil algae workers have a difficult time observing pigmentation, food storage products, cytologic and developmental characteristics, or any of the other nonpreservable features in fossil material, that algologists consider diagnostic in living algae. Adding to this problem is the disagreement among algologists about the relative diagnostic value of different nonpreservable characteristics.

Assuming that the taxonomy of the algae is a bit messy, and that botanists are not agreed on either a classification or the criteria for identification of algae, then what should the paleoalgologists do? To understand the nature and significance of calcareous algae, both fossil and living forms, they should consult Wray's book, *Calcareous Algae* (1977), and to become more familiar with living algae they should read a good general introductory treatment such as Morris's *An Intro-*

duction to the Algae (1968) or Doyle's *Nonseed Plants: Form and Function* (1970).

The specialist requires a more substantial background and should become familiar with the publications that have resulted from the international fossil algae symposia. These compilations contain short papers on a great variety of calcareous algal topics by paleontologists and geologists from all over the world, and include Flügel (1977), Poignant and Deloffre (1979), and Toomey and Nitecki (1985).

For a better grounding in the biology of the algae, the specialist should read a modern text such as Bold and Wynne (1985) or Trainor (1978). Some older books (e.g., Fritsch 1945; Smith 1950, 1951) are valuable to fossil algae specialists because their organization is taxonomic and they emphasize vegetative morphology to a greater degree than do most recent texts. Dawson (1966) provides a valuable introduction to marine algae including the calcareous algae.

Choosing a classification system is a subjective exercise and is not as critical as is the consistent use of a single system. The classifications of Papenfuss (1955), Christensen (1964), and Levring (1969) are among the most popular general algal classifications. But for criteria for the identification of specific fossil algae, the specialist must go to the more detailed algal literature.

Blue-Green Algae or Cyanobacteria

These are prokaryotic organisms with mucopeptide walls that possess only chlorophyll *a*. The blue-green algae are morphologically simple and typically form unflagellated unicells or filamentous forms, either branched or unbranched. The iconoclastic taxonomy of Drouet (1968, 1973) has resulted in a monumental lumping of low-level taxa. Drouet's fundamental taxonomic concept (see Forest and Khan 1972) is that blue-green algae are highly polymorphic and that morphologic variants tend to be ecologically induced. This concept has not gone unchallenged (Monty 1967; Golubic 1969), and it flies in the face of the time-honored practice of assigning species of fossil algae on the basis of minor size variations in the filaments (trichomes). However, Drouet's taxonomy is compatible with current thinking about blue-green algal polymorphism and evolutionary conservatism. In addition, important studies of fossil blue-green algae (e.g., Schopf and Blacic 1971) have used this classification successfully. Among the critical diagnostic morphologic features of genera and species of blue-green algae, as defined by Drouet (1973), are (1) shape of the trichome, (2) shape and character of terminal cells, (3) nature of branching, (4) shape and location of heterocysts or akinetes, and (5) attenuation of the filament. Obviously, only well-preserved fossil material would possess such morphologic features

(e.g., Schopf and Blacic 1971; Riding 1977a). It was because of the problem of preservation of these diagnostic features that Pia (1927) established the group Porostromata for tubular microfossils that he thought might have been blue-green algae (see Wray 1977; Riding 1977b).

Green Algae

These are eukaryotic, chlorophyll-bearing plants that include two important families of calcareous algae: the codiaceans and the dasycladaceans. Both of these families belong to the green algal complex of siphonous chlorophytes. The siphonaceous algae are macroscopic and coenocytic. There are no cross walls in the vegetative thallus (plant body). This means that these algae consist of hollow tubes in life and, depending on the vagaries of calcification and fossilization, they could be preserved as fossils that would be difficult to differentiate from any other tubular form, including blue-green algae that have lost their cross walls.

Luckily, the Dasycladaceae are usually found as distinctive forms that consist of cylinders perforated by radially symmetric pores. Dasycladaceans are among the best known and least problematic of the fossil calcareous algae, although their species can be difficult to differentiate. Valet's monographic treatment (1968, 1969) of the morphology, cytology, reproduction, and systematics of living dasycladaceans is the standard reference for the biology of this group. The diagnostic characteristics for the identification of living dasycladaceans include (1) nature of branching system, (2) nature and position of reproductive structures, (3) degree and distribution of calcification, (4) size and shape of thallus, and (5) number of whorls (branching levels).

Living codiacean green algae consist of a morphologically diverse group of erect, benthic marine plants. The internal morphology of these plants typically consists of a central network of relatively large, loosely interwoven tubular filaments (medulla) surrounded by a peripheral meshwork of smaller branching filaments (cortex). Unlike the dasycladacean green algae, codiacean fossils are subject to varying identifications and interpretations, as illustrated by the controversy over the codiacean affinities of a group of encrusting fossil genera including *Ortonella, Cayeuxia,* and *Hedstroemia* (Riding 1975; Wray 1977). Fossil algal specialists interested in the codiacean algae should become familiar with Hillis-Colinvaux (1980), which monograph includes information on most of the living calcareous codiacean algae. Among the vegetative morphologic features thought to be diagnostic in the identification of codiacean algae are (1) general character of the thallus, (2) character of the blade (including siphonaceous filament diameter), (3) presence (or absence) and nature of cortex, (4) presence or absence of segmentation, and (5) shape of utricles.

Red Algae

Modern red algae are almost exclusively marine and tend to be more complex than the green algae in terms of their morphology and reproductive cycles. The Corallinaceae are generally very well calcified, so that they are typically better preserved in fossil form than most other calcareous algae. In addition, the coralline red algae are geologically significant, because they encrust hard substrates and are the glue that binds together coral reefs to form wave-resistant ramparts.

Coralline red algae have received more attention from algologists than other calcareous algae, and numerous fossil taxa have been assigned to this group through the years. Besides the comprehensive treatment of the coralline algae in Johansen (1981), two other important contributions are Adey and Johansen (1972) and Adey and MacIntyre (1973). Perhaps because there is a good deal of interest in this group, the taxonomy of the coralline algae is even more chaotic than many other calcareous algal groups (Johansen 1981; Adey 1965). One of the current algological taxonomic controversies concerns which feature is of greatest significance for subdividing the Corallinaceae: secondary cellular pit connections or the presence (or absence) of genicula (uncalcified flexible tissue separating calcified segments). Further, there is disagreement concerning the most natural suprageneric classification scheme for the corallines. Algologists do agree on one point—the archaic nature of the classification system in which the Corallinaceae were subdivided into just two subfamilies, the Melobesiodeae or crustose corallines and the Corallinoideae or articulated corallines. Archaic or not, this subdivision remains the best suprageneric classification system for fossil corallines.

As in other algal groups, the dichotomy between the taxonomy of living and fossil coralline algal forms is large and growing. Algologists recognize the futility of trying to apply many of the diagnostic, but nonpreservable, criteria for the identification of living algae to fossil algae, but they continue to criticize paleoalgologists for continuing to use only preservable morphologic features in their identifications (e.g., Cabioch 1972; Johansen 1981).

The following morphologic features are being used to characterize the living genera of red algae (Johansen 1981): (1) habitat and substrate (variable fossil recognition), (2) nature of hypothallus, (3) nature of perithallus, (4) nature of epithallus (nonpreservable), (5) character of male conceptacles (nonpreservable?), (6) nature of megacells, (7) character of branching (articulated forms), (8) na-

ture of intergenicular medulla (articulated forms), and (9) nature of fusion cells (nonpreservable).

Recognize the nature of the samples

Two factors distinguish calcareous algal fossil samples from most other fossils and create some problems for paleoalgologists not faced by most other paleontologists. Since fossil calcareous algae are studied in petrographic thin sections, they appear as two-dimensional random cuts. In addition, the fossil algae occur as one of many different components in carbonate depositional fabrics.

Random, two-dimensional cuts of fossils that are sinuous, branching tubes or irregular cellular mats or straight cylinders result in a bewildering array of shapes and sizes. The job of the specialist is to reconstruct the full three-dimensional appearance of the fossil from the two-dimensional sample, as well as to recognize that a variety of shapes and sizes may result from intersecting a single three-dimensional form with a plane at different angles. This phenomenon has resulted in the creation of many synonymous taxa and in the misidentification of a variety of fossil organisms.

The other sample factor that affects the manner in which calcareous algal fossils are observed and perceived is the rock context. Fossil algae are typically found in carbonate lithologies and are identified from petrographic thin sections of these lithologies. This fact presents both advantages and problems. It ensures that the geologic framework of the fossil is readily available for observation, but it also means that the fossil cannot be isolated from the other carbonate components easily. This intimate association may lead to the misidentification of any unfamiliar component as algal ("algal by default") or to the confusion of organic, perhaps algal, components with inorganic components such as cements or other diagenetic products.

Recognize the nature of the training and prejudices of the observer-interpreter

The algae are being used as a taxonomic wastebasket for the classification of unknown, equivocal, or odd-looking fossils. This has created a problem because calcareous algae are geologically and paleontologically important fossils. Incorrect assignment of nonalgal forms to the algae confuses algal systematics and decreases the value of true algae for paleoenvironmental interpretations by making classifications suspect. This situation could be improved dramatically if paleontologists and geologists who work with fossil calcareous algae would be more careful in applying the algal label to fossils of questionable affinities.

Another important step forward would be the increased use of the problematic category instead of the algal classification. A useful guideline is: if in doubt, label it problematic (see Wray 1977). Finally, it would be most helpful if everyone who deals with Paleozoic problematica of potential algal affinities would provide detailed information on the nonmorphologic characteristics of the unknowns being studied (e.g., stratigraphic range, facies distribution, associated components, and diagenetic aspects).

In order to recognize and improve the nature of observations and interpretations of problematica thought to be algal, a flowchart (Fig. 4) is provided.

Step 1:

"Describe morphology." This is the fundamental step for evaluating any unknown, whether the unknown has never been observed before by anyone or is a much-discussed species whose affinities remain enigmatic. The morphologic description should include gross features (habit and gross morphology), internal structure, and microstructure (wall character, etc.). Nonspecialists can describe morphology as well as specialists, because basic geometric terminology is required, not specialized nomenclature that may be appropriate only for a single taxonomic category. The specialist should compare the unknown with the holotype specimen of the type species, if the unknown has been identified as a fossil taxon. This is necessary to ensure that the correct taxon is being studied and compared.

Step 2:

"Compare unknown with morphology of living analogs." Whether a thorough morphologic analysis supports an organic nature for the unknown or not, it is necessary to compare it with the diagnostic morphologic features of living forms that might be analogous. This step should be taken by specialists, because they are familiar with one or more taxonomic groups and with the procedures required to evaluate the unknown. Perhaps the best approach would be to make the assumption that the unknown does not belong to a specific taxonomic category and then to try to accumulate the evidence that documents this assumption. If careful objective evaluations are made in step 2, then the question, "Is unknown algal?" will be made much easier to answer accurately.

Step 3:

"List/evaluate further evidence to make decision." If comparative morphologic studies fail to document the algal nature of the unknown, then it is necessary to gather circumstantial evidence. This kind of documentation might indicate whether an algal classification was compatible with other characteristics of the unknown, such as (1) comparison of morphology of unknown with extinct fossil forms thought to be of algal affinities, (2) comparisons of morphology of unknown with nonalgal taxa that are potentially analogous, (3) geologic characteristics including the stratigraphic range, facies distribution, and nature and quality of preservation, (4) relict or ghost morphologic features, and (5) relationships with associated components (in carbonate lithologies).

Step 4:

"List/evaluate further evidence required to make decision." Although the command in step 4 is the same as that in step 3, the evidence required to make a decision about the organic affinities of an unknown are somewhat different from that needed to assign an organism to the algae. In fact, documenting the organic affinities of a morphologically simple fossil problematicum is one of the most difficult tasks that a paleontologist can face. What criteria are diagnostic of life in an inanimate form now preserved in a fragment of limestone? This is the dilemma facing those of us who work with Permian carbonate buildups containing the laminated form *Archaeolithoporella*. When multiple layers of the dark–light laminated couplets are found encrusting a sponge or bryozoan, then the organic, and probably red algal, nature of *Archaeolithoporella* appears to be irrefutable. However, when there is merely a single dark lamination, then it becomes impossible to differentiate *Archaeolithoporella* from a micritic cement or a diagenetic dust film that marks a hiatus in cementation or other diagenetic process.

Other problematic forms of questionable organic affinities also require further documentation and objective analysis (e.g., *Stromatactis* and *Microcodium*). Some of the nonmorphologic characteristics that could help to determine their organic nature include the following:

1. *Distribution in time and space.* Stratigraphic distribution—restricted or not? Facies distribution, geographic distribution.

2. *Spatial–geometric relationships.* Does the unknown form obey or defy natural forces such as gravity, capillarity, crystallization, or hydrodynamics? For example, a single layer of isopachous inorganic cement will entirely line a pore with a crystalline crust that is of approximately equal thickness everywhere. Succeeding cement layers will be of different thicknesses. On the other hand, *Archaeolithoporella* encrusts organic frameworks that may form constructional cavities, but the crusts are discontinuous—they do not continuously line the cavities. In addition, the *Archaeolithoporella* crusts vary little in thickness, no matter how many layers thick. The spatial–geometric relationships exhibited by *Archaeolithoporella* are compatible with those of an organism, not of a cement.

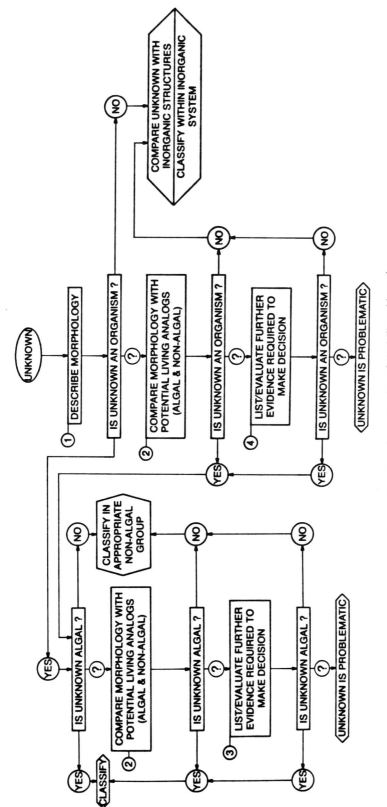

Fig. 4. Flowchart for evaluation of algal (?) problematica.

23

3. *Relationships with associated compo-nents.* Are there recurrent associations? What is the order of occurrence of the components?

CONCLUSIONS

Fossil taxa of questionable affinities found in the Capitan Limestone of the Permian Reef Complex offer a unique opportunity to observe and analyze a variety of problematica, most of which have been assigned previously to the algae—one of the great taxonomic wastebaskets. The engimatic forms of the Capitan provide concrete examples that illustrate the puzzle of algalike problematica and the sources of the algal taxonomic wastebasket problem.

By carefully examining and objectively describing morphologic and nonmorphologic characteristics, the affinities of equivocal fossils, including those that are truly problematic, can be evaluated more accurately. The proper classification of equivocal fossils thought to be algal requires that more attention be paid to the diagnostic morphologic criteria found in living algae and as defined in the botanic literature. Analogy with modern forms is still the accepted technique for the assignment of biologic affinities to fossils, so it is critical that paleontologists who deal with fossil algae be familiar with the rudiments of algal botany and with the pertinent algal literature.

Utilitarian classification systems are valid and should continue to be employed by specialists and nonspecialists alike. However, much of the confusion that has grown in the field of fossil algae is a result of the use of informal taxa in ways that should be restricted to more formal systematic classifications. The scientific study of modern calcareous algae is in its youth, and some of the questions posed by apparently unusual alga-like fossils may yet be answered. In the meantime, the best approach to the study of alga-like problematic fossils is to describe them thoroughly and carefully, including their paleoenvironmental setting, and to separate clearly interpretations from observations—exactly as all good systematic analyses are done.

Acknowledgments

I am indebted to Jack Wray and Bill Woelkerling for providing me with the opportunity to learn about both fossil and living algae. Jack Wray's and Rob Riding's thoughtful essays and discussions about critical problems in the systematics of fossil algae provided the inspiration for this paper. I thank Allen Ormiston, Paul Brenckle, and Laurel Babcock for reviewing the manuscript, and Amoco Production Company for permission to publish it.

REFERENCES

Adey, W. H. 1965. The Lithophylleae. *In* Johnson, J. H. and Adey, W. H. (eds.). Studies of *Lithophyllum* and related algal genera. *Q. Colo. School Mines* **60,** 71–102.

———— and Johansen, J. W. 1972. Morphology and taxonomy of Corallinaceae with special reference to *Clathromorphum, Mesophyllum* and *Neopolyporolithon* gen. nov. (Rhodophyceae, Cryptonemiales). *Phycologia* **11,** 159–180.

———— and MacIntyre, I. 1973. Crustose coralline algae: A reevaluation in the geological sciences. *Geol. Soc. Am. Bull.* **84,** 883–904.

Babcock, J. A. 1974. The role of algae in the formation of the Capitan Limestone (Permian, Guadalupian), Guadalupe Mountains, West Texas–New Mexico. Unpubl. Ph.D. dissert., Univ. of Wisconsin, Madison, 241 pp.

————. 1977. Calcareous algae, organic boundstones, and the genesis of the Upper Capitan Limestone (Permian, Guadalupian) Guadalupe Mountains, West Texas and New Mexico. *In* Hileman, M. E. and Mazzullo, S. J. (eds.). *Upper Guadalupian Facies, Permian Reef Complex, Guadalupe Mountains, New Mexico and West Texas,* vol. 1. Soc. Econ. Paleont. Mineral., Permian Basin Sect., 77-16, pp. 3–44.

————. 1979. Calcareous algae and algal problematica of the Capitan Reef (Permian), Guadalupe Mountains, West Texas and New Mexico, U.S.A. *Bull. Cent. Rech. Explor.-Prod. Elf-Aquitaine* **3**(2), 419–428.

————, Pray, L. C., and Yurewicz, D. A. 1977. Upper Capitan—massive, mouth of Walnut Canyon. Stop I. *In* Pray, L. C. and Esteban, M. (eds.). *Road Logs and Locality Guides,* vol. 2, *Upper Guadalupian Facies, Permian Reef Complex, Guadalupe Mountains, New Mexico and West Texas.* Soc. Econ. Paleont. Mineral., Permian Basin Sect. 77-16, pp. G17–G39.

Bold, H. C. and Wynne, M. 1985. *Introduction to the Algae.* 2nd ed. Prentice-Hall, Englewood Cliffs, N.J., 848 pp.

Braithwaite, C. J. R. 1973. Reefs: just a problem of semantics? *Bull. Am. Assoc. Petrol. Geol.* **57,** 1100–1116.

Cabioch, J. 1972. Etude sur les Corallinacées. II. La Morphogenèse; Consequences Systématiques et Phylogénetiques. *Cahiers Biol. Mar.* **13,** 137–288.

Christensen, T. 1964. The gross classification of algae. *In* Jackson, D. F. (ed.). *Algae and Man.* Plenum, New York, pp. 59–64.

Croneis, C. and Toomey, D. F. 1965. Gunsight (Virgilian) wewokellid sponges and their depositional environment. *J. Paleont.* **39,** 1–16.

Cronoble, J. M. 1974. Biotic constituents and origin of facies in Capitan reef, New Mexico and Texas. *Mountain Geol.* **11,** 95–108.

Dawson, E. Y. 1966. *Marine Botany: An Introduction.* Holt, Rinehart and Winston, New York, 371 pp.

Doyle, W. T. 1970. *Nonseed Plants: Form and Function.* 2nd ed. Wadsworth, Belmont, Calif., 240 pp.

Drouet, F. 1968. Revision of the classification of the Oscillatoriaceae. *Acad. Nat. Sci. Phila. Monogr.* **15**, 370 pp.

————. 1973. *Revision of the Nostocaceae with Cylindrical Trichomes.* Hafner, New York, 292 pp.

Endo, R. 1959. Stratigraphical and paleontological studies of the later Paleozoic calcareous algae in Japan, XIV: Fossil algae from the Nyugawa Valley in the Hida Massif. *Saitama Univ. Sci. Rep., Ser. B* **3**, 177–217.

Flügel, E. 1966. Algen aus dem Perm der Karnischen Alpen. *Carinthia II* **25**, 3–76.

————. (ed.). 1977. *Fossil Algae: Recent Results and Developments.* Springer, Berlin. 375 pp.

————. 1981a. *Tubiphytes—Archaeolithoporella* buildups in the southern Alps (Austria and Italy). *In* Toomey, D. F. (ed.). *European Fossil Reef Models.* Soc. Econ. Paleont. Mineral., Spec. Pub. 30, pp. 143–160.

————. 1981b. "Tubiphyten" aus dem Frankischen Malm. *Geol. Bl. NO-Bayern* **31**, 126–142.

————. 1983. *Tubiphytes:* Paleoecology and interpretation of an enigmatic alga/algal consortium from Paleozoic and Mesozoic carbonates (Abstr.). *3rd Int. Fossil Algae Symp., (Golden, Colo.,) Progr. Abstr.*, p. 10.

———— and Flügel-Kahler, E., 1980. Algen aus den Kalken der Trogkofel-Schichten der Karnischen Alpen. *Carinthia II* **36**, 113–182.

Forest, H. S. and Khan, K. R. 1972. The blue-green algae: a program of evaluation of Francis Drouet's taxonomy. *In* Desikachary, T. V. (ed.). *Taxonomy and Biology of Blue-Green Algae.* Univ. of Madras, Madras, India, pp. 128–138.

Fritsch, F. E. 1945. *The Structure and Reproduction of the Algae.* vols. 1 and 2. Cambridge Univ. Press, Cambridge, 791 pp. (vol. 1); 939 pp. (vol. 2).

Golubic, S. 1969. Tradition and revision in the system of the Cyanophyta. *Verh. Int. Verein. Limnolog.* **17**, 752–756.

Gould, S. J. 1983. Nature's great experiments. *Nat. Hist.* July, pp. 12–22.

————. 1984. The Ediacaran experiment. *Nat. Hist.* February, pp. 14–23.

Hillis-Colinvaux, L. 1980. Ecology and taxonomy of *Halimeda:* primary producer of coral reefs. *Adv. Mar. Biol.* **17**, 1–327.

Homan, W. 1972. Unter und tief-mittel permische Kalkalgen aus den Rattendorfer-Schichten, dem Trogkofel-Kalk und dem Tress-

dorfer Kalk der Karnischen Alpen (Österreich). *Senckenb. Leth.* **53**, 135–313.

Johansen, J. W. 1981. *Coralline Algae, A First Synthesis.* CRC Press, Boca Raton, Fla., 239 pp.

Johnson, J. H. 1942. Permian lime-secreting algae from the Guadalupe Mountains, New Mexico. *Geol. Soc. Am. Bull.* **53**, 195–216.

————. 1951. Permian calcareous algae from the Apache Mountains, Texas. *J. Paleont.* **25**, 21–30.

————. 1961. *Limestone-Building Algae and Algal Limestones.* Colorado School of Mines, Boulder. 297 pp.

Klement, K. H. 1966. Studies on the ecological distribution of lime-secreting and sediment-trapping algae in reefs and associated environments. *N. Jb. Geol. Paläont. Abh.* **125**, 363–381.

Konishi, K. 1959. Identity of alga *Tubiphytes* Maslov, 1956, and hydrozoan genus *Nigriporella* Rigby, 1958. *Trans. Proc. Palaeont. Soc. Jpn.* **35**, 142.

Levring, T. 1969. Classification of the algae. *In* Levring, T., Hoppe, H. A., and Schmid, O. J. (eds.). *Marine Algae: A Survey of Research and Utilization.* Cram, de Gruyter and Co., Hamburg, pp. 47–125.

Maslov, V. P. 1956. Fossil calcareous algae of the U.S.S.R. *Trudy Geol. Inst. AN SSSR* **160**, 301 pp.

Mazzullo, S. J. and Cys, J. M. 1978. *Archaeolithoporella*—boundstones and marine aragonite cements, Permian Capitan reef, New Mexico and Texas, U.S.A. *N. Jb. Geol. Paläont. Mh.* **1978**, pp. 600–611.

Monty, C. L. V. 1967. Distribution and structure of recent stromatolitic algal mats, eastern Andros Island, Bahamas. *Soc. Geol. Belg. Ann.* **90**, 57–100.

Morris, I. 1968. *An Introduction to the Algae.* 2nd ed. Hutchinson Univ. Library, London, 189 pp.

Newell, N. D. 1955. Depositional fabric in Permian Reef limestones. *J. Geol.* **63**, 301–309.

————, Rigby, J. K., Fischer, A. G., Whiteman, A. J., Hickox, J. E., and Bradley, J. S. 1953. *The Permian Reef Complex of the Guadalupe Mountains Region, Texas and New Mexico, A study in Paleoecology.* Freeman, San Francisco, 236 pp.

Papenfuss, G. F. 1955. Classification of the algae. *In* Kessel, E. D. (ed.). *A Century of Progress in the Natural Sciences, 1853–1953.* California Academy of Sciences, San Francisco, pp. 115–224.

Pia, E. 1927. Thallophyta. *In* Hirmer, M. (ed.). *Handbuch der Paläobotanik*, vol. 1. Oldenburg, München, pp. 31–136.

Pisera, A. and Zawidzka, K. 1981. *Archaeolithoporella* from the Upper Permian reef limestones of the northern Caucasus. *Bull. Acad. Polon. Sci., Ser. Sci. Geol. Geogr.* **29**, 233–238.

Poignant, A. F. and Deloffre, R. (eds.). 1979. Deuxième Symposium International sur les Algues Fossiles. *Bull. Cent. Rech. Explor.-Prod. Elf-Aquitaine (Pau)* **3**(2), 385–885.

Rauser-Chernoussova, D. M., in Elias, M. K. 1959. Facies of Upper Carboniferous and Artinskian deposits in the Sterlitamak–Ishimbaevo region of the pre-Urals based on a study of fusulinids. *Int. Geol. Rev.* **1**(2), 39–88.

Riding, R. L. 1975. *Girvanella* and other algae as depth indicators. *Lethaia* **8**, 173–179.

––––––. 1977a. Calcified *Plectonema* (bluegreen algae), a recent example of *Girvanella* from Aldabra atoll. *Palaeontology* **20**, 33–46.

––––––. 1977b. Problems of affinity in Paleozoic calcareous algae. *In* Flügel, E. (ed.). *Fossil Algae.* Springer, Berlin, pp. 202–211.

Rigby, J. K. 1957. Relationships between *Acanthocladia guadalupensis* and *Solenopora texana* and the bryozoan–algal consortium hypothesis. *J. Paleont.* **31**, 603–606.

––––––. 1958. Two new Upper Paleozoic hydrozoans. *J. Paleont.* **32**, 583–586.

Schmidt, V. 1977. Inorganic and organic reef growth and subsequent diagenesis in the Permian Capitan Reef Complex, Guadalupe Mountains, Texas, New Mexico. *In* Hileman, M. E. and Mazzullo, S. J. (eds.). *Upper Guadalupian Facies, Permian Reef Complex, Guadalupe Mountains, New Mexico and West Texas,* vol. 1. Soc. Econ. Paleont. Mineral., Permian Basin Sect. 77-16, pp. 93–131.

Schopf, J. W. and Blacic, J. M. 1971. New microorganisms from the Bitter Springs Formation (Late Precambrian) of the north-central Amadeus Basin, Australia, *J. Paleont.* **45**, 925–960.

Siesser, W. G. 1972. Relict algal nodules (Rhodolites) from the South African continental shelf. *J. Geol.* **80,** 611–616.

Smith, G. M. 1950. *The Freshwater Algae of the United States.* 2nd ed. McGraw-Hill, New York, 719 pp.

––––––. 1951. *Manual of Phycology.* Chronica Botanica Co., Waltham, Mass., 375 pp.

Tappan, H. 1980. *The Paleobiology of Plant Protists.* Freeman, San Francisco, 1028 pp. (chap. 3, Acritarcha or Hystrichophyta).

Toomey, D. F. 1969. The biota of the Pennsylvanian (Virgilian) Leavenworth limestone, Midcontinent region, part 2, distribution of the algae. *J. Paleont.* **43**, 1313–1330.

–––––– and Nitecki, M. H. (eds.). 1985. *Paleoalgology: Contemporary Research and Applications.* Springer, Berlin.

Trainor, F. R. 1978. *Introductory Phycology.* Wiley, New York.

Valet, G. 1968. Contribution a l'etude des Dasycladales: morphogenesis. *Nova Hedwigia* **16,** 21–82.

––––––. 1969. Contribution a l'etude des Dasycladales: Cytologie et reproduction, revision systematique. *Nova Hedwigia* **17**, 551–644.

Wolf, K. H. 1965. "Grain-diminution" of algal colonies to micrite. *J. Sed. Petrol.* **35,** 420–427.

Wray, J. L. 1977. *Calcareous Algae.* Elsevier, Amsterdam, 185 pp.

3. RECEPTACULITIDS AND THEIR RELATIONSHIP TO OTHER PROBLEMATIC FOSSILS

MATTHEW H. NITECKI

It is characteristic of human nature that the unknown and the uncertain attract curiosity. In paleontology, no organisms are more problematic than receptaculitids. Receptaculitids are those organisms included in the order Receptaculitales James 1885, as used by Nitecki and Toomey (1979) and Fisher and Nitecki (1982a). Receptaculitids are cosmopolitan fossils found almost exclusively in carbonate rocks, frequently forming a major component of organic buildups and found on all continents except Antarctica. They range from Lower Ordovician to Permian; however, Upper Devonian receptaculitids are not common, and Carboniferous and Permian forms are known from single localities only.

The receptaculitid skeleton (Figs. 1 and 2) is structurally simple and is composed of meroms arranged in whorls or in circlets around the central axis. Meroms consist of shafts and heads. Heads always have four-ribbed stellate structures and plates. There is an apical lacuna. In adults the plates in the lower (nuclear) parts of thalli are almost always fused.

Receptaculitids are not "good" algae. This means that they are not acceptable as algae to paleoalgologists, who are not able to place them in any known group of Recent or fossil thallophytes. Neither can they be placed among any group of known living or fossil animals. What are they, then?

Before any definitive conclusions can be made on the taxonomic placement of receptaculitids in relation to living organisms, it is necessary first to consider their relationship to other fossil groups. Such relationships have been discussed many times (see Fisher and Nitecki 1982a). This paper is concerned only with the most recent ideas dealing with the relationships of receptaculitids to two extinct taxa, archaeocyathids and the problematic cyclocrinitids. I hope that this discussion will place some limits on the consideration of the true affinity of receptaculitids.

RECEPTACULITIDS AND ARCHAEOCYATHIDS

There are many groups of fossil organisms that are difficult to place taxonomically, and whose nature is poorly understood. Two typical groups of such organisms are receptaculitids and archaeocyathids. Neither group has any close living relatives, and most of their morphologic structures have no exact extant counterparts. The problematic nature of receptaculitids has been recognized many times. Rauff (1892) and the German school of paleontologists were greatly concerned with the affinities of receptaculitids. American and British paleontologists joined the discussion later, and Soviet specialists on receptaculitids are just now emerging. Receptaculitids have been considered to be calcareous or siliceous sponges, algae, addenda to coelenterates, corals, echinoderms, an independent taxon, and other things. (For various assignments see Nitecki et al., in press.) This list is enormous, but it appears that the prevailing opinion, though far from unanimous, is that receptaculitids can best be placed among the algae.

Archaeocyathids, on the other hand, from the time of their recognition by Billings (1861) and until about World War II, have been considered a problematic group of sponges. At that time the difficulties of placing them among sponges became apparent, and archaeocyathids were removed from the Porifera and named an independent phylum of animals, taxonomically somewhere between sponges and coelenterates (Okulitch and Laubenfels 1953).

The similarities between archaeocyathids and receptaculitids have been noted many times in the past, but only in passing, and usually in a very broad and general discussion of their nature. Until recently there was no rigorous discussion of morphologic or functional similarities between these two groups. Zhuravleva (1970) proposed that receptaculitids, soanitids, archaeocyathids, and sphinctozoans represent a major subdivision of animals, which she named Archaeozoa. Zhuravleva and Miagkova (1970) added aphrosalpingidids and squamiferidids to that taxon and renamed it Archaeata. Aphrosalpingidids, at that time rare calcareous Silurian fossils, are now found in Devonian rocks of North America, and are being redescribed as sponges by Rigby, Nitecki, and Blodgett. I will not be concerned any more with their nature. Sushkin's squamiferidids (1962) is an unfortunate name, equivalent to receptaculitids, and should be dropped from paleontologic literature. Subsequently, Zhuravleva and Miagkova (1979) excluded sphinctozoans from Archaeata and, following Nitecki and De-

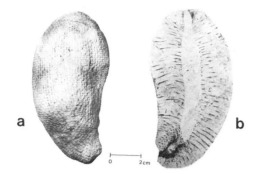

Fig. 1. *Tettragonis murchisoni* Eichwald 1842. Department of Historical Geology. Leningrad University, Holotype No. 4/26. Ordovician, Estonia. A. Lateral view of the surface of the thallus. B. Cross section of thallus.

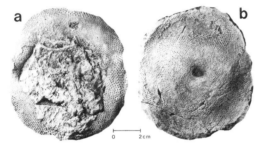

Fig. 2. Nuclear hemisphere of *Fisherites orbis* (Eichwald 1840). Mining Museum, Leningrad, Holotype No. 15/107. Ordovician, Estonia. A. Outer surface. B. Inner surface.

Fig. 3. Abaxial view of stellate structures and plate of *Ischadites tenuis* Nitecki and Dapples 1975. Field Museum of Natural History, Chicago, PP 18309. Silurian, Laurel Limestone, Indiana.

brenne (1979), added radiocyathids (see Chapter 4 by Zhuravlev, this volume). By this time they had fully developed their concept of Archaeata, which they elevated to the kingdom rank, placed between plants and animals. The origin of their kingdom was left in doubt, but recently Zhuravleva and Miagkova (1983) suggested that their

kingdom originated by symbiosis of various prokaryotic and eukaryotic cells. Their definition of the kingdom is based on two major morphologic characters: (1) presence of skeletons consisting of two walls and (2) presence of intramural (intervallum) spaces filled with skeletal elements. However, the double wall, as defined by them, is also present in other taxa (e.g., sponges, cyclocrinitids, dasycladaceous algae). Even more important, the presence of the intervallum skeleton is not an independent morphologic character but a necessary consequence of two-walled construction. In organisms of such a low-level organization as archaeocyathids, which lacked muscles, the support of one skeletal cup within another is functionally and structurally impossible without an intervallum skeleton. Thus, the second diagnostic character, the presence of skeletal elements between the true independent walls, is only a natural extension of two-walled construction. According to Zhuravleva and Miagkova, three groups of Archaeata, the archaeocyathids, aphrosalpingidids, and soanitids (but not receptaculitids), also possess another character, the exoskeletal and endoskeletal "overgrowth."

The case for the affinity between receptaculitids and archaeocyathids was presented by Nitecki et al. (1981). It was based on the analysis of the comparative morphology of ,*Soanites bimuralis* Miagkova 1965, an upper Lower Ordovician receptaculitid from eastern Siberia, and *Dokidocyathus lenaicus* Rozanov 1964, a lower Lower Cambrian archaeocyathid also from eastern Siberia. I have restudied these two taxa, and because I have reached conclusions somewhat contrary to those of Nitecki et al., I now wish to reexamine their arguments in detail.

Nitecki and co-workers believed that the following morphologic characters are shared by soanitid receptaculitids and dokidocyathids:

1. The shape of the body is an inverted cone consisting of a porous double wall connected with rods.

2. The pores are regularly arranged in the outer and inner walls.

3. The attachment to the substrate is by an extension from the exterior of the lower part of the body.

4. All external projections are similar and constitute parts of the original skeleton.

5. The connecting rods are of approximately the same thickness in corresponding parts of bodies.

6. Thicknesses of the two walls relative to the distances in between are identical.

7. The ecology and biologic adaptations appear similar.

Nitecki et al. were well aware that the morphology and distribution of rods, body dimensions, wall thickness relative to body size, and pore diameters and distributions were not the same in these two taxa. They considered these differences, however, of minor significance.

Since receptaculitids and archaeocyathids are represented by numerous taxa, any conclusions on the relationships of these two groups should be based on comparisons of more than just two species, as was done by Nitecki et al. It is in this light that before I proceed with point-by-point analysis of the arguments offered by Nitecki et al., I wish to comment on soanitids. I believe that naming the genus *Soanites* and creating a higher taxon of receptaculitids just for a single species (Miagkova 1965), based on poorly preserved specimens and without any documented comparisons with other receptaculitids, may have been premature. What has been described as *Soanites* is the well-known receptaculitid genus *Calathium,* described and figured by Billings (1865). Various species of *Calathium* and *Calathium*-like fossils have been described, and numerous papers describing the distribution of *Calathium* have been published (Bassler 1915; Foster 1973; Alberstadt and Walker 1976; Nitecki et al., in press). They are known to occur in numerous localities in North America, in a few localities in the United Kingdom, in Spitsbergen, Manchuria, and parts of Australia. Their receptaculitid nature and anatomy is fairly well understood. *Soanites* is identifiable by all morphologic characters known in all receptaculitids, that is, by globose body shape with apical lacuna, by the presence of central axis with meroms consisting of shafts, by four-ribbed stellate structures and plates (cf. Figs. 1–3), and by their growth pattern. For further discussion on why *Soanites* is *Calathium,* and why *Calathium* is a receptaculitid, see Fisher and Nitecki (1982a), and Miagkova (1984). [Miagkova (1984) now accepts *Soanites* as a junior synonym of *Calathium,* but still recognizes it as a base of an independent receptaculitid class Soanitida.]

Body Shape

For detailed illustrations of body shapes of receptaculitids see Rietschel (1969) and Fisher and Nitecki (1982a). It suffices to say here that among many various body shapes, whether elongate or not, all receptaculitids were globose, most of them ovoid, and none (when completely preserved) conical. I have collected in the Ordovician of Nevada an adult and a juvenile calathid that may be complete; they are ovoid and entirely free of any overgrowths.

Double Wall

Although many authors refer to the inner and outer receptaculitid walls, and by analogy with archaeocyathids to the area between them as an intervallum, it must be emphatically stated that such walls in receptaculitids are different from archaeocyathid walls. The outer "wall" in receptaculitids is found only in certain taxa, and in older, calcified parts of the body. In younger parts, or in younger, less frequently preserved parts of older individuals, the plates were not fused and may have been free to move to a certain extent. Receptaculitid inner "walls" represent calcification of merom feet and are found in many, but not all, receptaculitids; and in at least half of all known receptaculitid taxa the central axis was not calcified, and hence no inner "wall" formed.

In archaeocyathids, there may or may not be an inner wall, but when present it always represents, just like the outer wall, a continuous (even if porous) wall that never was composed of individual skeletal elements, never was subject to movement, and always consisted of a solid (though porous) continuous structure.

Porosity

In all other receptaculitids with the exception of calathids, the plates in the outer "wall" show no evidence of pores. As noted by Rietschel and Nitecki (1982), the overgrowth by epilithic organisms may have drastically altered the morphology of calathids by either entirely removing the plates or making pores in plates in a regular manner. These pores are always found in zones of smallest thickness, and only at the edges of plates. Other patterns of alterations are frequently found. For example, stellate structures may be enlarged or removed; exostructures formed entirely of sponge skeletons, reminiscent of archaeocyathid holdfast, may develop; and body "walls" may thicken. In addition, because most plates were not fused, there was no need for pores in those parts of the body, for I suspect they may have been somewhat free to move. However, there appear to have been openings in corners of the "feet" in some species, suggesting that the inner "wall" may have been porous. On the other hand, all archaeocyathids had pores in a manner very suggestive of sponge pores, and perhaps with the exception of the nuclear area, all archaeocyathid cups were porous.

Holdfast

Among over 20,000 receptaculitid specimens in numerous institutions, only one taxon, *Calathium,* possesses structures that can be interpreted as "holdfasts." All suggestions of attachment in receptaculitids (Nitecki et al. 1981, and references therein) are based on the presence of "overgrowths" in calathids. Many calathids, though certainly not all, were overgrown with numerous other organisms, particularly sponges and algae, while the so-called holdfasts (and the majority of other external overgrowths) were made entirely by sponges. The overgrowths occurred during the life of the *Calathium,* because sponges cemented the meroms together, but these overgrowths are present only in lower, older parts of the thalli. The formation of thick walls in *Calathium* (and also in *Hexabactron*) was also caused by epilithic organisms colonizing the living receptaculitids. In large and diverse collections representing localities with varied ecologic conditions, many calathids are preserved without any "holdfast" or any overgrowth. There are localities in which all calathids are free of overgrowth. All disarticulated and younger (i.e., actively metabolizing) parts of skeletons are free of overgrowth. While the erroneous interpretation of calathid morphology may mislead conclusions on receptaculitid relationship to archaeocyathids, no other receptaculitid taxa show such structures and such supposed similarities to archaeocyathids. All other receptaculitids were without any holdfast and either simply sat on the substrate or were attached only by a point attachment. Archaeocyathids, on the other hand, had such structures, and the overgrowth by epilithic organisms is a characteristic mode of their occurrence.

Meroms

The receptaculitid merom has absolutely no counterpart in archaeocyathids. The dokidocyathid rod is a simple structure, while in receptaculitids, including calathids, the merom is a separate, individual unit consisting of a shaft, a four-ribbed stellate structure, and a plate (Fig. 3). The fusion of plates (and thus formation of "walls") in receptaculitids is found only in older individuals, and only in lower parts of the body; and the "walls" thus made of plates are not comparable with archaeocyathid walls, which are continuous structures. The receptaculitid merom could be homologous with the archaeocyathid rod and wall only if it could be demonstrated either that the merom could evolve into a rod (or septum, or other structure) and wall, or, conversely, that the archaeocyathid rod and wall could evolve into the receptaculitid merom. However, there is presently no evidence for such speculation, and conclusions on homology between these skeletal parts are not supported.

Nucleus and Growth

Another fundamental difference between archaeocyathids and receptaculitids concerns the growth of these forms. All well-preserved receptaculitids have a basal nucleus surrounded by a single whorl of meroms. No such structure has ever been observed in archaeocyathids, but instead, a single solid small cone is present. Growth in receptaculitids is a process consisting of an increase in size of individual meroms, and of an apical addition of new meroms in circlets, spirals, or possibly one at a time. The growth pattern of individual plates has been illustrated a number of times (see Rietschel 1969), but no such growth is present in archaeocyathids. Thus the growth in receptaculitids is compound and proceeds outward from the cen-

tral axis. In archaeocyathids, however, the growth is simple, apical, and proceeds inward from the entire wall. No greater difference in growth pattern, as between these two groups, can be conceived. No evidence exists that new elements are inserted in any receptaculitid thallus, as is evident in archaeocyathids, where skeletal entities are constantly being added to the archaeocyathid cup. Receptaculitids generate new elements at their generative poles, the apices, and fill their outer surfaces with plates. This produces an intercalated outer surface. Archaeocyathids grow in a radial manner, producing taller and taller cups, adding new skeletal elements at the outer edge of the outer wall, and inserting new septa throughout the entire adult portion of the cup.

In summary, enough differences have been demonstrated between receptaculitids and archaeocyathids that it seems unnecessary to proceed further. It should be added, however, that other important considerations, such as anatomy of intercalation, gradients of number of skeletal elements per whorl (characters discussed in detail in Fisher and Nitecki 1982a) must be considered and shown to be similar before any positive conclusions on the relationships of such dissimilar fossils as receptaculitids and archaeocyathids can be made.

Although I have been concerned in detail only with a single publication (Nitecki et al. 1981), dealing only with one species in each group, I am by implication of that paper (and the references therein) concerned with the concept of Archaeata. Archaeata, a kingdom-level group, was erected in order to remove the difficulties of interpreting anatomic structures within existing systems of classification. All organisms included in Archaeata are problematic, and nothing but speculation is known about the soft parts of all these. Their physiology and their mode of life is conjectural, and their relationship to other organisms is highly inferential. These should be not bases upon which kingdoms of organisms are erected. Kingdoms represent very fundamental segments of life that should not be erected without great care and, for the same reason, should not be easily wiped out. Under some circumstances, however, the conditions may occur where kingdoms may have to be swept away. Receptaculitids and archaeocyathids are not related, and therefore Archaeata, created for their reception, is invalid.

RECEPTACULITIDS AND CYCLOCRINITIDS

Although it has been demonstrated that receptaculitids and archaeocythids are different organisms, and none of their morphologic features appear homologous, there are other groups that different authors at different times have compared with receptaculitids (see Chapter 4 by Zhuravlev, this volume). One such group is cyclocrinitids (see Nitecki and Toomey 1979; Fisher and Nitecki 1982b). The relationship of cyclocrinitids to receptaculitids has been claimed to be very close (Nitecki 1970), or it has been disputed (Campbell et al. 1974). I believe that the morphology of these two groups, as judged after the general structural body plan, body shape, and particularly the presence and shape of the central axis and meroms and their arrangement, does indeed appear very close. However, it is very important to assess which, if any, of the skeletal elements are homologous structures, what is the significance and the importance of the differences between these two groups, and how these differences help—if they do—in solving the problem of the taxonomic position of receptaculitids.

Cyclocrinitids

Cyclocrinitids are a group of small problematic fossils, abundant in North America, especially in the Siluro-Ordovician of Quebec, in the Silurian of the Midwest, and in the Ordovician of the Appalachian region. They are perhaps best preserved and certainly best known from the Ordovician of the Baltic area (Stolley 1896), and are found in the Ordovician and Silurian of Great Britain and Norway. They have been described in detail from Kazakhstan (Gnilovskaya 1972). As with many other problematic fossils, they have been moved from taxon to taxon and have been placed among protozoans, sponges, corals, bryozoans, mollusks, algae, receptaculitids, and the "unknown" organisms. The question of whether they are plants or animals has been asked many times; however, many invertebrate paleontologists still consider them to be animals of unknown phylum, or sponges, while most paleoalgologists assign them to Dasycladaceae. Many cyclocrinitid genera have been placed as an appendix to Porifera by Laubenfels (1955) and Sushkin (1962). On the other hand, Stolley (1896) monographed the Baltic cyclocrinitids as algae. It is a great puzzle why this work was virtually ignored by (particularly American) invertebrate paleontologists. The influential paper by Pia (1927) also placed cyclocrinitids in Dasycladaceae. The general outline of Pia's classification was followed in, among others, the Soviet treatise on algae (Maslov 1963), the revision of North American cyclocrinitids (Nitecki 1970), and the monograph on the Ordovician calcareous algae of Kazakhstan (Gnilovskaya 1972).

Taxonomic Composition

Many genera have been assigned by various authors to cyclocrinitids, and their stratigraphic range has been claimed for the entire Paleozoic, or even more. However, only the following genera are here recognized as cyclocrinitids: *Cyclocrinites, Coelosphaeridium, Mastopora, Lunulites, Apidium, Pasceolus,* and *Nidulites.* Under such restrictive definition, cyclocrinitids ranged from the lower Middle Ordovician to the upper Middle Silurian. They appear to have inhabited shallow, marine, carbonate environments, with only rare occurrences in shaly rocks.

Morphology

The cyclocrinitid body is spherical, sometimes elongated, club shaped, or, rarely, flattened and disc shaped. The last may be a result of postmortem compaction. Cyclocrinitids were benthic; their means of attachment to the substrate (known only in a few taxa) is by a distinct stemlike extension of the main axis. The cyclocrinitid skeleton consists of a generally thin, sometimes terminally bulbous but generally clublike central axis that bears branches (meroms) arranged apparently in whorls. The axis is not often preserved, hence I assume not often calcified. When preserved, it appears to have been rather short, seldom reaching two-thirds the height of the thallus. In *Coelosphaeridium,* it is always terminally bulbous, and in *Lunulites dactioloides* the main axis is very thin. A relatively thin axis is found in *Cyclocrinites.* Irrespective of the frequency of the preservation of the main axis, its presence is unequivocal. The meroms are packed closely together, presumably in whorls, or in a spiral. The shaft of meroms may be conical or clublike, but in the majority of taxa the shafts are thin and long. The meroms terminate with an enlarged head, which may or may not have a distal lid. The lid may be sculptured and may or may not be perforate. In some species stellate structures are present. These are relatively thin and long and consist of more than four ribs. The ribs of stellate structures are just below the head and are probably perpendicular to the axis of the meroms. The main axis originates at the nucleus, which, when preserved, is recognized on the surface by the characteristic rosettelike arrangement of merom heads. The presence of a lacunal opening at the upper end of the body has not been conclusively demonstrated.

The growth pattern of cyclocrinitids has not yet been investigated. In those cyclocrinitids in which a well-preserved surficial pattern is present, the preliminary examination reveals that the meroms are either single, or occasionally, branched. The branched meroms have been seen in sections of one species only (Ordovician *Apidium rotundum* from Norway). The surficial pattern has not yet been mapped. It is this occurrence of branched and unbranched meroms within a single thallus that makes the interpretation of the growth pattern difficult.

Relation to dasyclads

Contrary to the prevailing consensus among paleoalgologists that cyclocrinitids are Dasycladales, I believe that they are still inadequately known and thus are best considered a problematic group of algae. They cannot be assigned to Dasycladales for the following reasons:

1. Dasyclads, as understood today and as represented by living forms, are present already in the Ordovician. The Upper Ordovician *Archaeobatophora* possesses a thallus that consists of a weakly calcified main axis and well-preserved noncalcareous branches (Nitecki 1976). The preservation of the branches is so good that the nature and position of laterals in whorls allow for unquestionable assignment of *Archaeobatophora* to the living tribe Batophoreae. It is hard to compare other Paleozoic "Dasycladales" with living Dasycladales. The Paleozoic forms are known from skeletal remains only, and neither their gametangia, nor the position of gametangia, is known. There are great gaps in their stratigraphic distribution, making phylogenetic reconstructions difficult. Except for the well-preserved *Archaeobatophora*, all Lower Paleozoic "dasyclad" algae are poorly known. The only fossil taxon to which *Archaeobatophora* may be related is *Chaetocladus;* however, the size and shape of the thalli and the nature and distribution of the branches of these two genera are not very similar. Thus, if cyclocrinitids were siphonous algae, they represent very primitive forms, which are as different from the Ordovician as they are from living Dasycladales.

2. The sizes of cyclocrinitid thalli and particularly their lateral branches (meroms) are at least five times larger than comparable parts of dasyclads. Since dasyclads are coenocythic, their size is restricted to the body size defined by septa formation. The size of cyclocrinitids makes it very difficult to consider them as plants with a single nucleus. In addition, no septa have ever been observed in cyclocrinitids.

3. Cyclocrinitids possess meroms with stellate structures, anatomic entities entirely absent in dasyclads, and difficult to interpret functionally.

4. The lid found in heads of certain cyclocrinitids in the Baltic area is not present, nor can it be easily explained in living or fossil Dasycladales. If the head of the lateral was an organ holding gametangia (a reasonable assumption for an alga of such large proportions), why would it have a skeletal, porous, ornamented lid? On the other hand, if the porous lid served the function of holding the photosynthetic hair, how was photosynthesis conducted by those cyclocrinitids that differ from those with porous lids only in having nonporous lids?

This is not, however, a claim that cyclocrinitids were not algae. They have the following distinct algal characters (see Chapter 2 by Babcock, this volume):

1. The main axis is distinctly algal; the upper part of the main axis holds the branches and when preserved, there appears to be a holdfast.

2. The morphology and arrangement of meroms can be easily made homologous with lateral branches of thallophytes.

3. It is easy to postulate that they were photosynthetic, while it is difficult to assume that they had a nonphotosynthetic mode of life. The photosynthetic pigments could have been distributed in a thallus in a manner found in living *Bornetella* (see Nitecki 1970, fig. 20).

4. They are morphologically very close to the Lower Paleozoic amphispongids, which have all "good" algal (but not necessarily dasyclad) characters.

In conclusion, cyclocrinitids may have been algae, but not Dasycladales.

Relation to receptaculitids

The comparison of cyclocrinitids with receptaculitids can be based on the same characters as were used for comparison with dasyclads.

Thallus Shape. While the body shapes of both groups appear similar, the general body shapes in spherical organisms are of little value in drawing analogies. It is now almost certain that all receptaculitids had an apical lacuna, while no opening has yet been found in cyclocrinitids.

Main Axis. The morphology of the main axis in both groups appears very similar. It is better known in cyclocrinitids, but this may be due to the different degree of calcification. Presence of the central axis has been questioned for receptaculitids (Foster 1973; Zhuravleva and Miagkova 1979), but many specimens show internal structures that are impossible to explain as anything other than the central axis (see Rauff 1892, pl. 6, figs. 10 and 10a).

Surficial Pattern. The arrangements of meroms on the main axis (as seen in the surficial pattern) is known for receptaculitids but has not been studied yet in cyclocrinitids. At this stage of investigation there is a weak suggestion that they may be similar.

Meroms. The general outline of meroms is similar; however, the details of meroms are very different, and it is in the anatomy of the merom that the two groups differ most. In receptaculitids, the stellate structures served a function of holding the skeleton together, and yet their interlocking may have allowed for a certain motion of meroms. It is not known whether receptaculitids were capable of active movement of meroms, or whether the movement was passive. Neither do we know whether one side of the thallus could "open" while the other "closed." In any case, the function of stellate structures can be seen as adding to the stability and rigidity of the organism while at the same time allowing some freedom of movement. Stellate structures of cyclocrinitids are very rare, but even if they were a universal occurrence, they could not (because of their length and extremely thin character) have served the same function as stellate structures in receptaculitids. The rigidity of the cyclocrinitid body was maintained by fusion of the heads of meroms. This cementation of heads produced better-preserved thalli but certainly was not as effective as the receptaculitid interlocking of stellate structures, which may explain the attainment of a much larger size in receptaculitids. Thus, functionally, the stellate structures of both groups appear very different. However, functionally different organs may indeed be homologous and developmentally identical.

The differences between the receptaculitid plate and the cyclocrinitid head are also great. In receptaculitids, the plate is always flat and invariably rectangular to hexagonal. In cyclocrinitids the head is bulbous and generally hexagonal. Furthermore, certain cyclocrinitids (as shown by Stolley 1896) possess perforate lids, structures indistinguishable from tops of plates in receptaculitids, (see *Receptaculites neptuni* in Rauff 1892, pl. 2, figs. 1–3). It is possible to regard meroms in both groups as homologous, and it is possible to visualize the evolution of the cyclocrinitid head into the receptaculitid plate and stellate structures. Nevertheless, there are significant differences between these structures, too.

Nucleus. The nuclear areas of cyclocrinitids and receptaculitids are very similar—almost identical. In receptaculitids it consists of 4 to 12 meroms, and in cyclocrinitids from 4 to 8. Thus, in both groups a single whorl of meroms forms the first circlet, the nucleus.

CONCLUSIONS

Both cyclocrinitids and receptaculitids can be viewed as assemblages of meroms arranged around the main axis; receptaculitids, in addition, appear to have an apical lacuna. If it could be shown (1) that these assemblages are arranged in the same way or (2) if these arrangements are not the same, that the meroms are the same and that cyclocrinitids have apical lacunas—then the relationships of these two groups, without specifying the level of relationship, could be considered as demonstrated. If these two groups are indeed related, then the cyclocrinitids' "simpler" morphology and their stratigraphic distribution would suggest that cyclocrinitids are a less specialized group that descended from their common ancestry with receptaculitids before the early Ordovician.

Acknowledgments

This paper was written in Moscow during the 1984 Exchange Scientist Program between the National Academy of Sciences and the USSR Academy of Sciences. I am grateful to these institutions and to the Field Museum of Natural History for their financial assistance, as well as to Inga F. Bluman, Daniel C. Fisher, Marina B. Gnilovskaya, Ralf Männil, Ekaterina S. Paretzkaya and Andrey Yu. Zhuravlev for helpful discussions and comments on the manuscript, and for granting permission to study the specimens under their care.

REFERENCES

Alberstadt, L. P. and Walker, K. R. 1976. A receptaculitid–echinoderm pioneer community in a Middle Ordovician reef. *Lethaia* **9**, 261–272.

Bassler, R. S. 1915. Bibliographic index of American Ordovician and Silurian fossils. *Bull. U.S. Natl. Mus.* **92**, (2 parts), 1521 pp.

Billings, E. 1861. On some new or little known species of Lower Silurian fossils from the Potsdam Group. *Rep. Geol. Vermont* **2**, 942–960.

———. 1865. *Palaeozoic Fossils*, vol. 1. Geological Survey of Canada, 426 pp.

Campbell, K. S. W., Holloway, D. J., and Smith, W. D. 1974. A new receptaculitid genus, *Hexabactron,* and the relationships of the Receptaculitaceae. *Palaeontographica A* **146**, 52–77.

Fisher, D. C. and Nitecki, M. H. 1982a. Standardization of the anatomical orientation of receptaculitids. *Paleontol. Soc. Mem.* **13**, 1–40.

———. 1982b. Problems in the analysis of receptaculitid affinities. *Proc. 3rd North Am. Paleont. Conv.* **1**, 181–186.

Foster, M. 1973. Ordovician receptaculitids from California and their significance. *Lethaia* **6**, 35–65.

Gnilovskaya, M. B. 1972. *The Calcareous Algae of the Middle and the Late Ordovician of Eastern Kazakhstan. Akad. Nauk.* SSSR, Leningrad, 195 pp. [in Russian].

Laubenfels, M. W. de. 1955. Porifera. *In* Moore, R. C. (ed.). *Treatise on Invertebrate Paleontology.* Part E. Geological Society of America and Univ. of Kansas Press, Lawrence, pp. E21–E112.

Maslov, V. P. (ed.). 1963. Phylum Chlorophyta. *In* Orlov, Y. A. (ed.). *Osnovy Paleontologii.* Nauka, Moscow, pp. 199–223 [in Russian].

Miagkova, E. I. 1965. Soanitids—a new group of organisms. *Paleont. Zh.* **1965**(3), 16–22 [in Russian].

———. 1984. Comparative morphology of *Calathium* Billings 1865, and of *Soanites* Miagkova 1965. *In* Sokolov, B. S. (ed.). *Problematics of the Paleozoic and Mesozoic* Nauka, Moscow, pp. 38–43 [in Russian].

Nitecki, M. H. 1970. North American cyclocrinitid algae. *Fieldiana: Geol.* **21**, 1–108.

———. 1976. Ordovician Batophoreae (Dasycladales) from Michigan. *Fieldiana: Geol.* **35**, 29–40.

————, Bradof, K. L., and Nitecki, D. V. In press. Annotated bibliography of receptaculitids, 1805–1980. *Fieldiana: Geol.*

———— and Debrenne, F. 1979. The nature of radiocyathids and their relationship to receptaculitids and archaeocyathids. *Géobios* **12**, 5–27.

———— and Toomey, D. F. 1979. Nature and classification of receptaculitids. *Bull. Cent. Rech. Explor.-Prod. Elf-Aquitaine* **3**, 725–732.

————, Zhuravleva, I. T., Miagkova, E. I., and Toomey, D. F. 1981. An affinity of *Soanites bimuralis* to archaeocyathids and receptaculitids. *Paleont. Zh.* **1981**(1), 5–9 [in Russian].

Okulitch, V. J. and Laubenfels, M. W. de. 1953. The systematic position of Archaeocyatha (Pleosponges). *J. Paleont.* **27**, 481–485.

Pia, J. 1927. Thallophyta. *In* Hirmer, M. (ed.). *Handbuch der Paläobotanik,* vol. 1. Oldenburg, München, pp. 31–136.

Rauff, H. 1892. Untersuchungen über die Organisation und systematische Stellung der Receptaculitiden. *Abh. Bayer. Akad. Wiss. Math.-Phys.* **17**(3), 645–722.

Rietschel, S. 1969. Die Receptaculitiden. *Senckenb. Leth.* **50,** 465–517.

———— and Nitecki, M. H. 1982. Concept of Kingdom Archaeata. *J. Paleont.* **56**(2, suppl.), 22.

Stolley, E. 1896. Untersuchungen über *Coelosphaeridium, Cyclocrinus, Mastopora* und verwandte Genera des Silur. *Arch. Anthropol. Geol. Schlesw.-Holst. Benach.* **1**(2), 177–282.

Sushkin, M. A. 1962. Class Squamiferida. *In* Orlov, Y. A. (ed.). *Osnovy Paleontologii: Sponges, Archaeocyathids, Coelenterates, Vermes.* Nauka, Moscow, pp. 81–83 [in Russian].

Zhuravleva, I. T. 1970. Porifera, Sphinctozoa, Archaeocyathi—their connections. *In* Fry, W. G. (ed.). *Biology of the Porifera.* Zool. Soc. London Symp. **25**, 41–59.

———— and Miagkova, E. I. 1970. Higher group Archaeata. *Inst. Geol. Geophys. Sib. Acad. Sci. USSR,* p. 2 [in Russian].

———— and ————. 1979. Comparison entre les Archaeata et les Porifera. *In* Levi, C. and Boury-Esnault, N. (eds.). *Biologie des Spongiaires,* CNRS, Paris, pp. 521–526.

———— and ————. 1983. The status of Archaeata in the development of the organic kingdom. *In Morphology and Systematics of Phanerozoic Invertebrates.* Academy of Sciences, USSR, Moscow, pp. 60–65 [in Russian].

4. RADIOCYATHIDS

ANDREY YU. ZHURAVLEV

Radiocyathids were first described by Bedford and Bedford (1934, 1936) from the Lower Cambrian of the Flinders Ranges, South Australia, some of them *(Heterocyathus)* as archaeocyathids, others *(Uranosphaera)* as sponges. Okulitch (1937) replaced the name *Heterocyathus* with *Radiocyathus,* and the entire group has been consequently called radiocyathids.

Debrenne et al. (1970, 1971) revised all radiocyathid taxa. They determined that, in addition to *Uranosphaera* and *Radiocyathus,* the presumed hexactinellid sponge *Girphanovella* from Tuva (Zhuravleva et al. 1967), *Kuraya* from Altai (Romanenko 1968), and *Gonamispongia* from southern Siberia (Korshunov 1968) should also be placed among radiocyathids. More recently, radiocyathids were discovered in Morocco (Debrenne 1977) and central Australia (Kruse and West 1980).

The following discussion of radiocyathids is based largely on new material from Mongolia, Siberia, Canada, and Antarctica. Vologdin (1940a) reported *Archaeocyathus* cf. *proskurjakovi* Toll from western Mongolia, which he later described (Vologdin 1940b) as *Archaeocyathus neoproskurjakovi* Vologdin. Vologdin's type specimens are lost, but his photomicrograph suggests that he described a radiocyathid similar to *Girphanovella.* This inference is strongly supported by the radiocyathid material from Vologdin's type locality.

Rozanov (in Zhuravleva et al. 1964) described from Transbaikal a problematic fossil, *?Dokidocyathina georgensis* Rozanov, presumably an archaeocyathid of the suborder Dokidocyathina. However, analysis of the additional material from the type locality strongly suggests that it is *Girphanovella.* A very rich assemblage of *Girphanovella* has also been collected in the Lower Cambrian of western and northwestern Mongolia. Another radiocyathid, *Gonamispongia,* was collected in southwestern Prianabarye in northern Siberia.

All the material discussed in this paper is housed at the Paleontological Institute (PIN) of the USSR Academy of Sciences (Collections: PIN N 3457, 3482, 3900, and 4016).

RADIOCYATHID MORPHOLOGY

Our material permits us to emend the original descriptions of *Gonamispongia, Girphanovella,* and *Kuraya,* and the diagnosis of *Radiocyathus.* The radiocyathid terminology proposed by Nitecki and Debrenne (1979) is used in this paper.

Gonamispongia Korshunov 1968

This is the oldest known radiocyathid genus. It occurs in the upper part of Tommotian and possibly also in the lower part of Atdabanian in southern (Korshunov 1968) and northern Siberia (Shishkin et al. 1982).

The body shape of *Gonamispongia* is pyriform, > 5 cm in diameter. The wall is single, built of nesasters, (i.e., dumbbell-like structures consisting usually each of two rosettes connected by a shaft). In *Gonamispongia,* however, each nesaster has only one rosette consisting, as in all radiocyathids, of a central knob that gives origin to several rays (Fig. 1A–E). A well-preserved rosette has 8–10 rays parallel to the wall. The rays have their ridgelike bases just under the surface of the central knob. Neighboring rays are interconnected by subrays. Rays exceed 3 mm in length and are five to six times longer than the diameter of the central knob. Neighboring rosettes are interconnected by rays, with a ray of one rosette running beneath and parallel to a ray of the other. Where rosettes are very close to each other, rays of one rosette cover the central knob of the other rosette, and thus some rosettes appear depressed. They may have been formed later than the uplifted rosettes. The sunken rosettes are irregularly arranged, thus suggesting that the nesasters may have been formed irregularly. Nesaster shafts may exceed 5 cm in length and run from each central knob toward the center of the skeleton (Fig. 1F).

Girphanovella Zhuravleva 1967

This genus is known from the Botomian and possibly also the upper part of Atdabanian of Tuva and Transbaikal, Mongolia, Canada (Fig. 2C), and perhaps Morocco (Debrenne 1977).

Girphanovella has a two-walled globular or pyriform skeleton, up to 12 but most commonly 2–5 cm in diameter. Both solitary and branched morphotypes are known. Each rosette of the outer wall is connected with a rosette of the inner wall by a shaft (Fig. 4C). In some instances, two shafts may originate from a single central knob of the inner rosette, each of them terminating with an

outer rosette. Thus, the number of rosettes is approximately equal in both the walls, and the shape of the inner wall reflects the outline of the outer wall (Fig. 2A, E). The outer rosettes vary in size. They have 8–12 rays each, every ray being two to three times longer than the diameter of the central knob. Primary rays bifurcate immediately outside the central knob, producing > 20 secondary rays in total. The latter are connected with primary rays and central knobs of the neighboring rosettes, forming a dense surficial network with irregular gaps giving the appearance of porosity in the wall. Primary rays are a little thicker and slightly inclined toward the vertical axis of the skeleton. Primary rays of the neighboring rosettes do not contact but intertwine with each other. Thus, two layers can be distinguished in the outer wall of *Girphanovella:* the outer layer formed by rays fused with each other (Fig. 3B), and the inner layer consisting of intertwining rays (Fig. 3A). In the outer wall of *Girphanovella,* the rosettes are regularly arranged, with smaller rosettes interspersed among larger ones (Fig. 3C). The inner wall differs in structure from the outer wall. The rays of its rosettes are approximately constant in thickness, and generally no more than eight per rosette. They diverge from the central knob in two or three planes. Rosettes of the inner wall are seldom preserved (Fig. 4A, B, E); if preserved, however, the inner wall appears as a solid spongy mass with vermicular canals (Fig. 4D).

Kuraya Romanenko 1968

Kuraya is known from only one single locality in the Botomian of Altai. It is subspherical in shape, with a single wall. All nesasters are constant in size, with six rays per rosette. The rays are slightly inclined toward the vertical axis of the skeleton. They usually intertwine among neighboring nesasters and occasionally even fuse; however, Romanenko (1968) and Debrenne et al. (1971) suggest that they may be free.

Radiocyathus Okulitch 1937

This genus is known from the Botomian of Australia (Bedford and Bedford 1934; Kruse and West 1980) and Antarctica. Its morphology was described in detail by Bedford and Bedford (1934, 1936) and Debrenne et al. (1970). Kruse and West (1980) demonstrated that all the fossils assigned

---

Fig. 1. Photomicrograph of *Gonamispongia ignorabilis* Korshunov; PIN N 3900/40. A–C. Fusion of rosettes of nesasters. A. Outer surface. B–C. Inner surface. D. Part of wall, outer surface. E. Rosette. F. Rosettes with shafts, lateral view. USSR, Siberian Platform, Prianabarye, Kotuyi River. Lower Cambrian, Tommotian Stage. Scale bars: A–D and F = 2 mm; E = 1 mm.

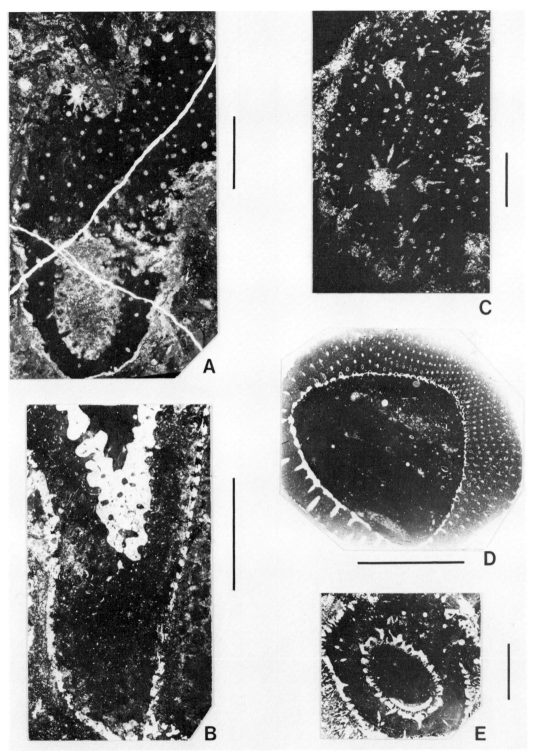

Fig. 2. A. *Girphanovella georgensis* (Rozanov); PIN N 3482/51. Oblique longitudinal section. Western Mongolia, Dzuhn-Artsa Mountain. Lower Cambrian, Botomian Stage. B. *Calathium bimuralis* (Miagkova); PIN N 3900/ 37. Longitudinal section. USSR, Siberian Platform, Moyero River. Lower Ordovician. C. *?Girphanovella* sp.; PIN N 4016/16 (91690 R. Handfield's collection). Tangential section of outer wall. Canada, Mackenzie Mountains. Lower Cambrian, Botomian Stage. D. *"Receptaculites"* sp. PIN N 3900/36. Longitudinal section. USSR, Kazakhstan. Silurian. E. *Girphanovella georgensis* (Rozanov); PIN N 3900/35. Cross section. USSR, Transbaikal. Lower Cambrian, Botomian Stage. Scale bars: A–C and E = 5 mm; D = 3 cm.

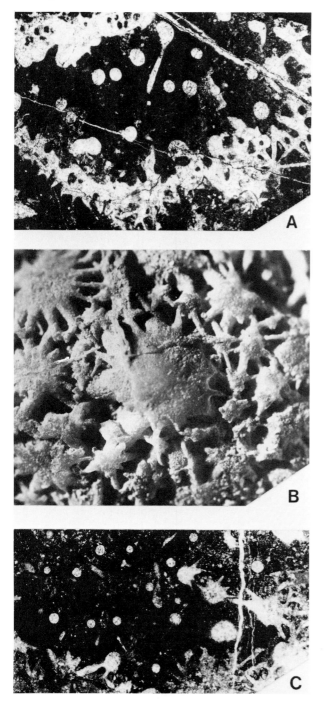

Fig. 3. *Girphanovella georgensis* (Rozanov).
A. PIN N 3482/55. Tangential section of
outer wall. Western Mongolia, Dzuhn-Artsa
Mountain. B. PIN N 3457/104. Tangential
section of outer wall. Northwestern Mongolia,
Khubsugul region. C. PIN N 3900/34. Etched
surface of outer wall. USSR, Tuva, Shivelig-
Khem River. Lower Cambrian, Botomian
Stage. Scale bars: A, B = 4 mm; C = 2 mm.

to this genus represent only a single species, and
that the three morphotypes discussed by earlier
authors reflect merely various modes of preservation (see also Nitecki and Debrenne 1979).
They determined that *Radiocyathus* has two layers in the outer wall, with rosettes anastomosing
in the inner layer and intertwining in the outer
layer. Each rosette in the outer wall has 6–12 primary rays. The rays and numerous small projections in between form a solid starlike plate at the
outer surface. Nesasters resemble *Gonamispongia*
and they are less variable in size than in *Girphanovella*. In all other details, however, *Radiocyathus*
is closer to the latter genus.

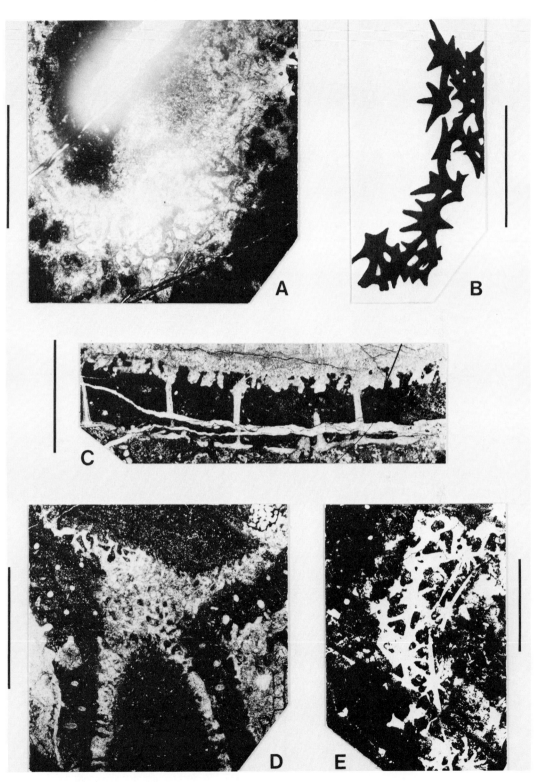

Fig. 4. *Girphanovella georgensis* (Rozanov); A, B, D. PIN N 3482/54. Tangential section of inner wall, nesasters. A. Thin section. B. Drawing of thin section. D. Fusion of rosettes of inner wall. C. PIN N 3482/52. Part of cross section. E. PIN N 3482/53. Tangential section of inner wall, nesasters. Western Mongolia, Dzuhn-Artsa Mountain. Lower Cambrian, Botomian Stage. Scale bar = 4 mm.

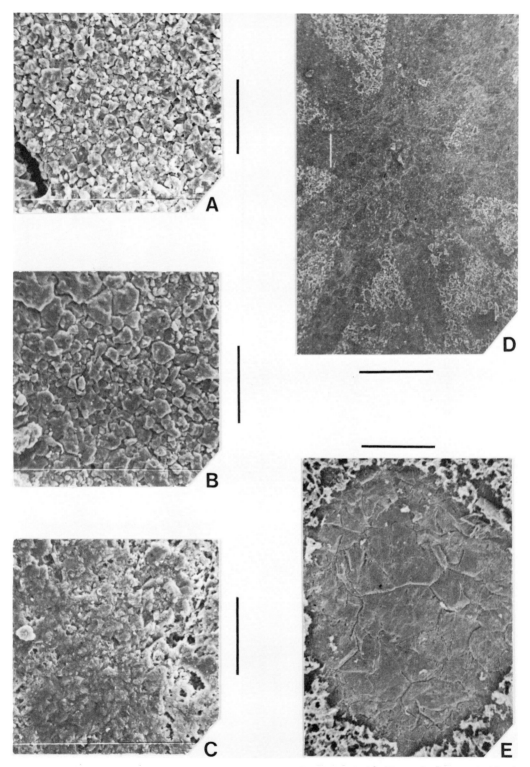

Fig. 5. *Calathium bimuralis* (Miagkova). PIN N 3900/37. Part of wall etched with trilon-B for 35 seconds. Siberian Platform, Moyero River. Lower Ordovician. B–D. *Girphanovella georgensis* (Rozanov); B. PIN N 3457/105. Part of wall etched with 10 percent HCl for 1 minute; C. PIN N 3457/106. Section of wall etched with 5 percent HCl for 1.5 minute. D. PIN N 3482/53. Tangential section of rosette etched with 5 percent HCl for 40 seconds. Western Mongolia, Dzuhn-Artsa Mountain. Lower Cambrian, Botomian Stage. E. *"Receptaculites"* sp. PIN N 3900/ 36. Part of shaft etched with 10 percent HCl for 4 minutes. Kazakhstan. Silurian. Scale bars: A–C = 20 μm; D = 200 μm; E = 100 μm.

Let us briefly summarize the anatomy of radiocyathids. Radiocyathids are radially symmetric, globular, ovoid or pyriform in shape, 20 mm wide and 50 mm high on the average (maximum height is 200 mm). The skeleton is built of nesasters each consisting of a shaft with rosettes on either end (Fig. 5). *Gonamispongia* has shafts with distal rosettes only; *Kuraya* and *Uranosphaera* may represent different modes of preservation of two-walled radiocyathids. Neighboring rosettes are interconnected by rays that form a solid wall. The only exception is *Blastasteria,* in which the rosettes are free (Debrenne et al. 1971); however, it is known from so few specimens that reliable conclusions about its morphology can hardly be drawn. At least two genera, *Girphanovella* and *Radiocyathus,* have two layers in the outer wall. The secondary rays form a dense network at the outer surface, whereas the inner surface consists of primary rays only and thus possesses more free space. The central cavity is delimited either by fused rosettes of the inner wall or by proximal ends of the shafts only. Presence of a lacuna is convincingly demonstrated exclusively in *Uranosphaera* (Nitecki and Debrenne 1979).

GROWTH PATTERN

Every large nesaster of *Girphanovella* is surrounded by smaller nesasters, each of which is surrounded by still smaller nesasters. This suggests that the nesasters were growing with the organism's growth. The spaces between neighboring nesasters were also increasing and filled by new, and consequently smaller, nesasters. Nesasters could originate at any place in the skeleton. This may also be true for *Gonamispongia,* where the position of nesasters relative to one another indicates their relative age. The center of growth of nesasters must have been near the central cavity, because shafts bifurcate away from the inner wall. In contrast, the growth pattern of *Radiocyathus,* with nesasters approximately constant in size and located all in the same plane (Nitecki and Debrenne 1979), must have been different.

MICROSTRUCTURE

When polished surfaces of the skeleton of *Girphanovella georgensis* are etched with a 5–10 percent solution of acetic acid or trilon-B, the microstructure appears to be granular and similar to the microstructure of archaeocyathids, certain Early Cambrian algae, and some other fossils. The granules are calcitic, isometric, angular, and 0.03–0.06 μm in size on the average (Fig. 6B). More frequently, however, the skeletons are recrystallized, and even large skeletal elements (e.g., individual rosettes of the inner wall) cannot be identified. Sometimes, separate granules can be recognized within amorphous structures (Fig. 6C). In other

Fig. 6. *Girphanovella georgensis* (Rozanov). Reconstruction of the skeleton.

cases, only big crystals of calcite are discernible; they are polygonal in shape and variable but up to 100 μm in size (Fig. 6D).

DISCUSSION

Some authors regard radiocyathids as an intermediate link between distinct groups of organisms, such as between archaeocyathids and sponges (Termier and Termier 1979) or between archaeocyathids and receptaculitids (Nitecki and Debrenne 1979). Although the present understanding of radiocyathid anatomy is inadequate to resolve their phylogenetic affinities completely, it allows for some speculation on their systematic position.

The radiocyathid body shape is comparable to certain sponges and receptaculitids, but it is different from the generally cuplike shape of archaeocyathids. Certain archaeocyathids (Capsulocyathina) have a globular body; in radiocyathids, however, the inner wall is connected with the outer wall by shafts only, while in globular archaeocyathids the eccentric inner wall is formed by parts of the outer wall being pushed down. The average radiocyathid body size (20–550 mm) is again comparable to certain receptaculitids and sponges, but it significantly exceeds the average size of archaeocyathid cups (5–25 mm). Radiocyathids and receptaculitids rarely divide, whereas archaeocyathids and sponges have many modes of division, and budding and colonial forms are frequent in the latter two groups.

As noted by Nitecki and Debrenne (1979) and Fisher and Nitecki (1982b), there are no solid continuous walls and platelike elements in radiocyathids. The skeleton is composed of numerous morphologically identical nesasters that form a wall-like structure only when their rosettes are fused. The nesasters have been considered homologous to receptaculitid meroms (Nitecki and Debrenne 1979) or to sponge spicules (Termier and Termier 1979). However, details of their morphology suggest a closer affinity to receptaculitids than to sponges, because the shaft of nesasters bifurcates as in receptaculitids (Fisher and Nitecki 1982a) but not in sponges and archaeocyathids. In radiocyathids, again as in receptaculitids but not in archaeocyathids, the central axis is delineated either by fused inner rosettes or by proximal ends of shafts.

The radiocyathid microstructure resembles the receptaculitid *Calathium bimuralis* (Miagkova) which has a similar granular microstructure (Figs. 2B and 6A). The microstructure described earlier for *C. bimuralis* by Zhuravleva and Miagkova (1981) is secondary. Such an altered microstructure of coarsely crystalline calcite occurs frequently in receptaculitids (Figs. 2D and 6E). It was studied by Van Iten and Fisher (1983), who noted that the coarsely crystalline calcitic microstructure was secondary, after the original high-magnesium calcite. It seems that the granular microstructure described by Zhuravleva and Miagkova (1981) was a diagenetically altered irregular aragonitic microstructure. Such diagenetic alterations have been shown for algae (Flajs 1977) and sponges (Wendt 1979). The microstructure of radiocyathids and *C. bimuralis* never consisted of spicules. Consequently, the radiocyathid nesasters cannot be homologous to hexactinellid spicules, although they may be comparable to the microstructure of nonspicular Sphinctozoa, whose fossils also have a granular microstructure (Boiko 1984). However, the gross morphology of Sphinctozoa is clearly different from the morphology of radiocyathids.

The growth pattern of radiocyathids appears to have been interstitial, as in many sponges, including some Cambrian forms (Rushton and Phillips 1973). This pattern distinguishes them from receptaculitids, which clearly had polar growth (Fisher and Nitecki 1982a). The radiocyathid and receptaculitid growth patterns are both clearly different from the archaeocyathid pattern, in which individual skeletal elements ceased growing once formed. Meroms in many receptaculitids and nesasters in radiocyathids were growing during the entire life span of the organism.

Thus, while the microstructure and morphology distinguish radiocyathids from sponges, they make a comparison with archaeocyathids even more difficult. The only group that can possibly be related to radiocyathids are receptaculitids (see Chapter 3 by Nitecki, this volume).

The main differences between the latter two groups include (1) construction of receptaculitid meroms (plate and stellate structure) and radiocyathid nesasters, (2) lack of gaps in the fused receptaculitid skeleton, (3) polar growth pattern in receptaculitids, and (4) growth pattern of individual skeletal elements. However, these are not fundamental differences. There are some receptaculitids with rosettelike elements under or in the plate—for example, *Calathium bimuralis* (see Miagkova 1984, pl. 26, fig. 2) and *Fisherites reticulatus* (see Nitecki and Debrenne 1979, pl. 3, fig. 5). Miagkova (1981) described "porous" outer walls in the Upper Ordovician *"Receptaculites" poelmi.* Its plates are regularly denticulate in shape, and the "pores" are actually gaps between denticles of neighboring plates. Rietschel and Nitecki (1982) suggested that such a "porosity" might result from activity of overgrowing boring organisms. In my opinion, however, the denticulate plates may rather represent shortened rays of rosettes. A similar trend to shorten the rays occurs in fact in radiocyathids. In the earliest known radiocyathid, the Tommotian *Gonamispongia,* the length of rays is five to six times the diameter of the central knob; in the Botomian *Girphanovella,* in turn, the rays are only two to three times the diameter of the central knob. The radiocyathid outer wall consists of two layers. A reduction in number of rays in the inner layer and a parallel increase in number of secondary rays in the outer layer could possibly lead to construction of the receptaculitid merom. The arrangement of rays of neighboring rosettes in *Gonamispongia* resembles the intercalation of neighboring stellate structures in receptaculitids as described by Fisher and Nitecki (1982a). It is difficult to explain the transformation from the interstitial growth pattern of radiocyathids to the polar growth pattern of receptaculitids. However, the earliest radiocyathids, *Gonamispongia* and *Girphanovella,* clearly had interstitial growth, whereas *Radiocyathus,* which is the latest representative of radiocyathids (it survived into the Tojonian; P. D. Kruse, personal communication 1984), had not.

The big stratigraphic gap between radiocyathids (Tojonian Stage of the Early Cambrian) and receptaculitids (Early Ordovician) might argue against a phylogenetic relationship between these two groups. But even greater gaps are known in the fossil record of many groups. For example, the Sphinctozoa, known previously from the Late Carboniferous to Late Cretaceous (Finks 1970), are now being found also in Recent oceans as well as in Eocene, Devonian, and Cambrian rocks (Vacelet 1979; Pickett and Jell 1983). Thus, stratigraphic gaps cannot serve as a valid criterion for taxonomy.

CONCLUSIONS

It is clear that radiocyathids are neither sponges nor archaeocyathids, and their relationship to receptaculitids appears somewhat tenuous. The microstructure of radiocyathids resembles protozoans, algae, receptaculitids, and sponges (Barskov 1984). However, their morphology and growth pattern rule out a relationship to protozoans.

On the other hand, the interpretation of Porifera as a single phylum has been questioned. Reiswig and Mackie (1983) placed hexactinellids in the subphylum Sinplasma, elevated to the rank of phylum by Barnes (1983). Paleontologic data are consistent with this assignment. The Early Tommotian *Protospongia* probably is a hexactinellid (Finks 1970; Sokolov and Zhuravleva 1983). This seems to be the case also with *Aulophycus*, described originally as an alga (Fenton and Fenton 1939). Spicules were identified in such forms, which were at that time described as the sponge *Multivasculatum* (Howell and Van Houten 1940). Astashkin (in Sokolov and Zhuravleva 1983) demonstrated recently that *Aulophycus* was a widespread frame-builder of organic buildups during the Cambrian. Demospongids, in turn, appeared at about the same time (Kruse 1983), and their isolated spicules are also found in the Tommotian (A. B. Fedorov, personal communication 1984). The first hexactinellids and the first demospongids are so different in morphology that their close phylogenetic relationship is doubtful. Moreover, the first sphinctozoans also are present in the Early Cambrian (Ordian, according to Pickett and Jell 1983). Sphinctozoans are not typical sponges because of the peculiar structure of their soft body (Vacelet 1979). Certain, though not all, sphinctozoans are very similar to archaeocyathids in their morphology, ontogeny, and microstructure (Zhuravlev in press). Archaeocyathids, traditionally regarded as a separate phylum of Metazoa, are now considered to possess sponge characteristics (Debrenne and Vacelet 1983).

Thus, "typical sponges" hexactinellids and demospongids, "probable sponges" archaeocyathids and sphinctozoans, and radiocyathids are all present in the Cambrian. I do not propose to erect a new kingdom for all these groups, but it may be necessary either to expand the concept of the phylum Porifera or to enlarge the composition of the superphylum Parazoa.

Acknowledgments

I would like to express my gratitude to Drs. I. T. Zhuravleva, E. I. Miagkova, and A. B. Fedorov for permitting me to study specimens from their collections, and to Dr. T. A. Sayutina for the data on microstructure.

REFERENCES

Barnes, R. D. 1983. Origins of the lower invertebrates. *Nature* **306,** 224–225.

Barskov, I. S. 1984. Paleontological aspects of biomineralization. *Proc. 27th Int. Geol. Cong.,* vol. 2, Nauka, Moscow, pp. 63–69.

Bedford, R. and Bedford, J. 1936. Further notes on *Cyathospongia* (Archaeocyathi) and other organisms from the Lower Cambrian of Beltana, South Australia. *Mem. Kyancutta Mus.* **3,** 21–26.

————— and Bedford, W. R. 1934. New species of Archaeocyathinae and other organisms from the Lower Cambrian of Beltana, South Australia. *Mem. Kyancutta Mus.* **1,** 1–7.

Boiko, E. V. 1984. The microstructure of Callovian Stromatoporata of Pamir. *In* Sokolov, B. S. (ed.). *Problematics of Paleozoic and Mesozoic.* Nauka, Moscow, pp. 67–72 [in Russian].

Debrenne, F. 1977. Archaeocyathes du Jbel Irhoud (Jebilets-Maroc). *Bull. Soc. Géol. Mineral. Bretagne C* **7**(2), 93–136.

—————, Termier, H., and Termier, G. 1970. Radiocyatha. Une nouvelle classe d'organismes primitifs du Cambrian inférieur. *Bull. Soc. Géol. Fr. Ser. 7* **12,** 120–125.

—————, —————, and —————. 1971. Sur les nouveaux représentants de la classe de Radiocyatha. Essai sur l'évolution des Métazoaires primitifs. *Bull. Soc. Géol. Fr. Ser. 7* **13,** 439–444.

————— and Vacelet, J. 1983. Archaeocyatha: is the sponge model consistent with their structural organization? *Abstr. 4th Int. Symp. Fossil Cnidaria, Washington, D.C.,* p. 4.

Fenton, C. and Fenton, M. 1939. Pre-Cambrian and Paleozoic algae. *Bull. Geol. Soc. Amer.* **50,** 89–126.

Finks, R. M. 1970. The evolution and ecologic history of sponges during Paleozoic times. *Symp. Zool. Soc. Lond.* **25,** 3–22.

Fisher, D. C. and Nitecki, M. H. 1982a. Standardization of the anatomical orientation of receptaculitids. *Paleont. Soc. Mem.* **13,** 1–40.

————— and —————. 1982b. Problems in the analysis of receptaculitid affinities. *Proc. 3rd North Am. Paleont. Conv.* **1,** 181–186.

Flajs, G. 1977. Die Ultrastrukturen des Kalkalgenskeletts. *Palaeontographica B* **160,** 69–128.

Howell, B. F. and Van Houten, F. B. 1940. A new sponge from the Cambrian of Wyoming. *Bull. Wagner Free Inst. Sci.* **15**(1), 1–8.

Korshunov, V. I. 1968. *Gonamispongia,* a new genus of sponges of family Chancelloriidae. *Paleont. Zh.* **1968** (3), 127–129 [in Russian].

Kruse, P. D. 1983. Middle Cambrian *"Archaeocyathus"* from the Georgina Basin is an anthaspidellid sponge. *Alcheringa* **7,** 49–58.

————— and West, P. W. 1980. Archaeocyatha of

the Amadeus and Georgina Basins. *BMR J. Aust. Geol. Geophys.* **5,** 165–181.

Miagkova, E. I. 1981. *Receptaculites poelmi* Miagkova sp. nov. *In* Sokolov, B. S. (ed.). *Problematics of the Phaneorozoic.* Nauka, Moscow, pp. 38–41 [in Russian].

————. 1984. Comparative morphology of *Calathium* Billings 1865, and of *Soanites* Miagkova 1965. In Sokolov, B. S. (ed.). *Problematics of the Paleozoic and Mesozoic.* Nauka, Moscow, pp. 38–43 [in Russian].

Nitecki, M. H. and Debrenne, F. 1979. The nature of radiocyathids and their relationship to receptaculitids and archaeocyathids. *Géobios* **12**(1), 5–27.

Okulitch, V. I. 1937. Some changes in nomenclature of Archaeocyathi (Cyathospongia). *J. Paleont.* **11,** 251–252.

Pickett, J. and Jell, P. A. 1983. Middle Cambrian Sphinctozoa (Porifera) from New South Wales. *Mem. Assoc. Aust. Paleont.* **1,** 85–92.

Reiswig, H. M. and Mackie, G. O. 1983. Studies on hexactinellid sponges. III. The taxonomic status of hexactinellida within the Porifera. *Philos. Trans. R. Soc. Lond. (B.)* **301,** 419–428.

Rietschel, S. and Nitecki, M. H. 1982. Concept of Kingdom Archaeata. *J. Paleont.* **56**(2, suppl), 22.

Romanenko, E. V. 1968. Cambrian sponges of the order Heteractinellidae of Altai. *Paleont. Zh.* **1968** (2), 134–136 [in Russian].

Rushton, A. W. A. and Phillips, W. E. A. 1973. A *Protospongia* from the Dalradian of Clare Island, County Mayo, Ireland. *Palaeontology* **16,** 231–237.

Shishkin, B. B., Fedorov, A. B., and Sundukov, V. M. 1982. Kotuyi archaeocyathid horizon of the southwestern Prianabarye. *In* Khomentovskiy, V. V. (ed.). *New Data on the Stratigraphy of Late Precambrian of Siberia.* Izdatelstvo AN SSSR, Novosibirsk, pp. 20–30 [in Russian].

Sokolov, B. S. and Zhuravleva, I. T. (eds.). 1983. *Lower Cambrian Stage Subdivision in Siberia.*

Atlas of Fossils. Nauka, Moscow, 216 pp. [in Russian].

Termier, G. and Termier, H. 1979. Hypothèse environmentale et symbiotique sur l'origine des spongiaires. *Coll. Int. C.N.R.S.* **291,** 513–520.

Vacelet, J. 1979. Description et affinités d'une éponge sphinctozoaire actuelle. *Coll. Int. C.N.R.S.* **291,** 483–493.

Van Iten, H. and Fisher, D. C. 1983. Microstructural and mineralogical analysis of the meroms of the receptaculitid *Fisherites reticulatus. Geol. Soc. Am. Abstr. Progr.* **15,** 710.

Vologdin, A. G. 1940a. *Archaeocyathids and algae of Cambrian limestones of Mongolia and Tuva,* part 1. Izdatelstvo AN SSSR, Moscow, 268 pp. [in Russian].

————. 1940b. Subphylum Archaeocyatha. *In* Vologdin, A. G. (ed.). *Atlas of Index Fossils of Faunas of the USSR. 1. Cambrian.* Gosgeoltekhizdat, Moscow, pp. 24–96. [in Russian].

Wendt, J. 1979. Development of skeletal formation, microstructure and mineralogy of rigid calcareous sponges from the Late Paleozoic to Recent. *Coll. Int. C.N.R.S.* **291,** 449–457.

Zhuravlev, A. Yu. In press. Recent archaeocyathids? *In* Sokolov, B. S. (ed.). *Problematics of Late Proterozoic and Paleozoic.* Nauka, Moscow, [in Russian].

Zhuravleva, I. T., Konyushkov, K. N., and Rozanov, A. Yu. 1964. *Archaeocyatha of Siberia. Two-walled Archaeocyatha.* Nauka, Moscow, 132 pp. [in Russian].

———— and Miagkova, E. I. 1981. Materials toward the study of Archaeata. In Sokolov, B. S. (ed.). *Problematics of the Phanerozoic.* Nauka, Moscow, pp. 41–74 [in Russian].

————, Zadorozhnaya, N. M., Osadchaya, D. V., Pokrovskaya, N. V., Rodionova, N. M., and Fonin, V. D. 1967. *Fauna of the Lower Cambrian of Tuva (the reference section of Shivelig-Khem River).* Nauka, Moscow. 181 pp. [in Russian].

5. CONSIDERATIONS ON SYSTEMATIC PLACEMENT OF THE STYLIOLINES (INCERTAE SEDIS: DEVONIAN)

ELLIS L. YOCHELSON AND RICHARD H. LINDEMANN

Styliolines are Devonian microfossils. Lardeaux (1969) reviewed several systematic schemes that include the styliolines and contributed his own arrangement. Students of Paleozoic calcareous tube-like fossils generally agree about the concept of the class Tentaculitoidea, but there is no consensus about to which phylum this class, which is based on extinct organisms, should be assigned. Much of the literature implies that it falls within the Mollusca, but this is disputed by some investigators of Paleozoic mollusks (Yochelson 1979).

Lardeaux distinguished three orders in the class: Tentaculitida, Homoctenida, and Dacryoconarida. Within the Dacryoconarida he recognized four families: Nowakiidae, Styliolinidae, Striatostyliolinidae, and Peneauiidae. The forms comprising three of the families are all annulated and have in common a laminated shell, as do the fossils in the other two orders. In contrast, the Styliolinidae have a smooth exterior and a fundamentally different shell structure. Only in their tiny size are styliolines and dacryoconarids superficially similar (Fig. 1A, B).

The family Styliolinidae contains only the genus *Styliolina* Karpinsky, and we have concentrated on this taxon as it is known in eastern North America. Much of our data are drawn from outcrops and samples in central and eastern New York State. Rickard (1975) gives the stratigraphic position of various formations and members mentioned below. The observations on morphology and local distribution of styliolines are supplemented by study of other areas in the Appalachian Basin where these fossils occur.

GENERAL MORPHOLOGY

An individual stylioline is about the size and shape of a pencil point and has even less characters; this is only a modest exaggeration. The stylioline hard parts consist of an extremely narrow cone, circular in transverse section, that expands at an angle of about 10° near the apex, but about 3° along most of the tube length, which rarely exceeds 3 mm (Figs. 1B, C, and 6). The cone is hollow, closed apically, and entirely open at the circular aperture. There are no internal protrusions, partitions, fillings, or septa in the cone.

A problem in describing extinct forms is that many descriptive terms carry some classificatory innuendo; for example, "protoconch" suggests mollusks, whereas "proloculus" suggest foraminifers. If one considers a "hard part" in its entirety, to some the term "tube" implies "worm," "test" indicates "foraminifer," and "shell" denotes "mollusk." "Shell" will be used in this paper only because it is shorter than "hard part."

The shell exterior may appear smooth (Fig. 1C), but all well-preserved specimens bear longitudinal ornament. This ornament is of incised striations, rather than raised lirae (Figs. 1D–F, and 2A). Study of many specimens from several localities suggests that the shell surface is readily affected by diagenetic change and the outermost striated surface commonly is lost.

Virtually the only other morphologic feature that styliolines possess is the apex. Some individuals have a slightly bulbous apex (Figs. 1B, C, and 2A), but other individuals expand nearly uniformly from essentially a bluntly rounded point (Fig. 1D, E). Even among more bulbous specimens, there is a large amount of individual variation with regard to the degree to which the cone is differentiated from the apex. Indeed, it has been impossible to quantify an average amount of apical inflation as a consistent feature. These variations are primary and are not artifacts of postmortem alteration or sample preparation; apical variation has been observed in thin sections as well as among shells and steinkerns (internal filling) removed from the matrix. A steinkern appears to show slightly less constriction near the apex than does an individual with shell preserved, but it is an optical illusion; broken specimens demonstrate that the shell is no thicker or thinner at the constriction than on other areas of the fossil.

The apex of styliolines is simple and uniformly curved, a point confirmed by both thin section and SEM examination. In contrast, some dacryoconarids have apical spines. A few dacryoconarids have a tube that is smooth for some distance from the apex before the first annulation is formed (Fig. 1A), and this has resulted in misassignment of species bearing apical spines to *Styliolina*.

Among the four genera that constitute the Striatostyliolinidae, *Striatostyliolina* Bouček and Prantl and *Metastyliolina* Bouček and Prantl are based on the presence of longitudinal costae, not

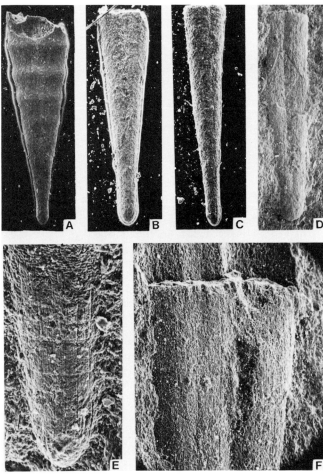

Fig. 1. A. Longitudinal view of a slightly flattened nowakiid, showing the apical area and the relatively long extent of the tube before the first annulations are formed. ×50. Wanakah Shale Member, Ludlowville Formation. USGS locality 10569-SD, USNM No. 398484. B. Longitudinal view of a stylioline slightly compressed in the apertural area. ×75. Wanakah Shale Member, Ludlowville Formation. USGS locality 10595-SD, USNM No. 398485. C. Longitudinal view of an undeformed stylioline; there is some difference in the degree of restriction at the "neck" and the sphericity of the apical area relative to the previous specimen. ×100. Wanakah Shale Member, Ludlowville Formation. USGS locality 10565-SD, USNM No. 398486. D. Crushed stylioline showing no "neck," but preserving longitudinal striations. ×100. Cashaqua Shale. USGS locality 10509-SD, USNM No. 398487. E. Apical area of an undeformed specimen showing the longitudinal striae; near the top of this specimen, pressure solution has obliterated the striations. ×300. Geneseo Shale Member, Genesee Formation. USGS locality 10565-SD, USNM No. 398488. F. Crushed specimen showing striations on the upper shell surface and corresponding ridges on the external mold below. ×150. Geneseo Shale Member, Genesee Formation. USGS locality 10565-SD, USNM No. 398489.

striae. Topotype suites in the National Museum of Natural History, identified by B. Bouček, include specimens with annulations on a shell layer lying atop that bearing the costae. We judge that both genera may be based on preservational effects, rather than biologic features. *Costulatostyliolina* Lardeaux also bears costae rather than striations; we have no data to show whether these occur on an inner shell layer. None of these three taxa can be related to the styliolines from the standpoint of ornament, regardless of whether its presence results from stripping away of an outer shell layer. In all three, what we deem to be the inner part of the shell is thicker than the shell of true styliolines. *Distritostylus* Lardeaux, the fourth genus currently placed in the Striatostyliolinidae, shows prominent growth lines and may be an early growth stage of *Coleolus* (cf. Yochelson, 1986).

SHELL COMPOSITION AND STRUCTURE

The stylioline shell is calcium carbonate. Rarely, individuals are secondarily replaced by pyrite, but

shell replacement is quite uncommon. In the Cashaqua Formation of western New York, concretions contain cephalopods, bivalves, and gastropods that are replaced by barite, but such alteration of stylioline shells has never been observed (W. Kirchgasser, oral communication, 1983). This is a strong indication that the original mineralogy of styliolines is quite distinct from that of mollusks.

Original mineralogy of fossils is difficult to study. Available observations indicate that the stylioline shell was composed of calcite (Lindemann and Yochelson 1984). Petrographic microscope examination shows that the shell consists of microcrystalline calcite prisms oriented with the C-axis normal to the shell surfaces. No inversion of aragonite to calcite has been observed. SEM observations support the conclusion that calcite is the original shell mineral.

The stylioline shell consists of a single homogeneous layer, uniformly about 10 μm thick (Figs. 1B, 2D, and 4). This basic point was noted by Bouček (1964), who nevertheless placed the styliolines in the same general taxonomic group as

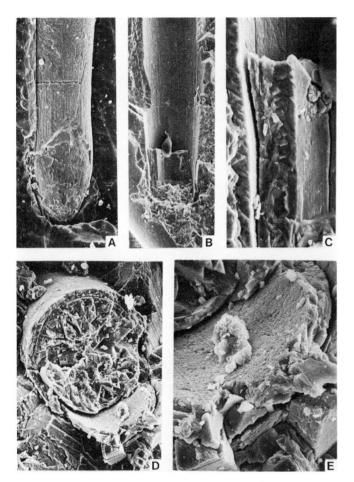

Fig. 2. A. Apical area of a specimen with most of the shell broken away from the steinkern; a homogeneous shell layer of constant thickness may be seen across the "neck" of the apical area, but higher up the shell contains internal cracks as a result of spalling. ×350. Genundewa Limestone Member, Genesee Shale. USGS locality 10561-SD, USNM No. 398490. B. External mold of specimen retaining a fragment of the shell near the apical area; a recent contaminant mars the photograph. ×50. Penn Yan Limestone. USGS locality 10506-SD, USNM No. 398491. C. Enlargement of B showing the shell fragment in cross section; the external mold is to the left, and recrystallization of the homogeneous shell is evident. ×750. D. Slightly oblique broken transverse section of specimen, showing the shell layer of uniform thickness around the aperture. × 500. Genundewa Limestone Member, Genesee Shale. USGS locality 10610-SD, USNM No. 398492. E. Enlargement of lower part of D showing inner surface of shell and the shell thickness. ×1500.

the nowakiids. Not only is the shell of a single layer, it is also imperforate. Not a single pore through the shell or hole on the shell surface has been observed.

Rarely, a broken specimen may give the illusion of more than one layer, but in every instance observed under the SEM, such multiple layers do not persist around the circumference (Fig. 2B–E); this differs markedly from the microstructure of the dacryoconarid *Nowakia,* which, under the SEM, shows a number of shell layers (Fig. 5C). Cursory observation of styliolines under the petrographic microscope may give an impression of two or three shell layers, but it is an illusion produced by Becke lines, spalling, calcite cleavage, or growth of syntaxial rims on the original shell.

Alberti (1975) reported three shell layers in a Lower Devonian specimen of *Styliolina* sp. He did not illustrate the entire fossil, but his figures show the layers clearly; they are not inbricated multiple shells. His observation has not been repeated by others and cannot be confirmed by us. We suspect that he may have dealt with unique material that had weathered or broken in a peculiar manner. Another possibility is that he exam-

ined a different organism whose early growth stages were smooth and mimicked *Styliolina.*

One unexpected aspect of our study of styliolines is that not a single growth line has been observed on the shell exterior. The only indication of growth interruption is seen on some steinkerns. These show a stepped appearance, rather than a smooth profile (Fig. 3A, B). The spacing of these "steps" varies among localities, but this may be a reflection of quality of preservation.

Even more surprising is the presence of small pits on the shell interior, represented by small elevations on the steinkern (Fig. 3C). These pits are associated with elongate low ridges. They show no consistent pattern and are uncommon, but their parity may be preservational. If the shell is not cleanly broken from the steinkern or the steinkern is not of the finest grained material, these features are not seen. What relation, if any, these pits have to the apparent "steps" is not known.

TAPHONOMY

The reason for the specific name of *Styliolina fissurella* (Hall) is obvious; most specimens in

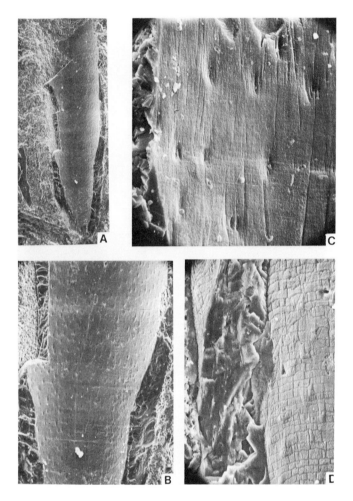

Fig. 3. A. Steinkern in longitudinal view, retaining a fragment of shell in contact with the matrix. ×75. Geneseo Shale Member, Genesee Shale. USGS locality 10565-SD, USNM No. 398493. B. Enlargement of A showing the shell to the upper left, and the "steps" and raised mounds on the steinkern. ×200. C. Enlargement of B showing drumlinoid shape of the mounds and faint elongate rounded ridges; the shell is to the left. ×750. D. Steinkern and shell in longitudinal view. Blocky texture from pressure solution is developed on the outer surface of the shell to the left and on the steinkern; some mounds are still apparent, but others have been nearly obliterated by the diagenetic effect. × 750. Geneseo Shale Member, Genesee Shale. USGS locality 10565-SD, USNM No. 398494.

shales have a longitudinal break caused by compaction (Figs. 1D, 1F, and 6C–E). Crushing may be complex, and some specimens show more than one longitudinal crack (Figs. 4 and 5D). Typically the early part of the shell is uncrushed, leading to misinterpretation that the apex is solid (Hall 1879). In units such as the Wanakah Shale Member of the Ludlowville Formation, the smaller part of the tube may be filled and the remainder broken during compaction, so that specimens are "short" compared to those from other beds. As we understand the compaction process, sediment enclosed in the shell is extruded through the aperture.

A single specimen noted by Lindemann and Yochelson (1984) has at the apical end a relatively large hole with tapering sides; this may have resulted from a radial fracture pattern because of the curvature of the apex. Occasionally, a similar hole is seen in thin section. Compression of the aperture and compaction of enclosed sediment may have caused the apical wall to pop out like the "frost plug" in a frozen automobile engine block.

In most shales, stylioline shells are dissolved, leaving a steinkern and an external impression. Many impressions preserve striations, and in so far as one can compare material from the various kinds of shales, ornament is similar and shell dimensions after crushing are consistent. The common occurrence of striae confirms that shell solution in shales is secondary and is not due to pressure solution during diagenesis.

One effect seldom commented upon is the imbrication of these tiny shells (Fig. 6E). This occurs locally in some limestones, but is common at the extreme base of the Geneseo Member of the Genesee Shale. The Geneseo exhibits all aspects of stylioline preservation from round to crushed.

Within limestones, secondary effects have been observed under SEM. As a result of minor-pressure solution, striations on the exterior commonly are lost (Fig. 1B, C). Where present on part of the shell, they may disappear abruptly on another part (Fig. 1E). Elongate pits of the shell interior, represented by drumlin shaped bumps on the steinkern (Fig. 3A–C), are readily obscured and obliterated by secondary blocky texture on

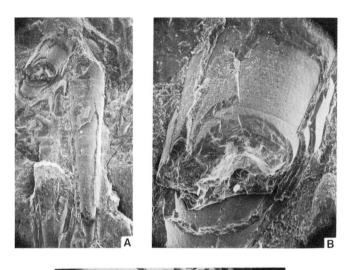

Fig. 4. A. Longitudinal view of several imbricated and oriented specimens. × 50. Genundewa Limestone Member, Genesee Shale. USGS locality 10565-SD, USNM No. 398495. B. Enlargement of A showing three imbricated styliolines, seen in the upper left of A, the central one possessing a steinkern. ×150. C. Enlargement of B showing the homogeneous shell of three specimens; below the middle shell layer is a prominent microstyliolite, demonstrating that the blocky exterior surface texture is produced by pressure solution. × 750.

the steinkern (Fig. 3D). This texture is also occasionally present on the shell exterior, where it can imitate growth lines (Figs. 2B and 4C). Examination of shell–shell contacts (Fig. 4C) and shell–matrix contacts (Fig. 3D) indicates that the blocky texture is a secondary feature produced by pressure solution. As with the striae, solution involves only a small part of the shell thickness.

STRATIGRAPHIC DISTRIBUTION

The oldest styliolines known in the eastern United States are in the early Eifelian Onondaga Limestone (Lindemann and Yochelson 1984). Attempts to find older ones in the eastern United States have been unsuccessful. To the best of our knowledge, no Silurian styliolines have been described. Pre-Eifelian styliolines, such as *S. glabra* Lardeaux, are reported from Europe (see Lardeaux 1969), but there is no information on shell structure of these forms; such occurrences should be reexamined to see if they are the early stages of taxa annulated during later growth.

Another source of confusion is with the early stages of smooth calcareous "worm" tubes, such as *Coleolus*. In the United States this has affected the interpretation of the highest part of the range of styliolines (Yochelson and Hlavin 1985), but there is no reason why it could not equally well lead to misinterpretations of the downward range. The Early Devonian (Pragian) *Styliolina elongata* Peneau, as illustrated by Lardeaux (1969), has a small spike at the apical end and is exceptionally large; it could be a *Coleolus*.

The youngest styliolines known in the eastern United States are in the Angola Member of the West Falls Formation (Late Frasnian) (Yochelson and Kirchgasser 1986). These fossils are at essentially the same stratigraphic interval as the youngest styliolines elsewhere in the world.

Styliolines underwent a population decline prior to extinction and seem to have disappeared before the Frasnian–Famennian "boundary event." We cannot shed any light on why such abundant and ubiquitous organisms died out. Perhaps they were victims of minor temperature or chemical changes that did not affect larger organisms. Whatever the cause of their extinction in eastern North America, it was not one that affected general physical conditions of deposition,

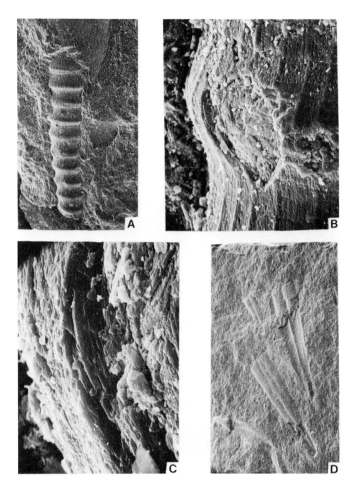

Fig. 5. A. A well-preserved specimen of *Nowakia gyrocanthus* (Hall), showing annulations and fine longitudinal threads. ×75. Geneseo Shale Member, Genesee Shale. USGS locality 10565-SD, USNM No. 398496.
B. Enlargement of A near central part of specimen where broken shell gives a fracture section. ×750. C. Enlargement of B to show the multiple, uniform and continuous wall laminae, quite unlike the spalling occasionally seen in a stylioline shell. ×3000. D. Several crushed styliolines in subparallel orientation, the lower one showing several fissures. ×25. Geneseo Shale Member, Genesee Formation. USGS locality 10567-SD, USNM No. 398497.

for Famennian rocks of post-*Styliolina* age appear no different from those slightly older that contain this fossil.

DIVERSITY AND SIZE

Overall, the squeezing and crushing of specimens, along with diagenetic effects in limestones, provide an Elysian field for the splitter. Fortunately most systematists have recognized and avoided these pitfalls. Only a single species, *S. fissurella* (Hall), is known throughout the range of the group in New York State (Yochelson and Kirchgasser 1986) despite determined efforts to divide specimens into taxa more biostratigraphically restricted. Only about a dozen species have been named, and, if the view expounded above is correct that some of these may be early growth stages of annulated genera, most named species may not be true *Styliolina*.

There may be few kinds of styliolines, because they have so few morphologic features. Yochelson (1978) discussed distinctiveness and diversity; the first indicates a basic morphologic type, whereas the second is a measure of what can be developed

from a basic plan. Apparently the stylioline ground plan was so simple that little diversity could be developed.

The relationship between morphologic complexity and taxonomic diversity is well illustrated by the work of Hall (1879) on *Tentaculites* and *Styliola* (later changed to *Styliolina*). *Tentaculites* is conical, annulated, and ranges in length from near 10 mm to > 10 cm; annulations vary in size, shape, and spacing, providing a number of features for study. Hall (1879) listed 19 species of Devonian tentaculitids. In contrast, he noted only *S. obtusa* and *S. fissurella*, with three varieties in the latter species. When artifacts of taphonomic alteration are considered, these three varieties and two species all constitute a single taxon.

In both physical and biologic realms, there are relationships between abundance and size. Sagan (1979) indicated that small particles produced by planetary collisions are two to three orders of magnitude more numerous than those 10 times larger. In rural northeastern America, hemlock is far less numerous than sumac, and sumac is far less numerous than golden rod. In the marine realm, the baleen whale is far rarer than the krill

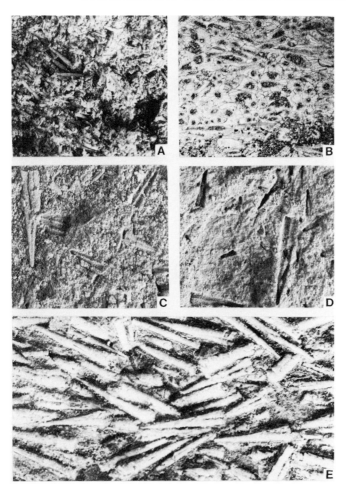

Fig. 6. A. Broken surface of a stylioline limestone. ×15. Genundewa Limestone Member, Genesee Shale. USGS locality 10506-SD, USNM No. 398498. B. Thin section of same rock. ×15. C. Slab showing numerous styliolines in random arrangement, preserved only as external molds. ×15. West River Shale. USGS locality 10562-SD, USNM No. 398499. D. Slab showing crushed styliolines, some with shell preserved having the characteristic "fissure," in random arrangement. ×15. Union Springs Member, Marcellus Shale. USGS locality 10613-SD, USNM No. 398500. E. Imbricated, oriented specimens. ×15. Geneseo Shale, Genesee Formation. USGS locality 10567-SD, USNM No. 398501.

upon which it feeds, and krill in turn is far rarer than the small plankton upon which it feeds. The smaller the organism, the more abundant. This smacks of being a truism, yet it is worthy of some consideration.

An exceptionally large stylioline usually is < 3 mm. If there is an "average" length it is slightly < 2 mm from apex to aperture. Interestingly enough, juvenile styliolines are unknown. There are no specimens that consist only of the bulbous part of the apex; invariably one-third or more of the cone is also present, except in those specimens obviously broken. Rarely is the largest stylioline one finds at any locality more than twice the length of the smallest.

ABUNDANCE

Even the largest stylioline is easily moved by quite gentle water motion, and thus there is no way of translating accumulations of specimens into former life populations. Allowing for all the necessary caution, one is still drawn to the conclusion that styliolines are fantastically abundant in

the Devonian of the eastern United States. An immediate second conclusion is that in the face of overwhelming regional abundance, the local occurrence of styliolines varies greatly. In the Marcellus Shale, as exposed in the southern part of the Massanutten Syncline of Virginia, lithology appears constant through several meters of very dark gray shale, yet the number of specimens ranges from zero to bedding planes white from shells present in profusion. Such change may take place in < 0.5 cm, about the minimum thickness one can sample.

In the Union Springs Member of the Marcellus Shale in New York, some bedding planes bear no specimens while others are completely covered by the specimens so that the surface texture of the plane itself is irregular. Commonly specimens are randomly arranged on shale bedding planes, suggesting extremely quiet water conditions; occasionally specimens in shales are water oriented. We know of no instance of a stylioline preserved apex upward or apex downward; invariably they are confined to a single bedding plane.

The gray Cashaqua Shale in western New York

may be about as unbiased a sample of a life population as might be obtained, for there is no evidence of current movement or concentration. Density of individuals is three per 10 cm^2 on a bedding plane. Using this small number, estimate of the number of bedding planes in the shale soon leads to individuals numbering in the millions at any one exposure. The numerous concretions in the gray shales contain scattered styliolines. A concretion is in a sense a bedding plane frozen in time by local geochemical changes; commonly decay of a cephalopod starts concretion growth, and it seems unlikely such localized decay could occur except in the absence of all currents. Because stylioline abundance in concretions approximates that on bedding planes given above, it provides some support that the density figure may be realistic.

In some areas, styliolines are abundant enough to be the predominant sedimentary component. The Genundewa Limestone, named by Clarke (1897) and referred to by him as the *Styliola* and later the *Styliolina* limestone, consists of a solid mass of styliolines with just enough additional calcium carbonate to cement them (Fig. 6B). At Buffalo, New York, the unit is about 15 cm thick, and thickens eastward so that about 75 km east in Livingston County, it is thicker than a meter. Further east, gray shales between the stylioline-rich beds thicken rapidly and through facies change, the Genundewa is not recognized east of Canandaigua Lake (W. Kirchgasser, written communication, 1983). Counts of pieces of the Genundewa from one locality show about 1500 to 6000 specimens/cm^3. A conservative estimate is that 500 billion specimens occur at a small stream or roadcut outcrop. If this formation is limited to a width of even 1 km, one can speak of quadrillions of specimens without exaggeration.

These figures may be compared to other Paleozoic invertebrates in North America. It is unlikely that more than a few hundreds of thousands of gastropods can be observed on all of the outcrops of Paleozoic rocks that contain them. Even with brachiopods, it is unlikely that more than a few million have been seen in all the outcrops in North America. Such figures pale beside those of stylioline abundance. Apart from fusulinid and endothyrid foraminifers, no Paleozoic fossils have such a profusion of individuals.

PALEOECOLOGY

A planktonic habitat for styliolines was accepted by Bouček (1964), Fisher (1962), and Lardeaux (1969). We agree with this concept, which is supported by their small size, worldwide distribution, and presence in many facies.

In stylioline limestones, rare cephalopods occur and are about the only other fossils ever found. In many gray shales, which might be considered stylioline shales, about the only other fossils present

are occasional specimens of a limited variety of pelecypods and brachiopods; faunal lists for many of these units are provided by Buehler and Tesmer (1963). These invertebrates are small, though far larger than the styliolines. Conventional wisdom (Ruedemann 1934) is that the shelled fossils may have been attached to floating plants. This seems a reasonable interpretation to evoke for many such assemblages, although a small size and few individuals equally favors the notion that bottom conditions may have been marginal for benthonic organisms. Thayer (1974) supported the "floating plant" concept by observations on the Geneseo Shale in which *Styliolina* occurs with locally large numbers of *Barroisella* and *Leiorhynchus*.

Even though epiplanktic rafting accounts for occurrences of small invertebrates in black shales, it also presupposes that the styliolines were characteristic of and limited to "deep water" or "basinal" facies. *Styliolina* and "basin" have become nearly synonymous. As a result, rocks such as the Genundewa are interpreted as deep-water deposits primarily because styliolines abound in them. We know of no data that demonstrate styliolines were epiplanktonic or restricted to pelagic habitats.

Distinction between planktonic and pelagic should be clarified. For decades, there have been arguments over the depth of epicontinental seas; a current tendency is again to suggest that black shales represent "deep water." Variations in lithology and abundance of styliolines that we have observed, in our view are better explained by changes in a shallow-water environment. Minor fluctuations of sediment source, water depth, and population densities can produce what appear to be major changes; in contrast, the greater the water depth the more likely the constancy of sedimentation. We would hold that the styliolines are planktonic, but definitely not pelagic. They swarmed in surface waters a few meters or at most a few tens of meters in depth.

The issue of absolute water depth is difficult to discuss. E. J. Anderson (oral communication, 1977) recognized three Punctuated Aggradational Cycles in the Nedrow Member of the Onondaga Limestone, at Cherry Valley, New York. Each cycle records an abrupt basal stylioline-rich matrix, followed by a gradual decrease through about 70 cm to the stylioline-rich base of the next overlying cycle. Compressed specimens indicate mechanical compaction of sediments prior to cementation (Lindemann and de Caprariis, 1984). Using the 50 percent thickness reduction suggested by Shinn and Robbin (1983), it appears that stylioline abundance increased significantly in water deepened as little as 1.5 m. This reinforces our view that styliolines inhabited extremely shallow water and that slight changes had major effects on their abundance and distribution. For the New York Devonian epicontinental seas, the terms "offshore" and "deep water" are not synonymous.

The characteristic occurrence of styliolines as the only fossil component in many shales has hampered interpretation of life habit. Fortunately, in the Onondaga Limestone they are associated with a rich, diverse fauna. Lindemann (1980) recognized nine paleocommunities in the formation and interpreted environments ranging from lagoon and shallow shelf, through bank and open shelf, to offshore. *Styliolina* occurs in each community, including coral reefs in the Edgecliff Member. Wolosz (1984) and Lindemann and Chisick (1984) demonstrated that these reefs were so shallow as to emerge during life; this places styliolines in the shallowest of waters. Beyond the reef-dominated shallow shelf area, the abundance pattern of styliolines parallels that of brachiopods and trilobites; they reach maximum abundance in strata interpreted as offshore.

Walker and Laporte (1970), studying the Early Devonian Manlius Formation, considered *Tentaculites* restricted to a tidal facies. In the Genesee Formation, Thayer (1974) found tentaculitids to be restricted to a limited number of localities; he made a clear distinction between the inshore tentaculitids and the offshore styliolines.

Throughout the eastern United States, annulated dacryoconarid cones, characteristically nowakiids, are far rarer than styliolines. Where dacryoconarids are common, larger invertebrates are also common, and styliolines are absent or exceedingly rare. If the apical spine of many dacryoconarids was for substrate attachment and if they were attached, at least in early life, differences in distribution between styliolines and annulated specimens becomes more meaningful. In addition to morphologic distinction, there was an ecologic separation between dacryoconarids and styliolines.

PRESUMED LIFE HABIT

The concept of styliolines as planktonic organisms provides a basis for speculation about the living animal. Apart from the bulbous apex in many individuals, there is no differentiation of developmental stages. A long larval stage cannot be ruled out, but it is unlikely, if one does not equate this apex with a molluscan protoconch. The total absence of growth lines implies rapid, uninterrupted growth and a short life span.

In fossils there is not necessarily any direct correlation between size of the hard part and bulk of soft parts. Radiographs of dacryoconarids from the Lower Devonian Hunsrück Shale indicate that only a small amount of material extended out from the aperture. Because styliolines are in the same size range, one *might* assume that styliolines also did not have extensive soft parts. The type of digestive system within the soft parts depends on how "advanced" the styliolines were. If they were metazoans, a blind gut would be logical. If they were "protists," an organized digestive system in that sense would be absent; the presence of symbiotic zooxanthellae within the tissue would not be out of the question.

Styliolines might have been suspension feeders, gathering varied smaller plankton on some sort of a curtain of soft parts that they swept through the water. If this speculation on diet is correct, it can be used to rationalize the presence of a shell. Shells are the rule rather than the exception among living zooplankton, yet little is specifically known of their function or significance, and a diversity of tiny organisms lack a shell but flourish in the plankton. If the stylioline shell was not used for defense or to anchor muscles, perhaps it was involved in the organism's buoyancy. One possibility is that the calcite served as ballast to offset the lower density of protoplasm; protoplasm, as it is commonly considered with enclosed gas vacuoles, has a specific gravity slightly less than that of water, about a third that of calcite. A heavy shell would ensure that the aperture pointed upward, the best position for feeding. It is far easier to catch food falling down into a funnel than when holding a funnel sideways or downward.

Modern plankton are involved in large diurnal vertical migrations. In some, the organism traps oil to assist upward movement, later to expel it and allow the organism to sink. Without the weight of the hard part, negative buoyancy would seem difficult. The pits on the interior of the stylioline shell may have served to increase contact between soft parts and shell. These pits extend close to the apex and if they held tissue firmly, the apex was not hollow during life. Were any oil present, it was in vacuoles within the tissue rather than in the apex.

One may also speculate as to the utilization of styliolines as prey, but this line of inquiry is fruitless. There is no indication that any organism higher on the food chain ate styliolines, or at least ate them as a principal dietary staple. Fish remains are quite rare in stylioline-rich shales. Accumulations that might be interpreted as coprolites have been searched for unsuccessfully. As noted, cephalopods do occur in some of the same beds as styliolines and could have eaten them as part of their diet. However, after the styliolines died out, cephalopods and fish persisted. Either styliolines formed a food source for an unknown, nonfossilized organism, or they were essentially ignored as a food.

POSSIBLE RELATIONSHIPS OF STYLIOLINES TO MOLLUSKS

More than a century ago, styliolines were assigned to the pteropods, a group of pelagic gastropods, based largely on general similarity of shell form to the living *Styliola*. Pelseneer (1888) removed all Paleozoic fossils from this group, and since Pelseneer's work, no student of the pteropods has seriously considered the styliolines to be related to them. Unfortunately, once styliolines were placed in the "pteropods," paleontologists accepted this assignment uncritically, even though we find no relationship between pteropods and styliolines that is worthy of serious consideration.

Styliola itself has a fossil record that extends back into the Tertiary; as a group, the pteropods do not predate the Cretaceous. There is a time gap of about 250 Ma between them and the Devonian styliolines. We consider this absence to be significant, for both styliolines and pteropods are extremely abundant.

Pteropods show a fair degree of variation in shell shape, but most taxa have the coiled form characteristic of typical gastropods. The very slightly arched, nearly straight shell of *Styliola* is an exception. Most modern specimens of *Styliola* are larger than the styliolines. In cross section they are subtriangular, not circular. Herman (1978) noted the presence of a longitudinal ridge on the dorsal side of the arched shell. Thus, the hard-part morphology of *Styliolina* and of *Styliola* is not particularly close. Pteropods employ more than one microstructural plan; some show three shell layers, although several apparently have a single homogeneous aragonitic layer (Herman 1978). Significantly, all modern pteropod shells are composed of aragonite, and none of the rare Cenozoic and late Mesozoic pteropod fossils have ever been reported to be calcitic.

Assignment of styliolines to cephalopods is a more recent idea, traced mainly to the work of Blind (1969). Apparently there are three bases for this interpretation. First is a presumed relationship of styliolines to tentaculitids as members of the same class, coupled to the assumption that tentaculitids are cephalopods. Second is interpretation of soft parts of nowakiids, seen in radiographs from the Lower Devonian Hunrück Slate, as those of cephalopods (Brassel et al. 1971; Blind and Stürmer 1977), again coupled to the assumption of a close relationship between these dacryoconarids and styliolines. Third is interpretation of the shell structure of *Styliolina* as being like that of cephalopods.

Blind (1969) considered the septa of some tentaculitids to indicate relationship to cephalopods. However, the septa are of a different construction, and this may not be a valid point for relationships. Regardless, the styliolines show no internal partitioning. Several tens of thousands of specimens in all states of preservation have been examined, and nothing in the nature of a septum has been observed; flattened specimens in particular would show if any septa were present, for the mode of crushing would be different. No internal structures have been observed in thin sections.

The second point is open to alternative interpretation. Yochelson (unpublished data) examined additional radiographs from the Hunsrück Slate. The replaced soft parts of authentic cephalopods from that deposit do not resemble those of associated tiny nowakiids. So far as the shell itself is concerned, Hunsrück nowakiids may show in radiographs what appears to be septa or may show what appears to be a siphuncle. The apparent septa result from variable pyritization along the shell length, and the apparent siphuncle is the fissure produced by postdeposition compression. No nowakiids show both siphuncle and septa in radiographs.

As to the third point, we have not been able to reproduce the microstructure results of Alberti (1975), discussed above. Whenever we have examined well-preserved shell structure of a stylioline, invariably it consists of a single homogeneous layer. The one specimen studied by Alberti is older (Early Devonian) and in his own words poorly preserved; it may not be a *Styliolina*. There is sufficient variation in diagenesis of material that we do not question his report of three shell layers, but these three layers are not like the shell structure of any cephalopod that has been illustrated. Horowitz and Potter (1971) and Majewske (1974) record three layers in cephalopods, but they are of aragonite and nacre, not calcite. The shell structure evidence does not support interpretation of styliolines as cephalopods, nor does it support Alberti's material as being part of the Cephalopoda.

We emphasize that we see no relationship between the styliolines and any extant or extinct class of mollusks.

POSSIBLE RELATIONSHIPS OF STYLIOLINES TO TENTACULITIDS

Two major classifications of Tentaculitoidea accepted a general relationship between annulate tubes and smooth styliolines, but placed different emphasis on features for classification. Fisher (1962) suggested that the bulbous apex or protoconch linked all forms, whereas Bouček (1964) emphasized the wall structure as a prime factor in classification. Our view is that wall structure is far more significant than apical features. The wall of Tentaculitida is laminated or foliated (Larsson, 1979). Majewske (1974) considered the tentaculitid wall microstructure as foliated in a similar manner to that of the bryozoans. The other two orders also have a laminated wall structure.

In contrast, the wall of the styliolines is a single, thin layer, not laminated and not accretional. We consider the difference in wall structure to be so fundamental that we would not judge these forms to be related at a class or phylum level. We cannot accept Bouček's placement of styliolines as a family within the class Tentaculitoidea, a point that heretofore has been followed by all subsequent workers.

Tentaculitids and styliolines have in common a closed apex and a slowly expanding radially symmetric conical shape. Recall that in older classifications brachiopods and pelecypods were classed together, for both groups had two valves. This is not meant as a frivolous remark, but an attempt to rank features of systematic importance. Even in the absence of soft parts, paleontologists would be able to see differences in the shell structure of the two kinds of valves and soon distinguish between them. We submit that styliolines and tentaculitids form a parallel case. A closed apex and radial symmetry need not by themselves indicate close relationship or indeed any more relationship than two valves relate brachiopods and pelecypods.

We do judge that annulations and laminated shell microstructure link the tentaculitids, homoctenids, and dacryoconarids into a major group that may be of class rank. It is not necessary to discuss their relationships in this paper, but we think they are "worms" rather than part of the more conventional phyla of Paleozoic invertebrates.

The significance that systematists attach to hard-part composition as a taxonomic character varies widely. For example, Porifera have spicules of organic matter, calcium carbonate, or silica, whereas in contrast, the Echinodermata have so concentrated on calcite in formation of hard parts, that to observe a cleavage face on a broken unknown fossil often provides sufficient information to assign it to the phylum. With exceptions, one can generalize about the way most individuals of a major group will appear when fossilized, for there is a characteristic "habitus" of mineralogy and microstructure. This observation may be important in classifying fossils of unknown affinities, for the hard part is a product of physiology that is preserved for study. If that logic is followed, the styliolines are not part of the Tentaculitoidea.

The Tentaculitoidea began at least in the Early Ordovician and died out in the Late Devonian, apparently persisting into the Early Famennian. One suggestion is that tentaculitids were benthonic, but during the mid-Paleozoic one branch produced smaller organisms, the dacryoconarids and homotcetids, which became pelagic in habitat. The styliolines began later than the class Tentaculitoidea and died out earlier. It should be no surprise that tiny annulated forms sometimes occur rarely with masses of styliolines, for the shells are hydrodynamic equivalents; this does not imply they are relatives.

POSSIBLE RELATIONSHIP OF STYLIOLINES TO PROTISTS

To the best of our knowledge, comparison of features of styliolines to those of the protists has never been seriously considered by previous workers, apart from Lardeaux (1969). The kingdom Protista is staggering in the diversity of taxa and number of individuals, relative to the kingdom Animalia. This generalization holds for both the Recent and the fossil record. The size of styliolines places them above the average size of autotrophic protists, though within the size range of zooplankton, particularly foraminifers (Table 1). In shape they are comparable to the tintinnids, and some unilocular foraminifers.

The notion of continuous rather than intermittent growth is to be expected in the protists. This style of development is particularly characteristic of many unilocular Foraminifera tests, for their continuous growth does not produce growth lines. In suggesting possible evolutionary events in chamber arrangement among foraminifers, Pokorny (1963) illustrated a straight, conical, uniloc-

TABLE 1. *General sizes and shapes of selected marine plankton*

Group	Size Range (μm)	Most Common Shapes
Phytoplankton		
Cyanophytes	< 25	Coccoid, ovoid, discoid, pyriform
Dinoflagellates	20–150	Subspherical, ovoid, discoid, rod
Silicoflagellates	20–100	Discoid, hemispherical
Chrysomonads	3–25	Subspherical
Diatoms	20–200	Pennate, centric, variable
Coccoliths	< 20	Spherical, ovoid
Calcispheres	200	Spherical
Zooplankton		
Tintinnids	20–2000	Cylindrical, conical, bell shaped
Calpionellids	45–150	Vaselike, conical, bell shaped
Radiolarians	100–2000	Spherical, conical, bell shaped
Foraminifers	500–3000	Extremely variable
Chitinozoans	30–1500	Flask or bell shaped
Pteropods	1000–3000	Coiled, conical

ular foraminifer, not unlike *Styliolina*. The fossil record shows that the Devonian was a time of rapid evolutionary change among calcareous foraminifers; several taxa of the Parathuramininacea and Endothyracea were unilocular, tubular, or conical in shape.

The possibility that styliolines and foraminifers may be closely related is intriguing, but shell structure data belies it. Members of the two Devonian superfamilies mentioned above construct the test wall from microgranular calcite of two layers, and sometimes include arenaceous material; the wall of the Endothyracea is perforate (Loeblich and Tappan 1964). Bilayered or multilayered shells are characteristic of foraminifers.

The styliorine shell microstructure bears striking similarity to that of calpionellids. Both construct homogeneous, thin, nonaccretionary walls of calcite having the C-axis of each crystallite oriented normal to the shell surface. Calpionellids range from Late Jurassic to Early Cretaceous, and their morphology is unlike that of the styliolines.

The overall shape and biostratigraphic distribution of the styliolines preclude them from inclusion in any known taxon of the Protista. However, when paleoecology, shell mineralogy, microstructure, and size are all taken into account, *Styliolina* falls squarely within the ground "plan" of the protists.

CONCLUSIONS

In working with any extinct fossil group, there is a very human tendency, and at least in the past an urgency, to look for descendants in the recent fauna and use them as the basis for interpretations (see Chapter 1 by Bengtson and Chapter 2 by Babcock, this volume). Thus *Nautilus* has provided a window into the past that has been of considerable utility in understanding a group of mollusks that otherwise would have seemed very strange. Where there are no obvious close living descendants, one hopes for them, as in the case of the Loch Ness monster as a presumptive *Plesiosaurus*. Between these two extremes one must reason by analogy. More than a century ago, both styliolines and pteropods were poorly known, and to our forebears'there was some basis for assuming a relationship.

The report by Pelseneer (1888) marked the beginning of new insights into these gastropods. In contrast, the work of Hall (1879) on the Devonian faunas was treated as effectively the final work on the fossil forms. Bouček (1964) outlined the history of study of Tentaculitoidea giving full credit to all workers, but it is fair to say that for three-quarters of a century virtually nothing of a biologic nature, and very little of a stratigraphic nature was done with the group.

Work by Fisher and Bouček in turn stimulated the investigations of Blind. His major investiga-

tion of tentaculitids led to his studies of nowakiids, and by inference the styliolines. We consider the data on shell mineralogy and microstructure to be features separating the styliolines from the tentaculitids, nowakiids, and all other annulated forms, and thus from the cephalopods. Others may argue the merits of classifying Tentaculitoidea as cephalopods; we find the assignment implausible.

We are impressed with both the vast number of specimens and the limited variety among the styliolines. A single microfossil species represented by billions upon billions of individuals is not at all typical of Devonian benthonic invertebrates. Indeed styliolines are so different that one is led to consider models based on the living protists, rather than on metazoans.

One author (E. L. Y.) is mainly interested in Paleozoic mollusks and has attempted to remove from that phylum taxa that seem alien to his view of that group. Within the Mollusca, Yochelson has proposed some high-level groups, including classes, based entirely on extinct forms. The other author (R. H. L.) encountered styliolines in thin sections while concerned with matters of carbonate petrology. Lindemann found that styliolines and mollusks are easily differentiated on attributes of microstructure and diagenetic alteration. Thus, there is some basis for our joint opinion that these fossils are not Mollusca. The styliolines are different.

One aspect of classification is embodied in the old proverb that the pitcher carried to the well too many times will be broken, sooner or later. Yochelson (1977) proposed a phylum based on a single extinct genus. This is as far out on a limb of the tree of classification as it is possible to go. Once is enough. We think the styliolines eventually will be judged to be a group of high-level rank, perhaps even a phylum, and that they will be placed in the kingdom Protista, but it is too soon to make a formal proposal. We leave them as phylum *Incertae Sedis*.

Acknowledgments

A U.S. Geological Survey G. K. Gilbert Fellowship to Yochelson and extensive collection and sample preparation partly supported by a Skidmore College Research Grant to Lindemann allowed the research reported on to be performed.

Thanks should begin with acknowledgment to the late B. Bouček, who shared his work in 1962. He and Dr. L. Marek guided Yochelson to collecting localities near Prague. In 1981 and 1982, Yochelson's debt to a number of persons increased for assistance provided in locating outcrops. In alphabetical order, these include Mr. Loren Babcock, Mr. Gerald Kloc, Dr. William Kirchgasser, Mr. Charles Mason, Dr. Eugene Radar, and Mr. John Way.

Types, figured specimens, and other material of

historical nature were lent by Drs. R. L. Batten, Ed Landing, and M. Nitecki. Dr. D. Fisher generously shared his ideas with us. As usual, Dr. G. A. Cooper was a source of specimens, data, and inspiration.

Others who contributed specimens, ideas, or both, include the late Dr. Alan Be, Mrs. Ruth Chilman, Dr. James Conkin, Dr. J. Dennison, Dr. R. Effington, Dr. K. Hasson, Dr. Donald Hoskins, Dr. A. Horowitz, Dr. L. Kent, Mrs. J. Lawless, Dr. Robert M. Linsley, Dr. J. Rosewater, Dr. D. Schumacher, and Dr. C. Thayer.

During a trip to Europe, Yochelson had the opportunity to exchange ideas with Prof. W. Blind, Prof. H. Lardeaux, Prof. K. Larsson, Dr. F. Lütke, and Prof. W. Stürmer. He thanks all for both stimulating scientific discussion and warm personal hospitality.

The staff of the SEM laboratory at the U.S. National Museum have been unfailingly cooperative and enthusiastic through many long sessions; their pictures have provided a wealth of new information. Additional photographic service were provided by Robert McKinney and Haru Makazaki of the U.S. Geological Survey. Finally, for a year, Mr. A. Casilio prepared swarms of samples and listened patiently as new interpretations flowered and withered in quick succession.

REFERENCES

Alberti, G. K. B. 1975. Zur Struktur der Gehäusewand von *Styliolina* (Dacryoconarida) aus dem Unter-Devon von Oberfranken. *Senckenb. Leth.* **55,** 505–511.

Blind, W. 1969. Die Systematische Stellung der Tentakuliten. *Palaeontographica A* **133,** 101–145.

———— and Stürmer, W. 1977. *Viriatellina fuchsi* Kutscher (Tentaculoidea) mit Sipho und Fangarmen. *N. Jb. Geol. Paläont. Mh.* **1977,** 513–522.

Bouček, B. 1964. *The Tentaculites of Bohemia.* Publishing House of Czechoslovak Academy of Sciences, Prague, 215 pp.

Brassel, G., F. Kutscher, and W. Stürmer. 1971. Erste Funde von Weichteilen und Fangarmen bei Tentaculiten. *Abh. Hess. I. A. Bodenforsch.* **60,** 44–50.

Buehler, E. H. and Tesmer, I. 1963. Geology of Erie County, New York. *Buffalo Soc. Nat. Sci. Bull.* **21**(3), 1–118.

Clarke, J. M. 1897. Stratigraphic and faunal relations of the Oneonta sandstones and shale, the Ithaca and the Portage groups in central New York. *N.Y. State Geol. 15th Ann. Rep.* pp. 27–81.

Fisher, D. W. 1962. Small conoidal shells of uncertain affinities. *In* Moore, R. C. (ed.) *Treatise on Invertebrate Paleontology,* vol. W, *Miscella-nea.* Geological Society of America and Univ. of Kansas Press, Lawrence, pp. W98–W143.

Hall, J. 1879. *Geological Survey of the State of New York. Paleontology,* vol. 5, part 2, *Containing Descriptions of the Gastropoda, Pteropoda, and Cephalopoda of the Upper Helderberg, Hamilton, Portage, and Chemung Groups.* Charles Van Benthuysen and Sons, Albany, N.Y., 492 pp., 120 pl. (2 vols.).

Herman, Y. 1978. Pteropoda. *In* Haq, B. U. and Boersma, A. (eds.). *Introduction to Marine Micropaleontology.* Elsevier, New York, pp. 151–159.

Horowitz, A. S. and Potter, P. E. 1971. *Introductory Petrography of Fossils.* Springer, New York, 302 pp.

Lardeaux, H. 1969. *Les Tentaculites d'Europe Occidentale et d'Afrique du Nord.* CNRS, Paris, 238 pp.

Larsson, K. 1979. Silurian tentaculitids from Gotland and Scania. *Fossils and Strata* **11,** 1–180.

Lindemann, R. H. 1980. Paleosynecology and paleoenvironments of the Onondaga Limestone in New York. Unpubl. PhD. dissert. Rensselaer Polytechnic Institute, Troy, N.Y., 121 pp.

———— and de Caprariis, P. 1984. *Styliolina fissurella* (Hall): evidence for the compaction of Devonian limestones. *Geol. Soc. Am. Abstr. Progr.* **16,** 47.

———— and Chisick, S. A. 1984. Cyanophyta and Rhodophyta from the Onondaga Formation (Lower Middle Devonian) of New York State. *Geol. Soc. Am. Abst. Progr.* **16,** 576.

———— and Yochelson, E. L. 1984. Styliolines from the Onondaga Limestone (Middle Devonian) of New York. *J. Paleont.* **58,** 251–259.

Loeblich, A. R., Jr. and Tappan, H. 1964. *In* Moore, R. C. (ed.). *Treatise on Invertebrate Paleontology,* part E, *Protista* 2. Geological Society of America and Univ. of Kansas Press, Lawrence, 900 pp.

Majewske, O. P. 1974. *Recognition of Invertebrate Fossil Fragments in Rocks and Thin Sections.* E. J. Brill, Leiden, 101 pp.

Pelseneer, P. 1888. Pteropoda: part 2. *Zool. Challenger Exped.* **23,** 1–97.

Pokorny, V. 1963. *Principles of Zoological Micropaleontology.* Macmillan, New York, 652 pp.

Rickard, L. V. 1975. Correlation of the Silurian and Devonian rocks in New York State. *N.Y. State Mus. Sci. Serv. Map Chart Ser.* **24.**

Ruedemann, R. 1934. Paleozoic plankton of North America. *Geol. Soc. Am. Mem.* **2,** 141 pp.

Sagan, C. 1979. *Broca's Brain.* Ballantine Books, New York, 398 pp.

Shinn, E. A. and Robbin, D. M. 1983. Mechanical and chemical compaction in fine-grained shallow-water limestones. *J. Sed. Petrol.* **53,** 595–618.

Thayer, C. W. 1974. Marine paleoecology in the

Upper Devonian of New York. *Lethaia* **7,** 121–155.

Walker, K. P. and Laporte, L. F. 1970. Congruent fossil communities from the Ordovician and Devonian carbonates of New York. *J. Paleont.* **44,** 928–944.

Wolosz, T. H. 1984. Evidence for an Onodaga sea level fluctuation derived from patch reefs. *Geol. Soc. Am. Abstr. Prog.* **16,** 72.

Yochelson, E. L. 1977. Agmata, a proposed extinct phylum of Cambrian age. *J. Paleont.* **51,** 437–454.

————. 1978. An alternative approach to the interpretation of the phylogeny of ancient mollusks. *Malacologia* **17,** 165–191.

————. 1979. Early radiation of Mollusca and Mollusc-like groups. *In* House, M. R. (ed.). *The Origin of Major Invertebrate Groups.* Academic Press, London, pp. 323–358.

————. 1986. Middle Devonian *Styliolina obtusa* (Hall) (Incertae Sedis) and *Styliolina spica* (Hall) ("Vermes") from western New York, reconsidered. *J. Paleont.,* in press.

———— and Hlavin, W. J. 1985. *Coleolus curvatus* Kindle ("VERMES") from the Cleveland Member of the Ohio Shale, Late Devonian (Famennian) of Ohio. *J. Paleont.* **59,** 1298–1304.

———— and Kirchgasser, W. T. 1986. The youngest styliolines and nowakiids (Late Devonian) currently known from New York. *J. Paleont.,* in press.

6. PRECAMBRIAN PROBLEMATIC ANIMALS: THEIR BODY PLAN AND PHYLOGENY

MIKHAIL A. FEDONKIN

Many, if not all, Precambrian metazoans can be considered as problematica. Attempts to place the Vendian or Late Precambrian animals within post-Vendian taxa usually were contrived. In this paper, I discuss the main body plans of Precambrian soft-bodied metazoan fossils in the perspective of their symmetry as the fundamental criterion of body plan analysis, or promorphology.

Body plan analysis appears indeed to be a very useful tool for the classification of Vendian fossils. The first reason is that the symmetry varies widely among lower metazoans, as for instance within the class Hydrozoa. The second reason is the small number of characteristic morphologic features that could be used for assessment of the systematic position of Vendian metazoans. The effectiveness of promorphologic analysis has, in fact, been well demonstrated by Beklemishev (1964).

VENDIAN RADIATA AND THE EARLY PHYLOGENY OF COELENTERATA

Radial symmetry predominates among Precambrian metazoan fossils (Fig. 1). Radially symmetric forms constitute approximately two-thirds of the soft-bodied metazoan specimens collected near Ediacara in South Australia (Glaessner 1972, 1979). They account for ≥ 70 percent of the total number of species described from the White Sea biota of the Vendian (Fedonkin 1981a,b, 1982). An even higher proportion of radially symmetric forms is found in the Vendian of the Ukraine (Palij et al. 1979; Fedonkin 1981a, 1983b), whereas no other fossils have been discovered in some other Precambrian localities, for example in the Avalon fauna of Newfoundland (Anderson and Misra 1968; Anderson 1976, 1978; Anderson and Conway Morris 1982) or in the Khatyspyt Formation of Olenek Uplift in northern Yakutya (Sokolov and Fedonkin 1984).

On the basis of their symmetry, all the Enterozoa are united into division Bilateria, while all the Coelenterata represent division Radiata. The wide diversity of radial symmetries in extant coelenterates may indicate that the evolution of this phylum led to realization of various kinds of radial symmetry. Precambrian Radiata also represent a considerable variety of symmetries, but the range of this diversity differs from later Radiata.

The Vendian fossils with radial symmetry can be subdivided into three main groups: (1) concentrically organized forms with a symmetry axis of infinitely high order, (2) radially arranged forms with a symmetry order changing in ontogeny, and (3) forms with a constant order of radial symmetry. These three groups correspond to some distinct levels of organization of Precambrian Radiata.

Concentric body plan

Coelenterates with monaxial heteropolar symmetry of infinitely high order are the most abundant group of Precambrian megascopic metazoans. Representatives of this group have in common the type of symmetry (i.e., the concentric body plan resulting from additive growth), and asexual reproduction, in particular simple longitudinal fission. Well-developed radial elements in their wide gastral cavity and radially arranged branching canals of the gastrovascular system are absent. All these traits suggest the primitive and archaic character of the group as a whole and its monophyletic origin. I consider this group of the Vendian Radiata as the class Cyclozoa within the Coelenterata (Fedonkin 1983a, 1984). Three taxa of lower rank are recognized within this class. They differ in body structure, mode of growth and reproduction, and ecology.

One group consists of sedentary organisms with, or sometimes even without, irregular concentric body plan (e.g., Nemiana, Beltanelliformis, Sekwia). The symmetry and the body plan of the simplest representatives of this group are strikingly similar to the hypothetical gastrula-like ancestor of all multicellular organisms, which is conventionally placed at the base of the phylogenetic tree of Metazoa. This body plan is the most primitive one among Vendian coelenterates and possibly in the entire history of the phylum. Such a symmetry is rare among Recent cnidarians; it occurs only at their larval stage and in adults of Protohydra.

Another group includes medusoids with prominent, regular concentric body plan, in which the number of concentric zones increases during ontogeny. Oligomerous forms in this group usually show very few wide, concentric rings embracing either the entire disclike body (Cyclomedusa dav-

Fig. 1.   Reconstruction of marine life in the Vendian. Drawing by Leonid Tolpygin.

idi Sprigg) or its central part only (*Cyclomedusa plana* Glaessner and Wade). Their concentric organization may be combined with radial structures, such as numerous radial furrows in *C. davidi* or irregular bosses that increase in size outward and form radially elongated lobes in *Mawsonites spriggi* (Glaessner and Wade 1966). Polymerous forms, in turn, have numerous, relatively narrow, concentric rings that are particularly narrow in the center but wider in the outer zone of the disc (e.g., *Kullingia, Eoporpita, Kaisalia*).

Some polymerous cycloidal Radiata combine concentric organization with radial symmetry, for example, *Eoporpita medusa,* with several concentric series of radially oriented polypids (Wade 1972a); others combine the concentric body plan with bilateral symmetry, for example, *Ovatoscu-*

tum concentricum and *Chondroplon bilobatum* (Glaessner and Wade 1966; Wade 1971). The latter phenomenon demonstrates the plausibility of derivation of bilaterally symmetric, segmented animals from forms with concentric body plan; for example, derivation of *Dickinsonia* from *Kullingia,* over *Ovatoscutum* and *Chondroplon* (Fedonkin 1983a,c). This unexpected discovery emphasizes the significance of Vendian fauna for the most fundamental problems in comparative anatomy of invertebrates.

The third group of Precambrian Radiata with a concentric body plan includes forms with small numbers of concentric rings that remain constant throughout the ontogeny. Typical representatives of this group include *Medusinites asteroides* Glaessner and Wade, *Paliella patelliformis* Fedonkin, *Nimbia occlusa* Fedonkin, *N. dniestri* Fedonkin, and *Irridinitus multiradiatus* Fedonkin, each with two concentric zones, and *Ediacaria flindersi* Sprigg and *Tirasiana disciformis* Palij, with three concentric zones each. Many of these forms tend to combine concentric organization with prominent radial elements, such as fine radial furrows in the outer ring of *Ediacaria* and *Irridinitus* or marginal tentacles in *Ediacaria* and *Nimbia* (Fedonkin 1983b).

The combination of concentric and radial symmetries may be indicative of a phylogenetic relationship of the Cyclozoa to more advanced Coelenterata that have a radial body plan. On the other hand, such a combination may also characterize very primitive forms.

Sedentary forms prevail among the Cyclozoa, while pleistonic and planktonic animals are rare. This may suggest the primary nature of polypoid stage in the evolution of cnidarian metagenesis. If so, the environmental anisotropy of the seafloor and a wide spectrum of benthic habitats could have promoted diversification of Precambrian coelenterates. Diversification within the framework of essentially concentric body plan is represented by a gradation among Vendian coelenterates with symmetry of infinitely high order. The most primitive step was the concentric zonation of the body. It was followed by polymerization of concentric rings and subsequent oligomerization and strong regulation of their number, in combination with either radial symmetry of variable order or bilateral symmetry.

Radial symmetry of variable order

Forms with monaxial heteropolar symmetry of variable but measurable order represent a large and diverse group of Vendian metazoan fossils. Their symmetry is reflected in the structure of the entire organism as well as in particular systems of organs: marginal tentacles (*Hiemalora stellaris* Fedonkin), channels of the gastrovascular system (*Rugoconites enigmaticus* Sprigg), gonads (*Hallidaya brueri* Wade, *Armillifera parva* Fedonkin, *Elasenia aseevae* Fedonkin), gastral cavity (*Bonata septata* Fedonkin), and other organs (Glaessner 1979; Fedonkin 1981a, 1983a).

The main peculiarity of this group, which is here considered as the class Inordozoa, consists in an increase in the order of radial symmetry during ontogeny; new antimeres are added while the organism is growing, but no order has been detected in their formation. Forms with this kind of symmetry were more widespread in the Vendian than in the Phanerozoic. Among living cnidarians, this kind of symmetry is believed to represent one of the lowest levels of organization (Beklemishev 1964).

Despite the apparently primitive mode of formation of antimeres, this group of Radiata seems to exhibit a higher organization than the Cyclozoa, characterized by a symmetry axis of infinitely high order. The Inordozoa usually have a well-organized system of tentacles and variable, often complicated systems of gastrovascular canals that branch several times toward the periphery and sometimes form a plexuslike network. The system of reproductive organs is well developed and seems to indicate prevalence of sexual reproduction; asexual reproduction appears to be rare in this group of Radiata.

Radial symmetry of constant order

Coelenterates with radial symmetry of constant order throughout the ontogeny are less numerous among the Vendian Radiata than representatives of groups with more primitive organization and archaic symmetry. The opposite is true for the Recent Cnidaria. This may indicate that the regulation of radial symmetry by stabilization of the number of antimeres was a relatively late phenomenon in the evolution of Coelenterata. The Vendian Radiata with radial symmetry of stable order are more similar to living cnidarians than to other Precambrian coelenterates. Nevertheless, they have some characters that are typical neither of living coelenterates nor of Recent metazoans in general.

One such peculiarity is the threefold symmetry. This kind of symmetry occurs only exceptionally in Recent cnidarians; in other metazoan groups, it is rare and always a secondary phenomenon. However, it was widespread among the Vendian medusae, such as *Skinnera brooksi* Wade, *Tribrachidium heraldicum* Glaessner, *Albumares brunsae* Fedonkin, and *Anfesta stankovskii* Fedonkin (Glaessner and Wade 1966; Wade 1969; Keller and Fedonkin 1977; Fedonkin 1981a, 1983a). This unique kind of radial symmetry makes this group of animals sufficiently homogeneous in the phylogenetic sense to allow us to consider it as a high-rank taxon, the class Trilobozoa (Fedonkin 1983a). I tentatively include in this class also the oldest known skeletal organisms, described as the Angustiochreidae (Valkov and Sysoev 1970) or Anabaritidae (Missarzhevsky 1974; Glaessner 1979) and attributed by Valkov (1982) to the subclass Angustiomedusae. They appeared at the end of the Vendian and became widespread in the early Cambrian (see Chapter 8 by Rozanov, this volume). Their relationship to the class Scyphozoa is questionable, because threefold symmetry is not typical of the latter class of cnidarians. The fourfold symmetry prevails in the Scyphozoa. Therefore, the Trilobozoa are here recognized as a separate class of coelenterates, similar in their organization to, but different in symmetry from, the Scyphozoa.

Forms with fourfold symmetry comprise the bulk of Precambrian Radiata with radial symmetry of constant order; they include *Conomedusites lobatus* Glaessner and Wade, *Kimberella quadrata* (Glaessner and Wade), *Ichnusina cocozzi* Debrenne and Naud, *Persimedusites chahgasensis* Hahn and Pflug, *Staurinidia crucicula* Fedonkin, and others (Glaessner and Wade 1966; Wade 1972a; Hahn and Pflug 1980; Debrenne and Naud 1981; Fedonkin 1983a, 1984). Their high level of organization and typically scyphozoan symmetry allow for attribution of these small jellyfish to the class Scyphozoa. Indirect evidence for such an interpretation is that one of these forms, the genus *Conomedusites,* has a theca. Glaessner (1979) placed this genus in the class Conulata, but the Conulata were originally considered as a subclass of the Scyphozoa (Moore and Harrington 1956).

Forms with radial symmetry of constant order are relatively rare among the Vendian Radiata, and they show a considerable variation in the body plan. The most abundant are forms with three- and fourfold symmetry. They have a wide variety of gastrovascular systems (single and paired, branching and nonbranching radial channels). In some forms, the umbrella is divided into narrow, branching lobes that may contain channels *(Anfesta).* Such an organization can be considered as primitive, for thin, branching channels of the gastrovascular system may have developed from such hollow lobes.

Unlike the Cyclozoa and Inordozoa, the Vendian coelenterates with radial symmetry of constant order do not show any indication of asexual reproduction. Their reproductive organs display some variation in shape and location within the body.

In summary, the body plan analysis of the Vendian Radiata gives new insights into the earliest stages of the morphologic evolution of coelenterates. The concentric body plan and the symmetry of infinitely high order indicate that the class Cyclozoa was an independent, large phylogenetic branch of early coelenterates. Presumably, this kind of symmetry was predominant at the initial stage of the evolution of the Radiata. In the Vendian, however, we already see another evolutionary tendency of the Coelenterata, one that later prevailed in the Phanerozoic; that tendency is development of radial organization.

Some Precambrian Radiata show co-occurrence of the concentric and the radial symmetries and thus suggest a historical linkage between the concentric and the radial body plans.

Radial symmetry of the order that changes in ontogeny is very rare among living cnidarians, and it is considered a primitive feature. In the Vendian Coelenterata, however, this kind of symmetry prevailed in many groups of relatively high organization. It may have represented an important, perhaps necessary, stage in the early evolution of many coelenterate groups (Fedonkin 1983a, 1984). Many structures and many systems of organs characteristic of the Recent Cnidaria appear to have first formed in the Precambrian Radiata with this kind of symmetry. It is logical to consider the Vendian Coelenterata with such a peculiar symmetry as a separate class Inordozoa, even though, on the other hand, it cannot be ruled out that this kind of symmetry evolved independently in various phylogenetic lineages of the Coelenterata.

The present analysis of the body plan of the Precambrian Radiata thus indicates that the Coelenterata evolved from the concentric to the radial body plan, and from a variable to a constant order of radial symmetry in its ontogeny.

VENDIAN BILATERIA AND THE ORIGIN OF METAMERISM

Among Recent metazoans, bilaterally symmetric animals are more numerous and more diverse than radially symmetric forms. However, the paleontologic evidence indicates predominance of the Radiata in the Vendian (Glaessner 1979, 1984; Fedonkin 1981a, 1983a). This observation supports Beklemishev's idea (1964) that the most primitive groups of Bilateria are related to the Coelenterata. The abundance, diversity, and large size of Vendian coelenterates all suggest that this is the oldest metazoan phylum. The composition of the animal biota of the Vendian seems to indicate that radially symmetric organisms prevailed completely at the earliest, possibly very short stage of the evolution of multicellular animals. This seemingly fantastic speculation is based merely on retrospective extrapolation, but it is supported by paleontologic data. One piece of relevant evidence is the Precambrian fauna of the Avalon Peninsula in Newfoundland, which consists exclusively of solitary and colonial coelenterates (Anderson and Conway Morris 1982).

It is noteworthy that comparative anatomists widely accept the notion of the primary nature of radial symmetry in ontogeny and phylogeny of the Eumetazoa. What is subject to debate, however, is how bilaterally symmetric organisms could evolve from radially symmetric ones (Beklemishev 1964; Clark 1964, and references therein). The only major exceptions to this consensus are the concepts of Jägersten (1955, 1959, 1972) and Hadži (1944, 1963). Jägersten's concept of bilaterogastrea derives Recent Radiata from bilaterally symmetric forms; Hadži, in turn, postulates the origin of Turbellaria by a cellularization of protists and the derivation of Coelenterata from turbellarians.

The apparent prevalence of Radiata in the Vendian fauna may be partially explained by ecologic and taphonomic peculiarities of Vendian fossil assemblages. However, the main reason for this disproportion may be that the Radiata diversified earlier than the Bilateria. Nevertheless, there is a large diversity of body plans among the Vendian

Bilateria, including fairly advanced and well-separated branches of this group (Fedonkin 1983a).

An overwhelming majority of the known Vendian bilaterally symmetric metazoans are flat bodied. Segmented animals predominate, including both polymerous and oligomerous forms, with homonymous and heteronymous metamerism as well as with imperfect segmentation and the symmetry of glide reflection. This predominance of segmented forms in the Vendian lends support to the idea that in many metazoan groups the processes of formation of bilateral symmetry and metamerism were related to each other or at least coeval (Fedonkin 1983a).

The bulk of Vendian Bilateria exhibit some peculiarities rarely observed in later multicellular animals. The most important of these is the symmetry of glide reflection. This kind of symmetry is characteristic of very primitive Bilateria, in particular polymerous forms with homonymous metamerism (Dickinsoniidae) and oligomerous forms with distinct signs of cephalization (Vendomiidae). The appearance of symmetry of glide reflection seems to represent a fundamental, perhaps necessary step in the evolution of metamerism in various groups of the Metazoa. It provides a paleontologic line of evidence to support derivation of the metamerism of segmented forms (Articulata) from radially symmetric animals. It is to be emphasized that the distribution of the symmetry of glide reflection among Vendian Bilateria may indicate the possibility of the early origin of bilaterally symmetric segmented forms from animals with radial symmetry of variable order.

Thin, sheet shaped Dickinsoniidae appear to have the lowest level of organization among segmented forms. At first, they were attributed to the coelenterate class Dipleurozoa (Harrington and Moore 1956), later on, to either annelids (Glaessner and Wade 1966; Wade 1972b; Glaessner 1979; Runnegar 1982) or flatworms (Fedonkin 1981a). Promorphologic analysis of this family and the comparison of its body plan to the architecture of coelenterates, turbellarians, and annelids suggest that the Dickinsoniidae are an independent branch of the primitive Bilateria (Fedonkin 1983a). This interpretation is supported by the plausibility of derivation of the dickinsoniid symmetry from the most archaic Radiata, either from those with concentric body plan (*Kullingia* to *Ovatoscutum* to *Chondroplon* to *Dickinsonia*) or from those with radial symmetry of variable order (*Irridinitus* to *Dickinsonia*). The Dickinsoniidae may, in fact, represent one of the branches of the Bilateria that separated from the Radiata earlier than others. However, since in addition to the Dickinsoniidae, other highly organized Bilateria existed in the Vendian, the Dickinsoniidae seem to have evolved as a blind alley that possibly ceased to exist during the Vendian (neither true dickinsoniids nor similar forms are known from the Late Vendian through Cambrian). Thus, the Dickinsoniidae can indeed be assigned to the class Dipleurozoa, though not within the Coelenterata but rather within primitive Bilateria of the phylum Proarticulata (Fedonkin 1983a).

Animals with a flat, oval body, a cephalic area with an axial lobe, and postcephalic segments, are a separate group of the Vendian Bilateria. Their left and right segments either appear opposite to each other or alternate. This group, represented by the family Vendomiidae (Keller and Fedonkin 1977), tends to develop a large cephalic area, apparently a progressive feature. But the symmetry of glide reflection, as in *Vendia sokolovi* Keller (see Keller 1969), and the flat, oval body without lateral appendages indicate a primitive character. This group seems to represent a blind alley in the evolution toward true Articulata, and it may be recognized as a separate class Vendiamorpha, of the phylum Proarticulata (Fedonkin 1983a).

Some peculiar groups of the Vendian Bilateria, including those aforementioned, are difficult to place within the existing classification of Metazoa. They are often characterized by co-occurrence of features that are typical of widely separated phyla. An example is the presence of shoeshaped heads, like the head shields of trilobites, and annelid bodies in the families Sprigginidae and Bomakellidae. Such groups may be transitional or ancestral to groups separated later on, in the course of evolution. Thus, the Sprigginidae and Bomakellidae may be interpreted as belonging to the class Paratrilobita, possibly of the phylum Arthropoda. Some other groups of the Vendian Bilateria can be conceived of as continuing into the Phanerozoic, for example, the arthropods Parvancorinidae (Glaessner 1980) and the Sabelliditidae, comparable to the Recent Pogonophora (Sokolov 1972).

The prevalence of segmentation among the oldest Bilateria sheds new light on the origin of bilateral symmetry and metamerism in animals. Some authors (Hyman 1940; Ivanov 1968, 1976) assume protaxonia of all the Bilateria and consider the origins of bilateral symmetry and metamerism as two separate problems. They claim that the nonsegmented Scolecida appeared first and the segmented Articulata second. A similar view is advocated by Clark (1964). The Vendian fossil record, however, does not lend support to this idea. Nonsegmented forms are very rare among the Vendian Bilateria (Glaessner 1979; Fedonkin 1983a), and their nature is questionable; a morphologic transition from these nonsegmented animals to the Vendian Articulata appears practically impossible.

Other authors derive the Bilateria from the Radiata through transformation of the oral into the ventral side and on the basis of plagiaxonia. In such a case, metamerism could be derived from cyclomerism (antimerism) of the Coelenterata (Sedgwick 1884; Melnikov 1971, 1977), or it could have evolved independently, by metamerization of the third pair of coelomes of the archicoelome (Masterman 1898; Remane 1950, 1952).

In my opinion, the diversity and morphology of the Vendian Bilateria suggest the simultaneous appearance of bilateral symmetry and metamerism. This concept is supported by at least three lines of paleontologic evidence: (1) prevalence of segmented over nonsegmented forms, (2) presence of primitive animals with a large number of metameres and with preserved traces of their radial arrangement, and (3) presence of segment alternation in both polymerous and oligomerous forms, which can be viewed as due to the peculiar structure of growth zones that "shifted" relative to each other.

Three major theories explaining the origin of metamerism refer to (1) metamerized differentiation and subsequent regulation of the position of organs, (2) suppression of cross division (strobilar theory), and (3) derivation from the cyclomerism of the Coelenterata. New data on the Vendian fossil metazoans allow for discussion of the latter theory, which was first proposed by Sedgwick (1884).

Sedgwick claimed that annelids had originated from forms similar to corals (Anthozoa). He derived the annelid mouth and anus from siphonoglyphs of the closed slit mouth; therefore, he considered the ventral part of the annelid body to be homologous to the oral side of anthozoans. According to Sedgwick, the nerve plexus of the oral disc of the polyp, having become compact, developed into the nerve ring, which resulted in several cross commissures after the primary slit shaped mouth became closed. A part of the ring anterior to the ultimate mouth developed into ganglion. The body became elongated, and the radial chambers of the gastral cavity closed and developed into coelomic pouches. The tentacles of the polyp produced extremities. The posterior growth zone of annelids is homologous to the only growth zone of such corals as *Ceriantharia*.

According to Beklemishev (1964), Sedgwick's theory can be partially applied to the ectodermal metamerism of the larval body but not to the metamerism of the coelome and its derivatives associated with the postlarval body. The differences in body organization are too great between coelenterates and coelomates. The failure of Sedgwick's simple theory, however, does not ultimately refute the derivation of metamerism from coelenterate cyclomerism, for it may be due to Sedgwick's attempt to derive annelids directly from highly specialized forms such as higher corals rather than from primitive coelenterates (Beklemishev 1964).

Remane (1950, 1952) may have avoided the difficulties of directly deriving coelomates from the Coelenterata, because he derived them from cnidarians with four gastral pouches, a number that does not correspond to the number of coelomic pouches of segmented organisms. According to Remane, an unpaired front gastral pouch and two lateral pouches became the protocoel and the mesocoels, respectively. This is preserved in oligomerous animals (hemichordates, echinoderms, pogonophores), although neither the protocoel nor the mesocoels are present in their adults.

Remane regarded the tetramerous polyp as the prototype of all the Bilateria. While crawling on the oral side, such a polyp stretches the gullet and thus creates a new longitudinal body axis perpendicular to the primary axis. When the architecture of Enteropneusta is derived from this prototype, it is assumed that the anterior gastral pouch forms an unpaired proboscic coelome, the lateral pouches produce collar coelomes, and the posterior pouch divides into two parts to form thoracic coelomes. According to Beklemishev (1964), this theory is inadequate for Deuterostomia, although it can be applied to trochophorous animals. The major deficiency of Remane's theory is an attempt to derive all the Bilateria from a single root, by reference to a single route of morphologic transformation, while the major stems of the Bilateria can be derived from at least three different prototypes at different levels of organization.

The third version of the theory of transformation of cyclomerism into metamerism was put forward by Lemche (1959, 1960). It is similar to the concept of Sedgwick, but although Lemche accepts the enterocoel theory, he does not consider the segmental subdivision of coelome as an important characteristic feature of metamerism. In his view, the crown of outer tentacles of the polyp is homologous to the mollusk ctenidia (primitively arranged around the foot, as in *Neopilina*), polychaete parapodia, trilobite branchiae, and arthropod limbs. In turn, the aboral pedal disc of the Cnidaria, which secretes the theca, corresponds to the aboral shell-secreting gland of the Mollusca.

The fourth variant of this theory was suggested by Melnikov (1977). In his opinion, the metamerization of cyclomerism during the evolution of Articulata (which are supposed to have originated from pelagic coelenterate ancestors that began to crawl on the blastoporal side of the body) was associated with formation of pairs of metamerized cyclomeres of completely segmented rings. While the blastoporal body became large, this process proceeded in such a way that the segmented rings closed behind the aboral part.

Current promorphologic analysis of the Vendian Metazoa provides paleontologic arguments to support the theories of Sedgwick and Melnikov. The Vendian irregularly metameric Bilateria, characterized by the symmetry of glide reflection, represent one line of the evidence. This kind of symmetry is a natural, and even necessary, step at the beginning of the formation of true metamerism in the Bilateria. It seems to be derived from less specialized and phylogenetically primitive forms of Radiata with symmetry of variable order, which were widespread in the Vendian. If the radial symmetry of infinitely high and/or variable order was indeed predominant at the early stages of metazoan evolution, as suggested by analysis of the Vendian fossils, the forms with irregularly arranged metameres (*Vendia, Dickinsonia,* etc.) may represent the morphologic transition to truely metameric Articulata.

Consequently, the grade of irregular metamerism of the symmetry of glide reflection could be achieved independently in various lines of the Bilateria. However, it could not persist for long. The alternating arrangement of metameres is of little advantage for actively moving animals, although it may be more adequate for forms that moved by peristaltic waves along the ventral side of the body rather than for organisms with appendages. Flat-bodied shapes and the absence of lateral appendages in many Vendian Bilateria suggest indeed the peristaltic mode of crawling. It is noteworthy that the Vendian trace fossils reflect chiefly this kind of locomotion, whereas traces of crawling with appendages are extremely rare in Vendian ichnocoenoses (Fedonkin 1981a).

Thus, the analysis of promorphology of the Precambrian metazoans lends new support to the theory that derives bilaterally symmetric animals directly from radially symmetric forms. It also seems to indicate that the processes of formation of bilateral symmetry and metamerism were closely related to each other at least in some lineages of the Metazoa.

CONCLUSIONS

The present promorphologic analysis of Vendian metazoans sheds much new light on some fundamental problems in comparative anatomy of invertebrates and provides new data for the general theory of morphologic evolution of animals. In particular, it seems that the earliest radiation of metazoans allowed for realization of considerable diversity in symmetry of the body plan. Various body plans appeared in the course of this radiation, some of which later became fixed as the fundamental features of higher taxa of the Metazoa.

The prevalence of Radiata in the Vendian fauna is here regarded as an indication of the phylogenetically primitive nature of radial symmetry in metazoan evolution. The most archaic kind of symmetry in the Metazoa is the monaxial heteropolar symmetry of infinitely high order, usually related to a concentric body plan that is characteristic of many primitive coelenterates in the Vendian. The next step is represented by Radiata, with radial symmetry variable and increasing in order during ontogeny. This kind of radial symmetry is here considered to be a necessary stage in the evolution of many lineages of Radiata, particularly those evolving toward the symmetry of stable order that is so typical of the Phanerozoic Cnidaria. The radial symmetry of stable order is represented in the Vendian by forms with three- and fourfold symmetry. These are the animals with the highest level of organization among the Vendian Radiata. Thus, it appears that the evolution of Radiata proceeded from forms with symmetry of infinitely high order, to those with symmetry of variable order, to those with the order of symmetry constant in ontogeny—and from animals with concentric organization to forms with a radial body plan.

In comparison to the Radiata, the remains of Vendian Bilateria constitute a minority. Many Precambrian bilaterally symmetric animals are characterized by a flat, segmented or metamerized body, but there is large variety of body plans in this group. The prevalence of segmented forms among the Vendian Bilateria suggests that in many branches of Metazoa the evolutionary formation of bilateral symmetry and metamerism was a concerted process that did not always, however, lead to coelomates. Among the most important peculiarities of the Vendian segmented animals is the symmetry of glide reflection, characteristic of the most primitive Bilateria. This kind of symmetry may indicate a very early origin of metameric Bilateria from Radiata with a symmetry of variable order. Perhaps the imperfect metamerism expressed by the symmetry of glide reflection represented a fundamental step in the evolution of some groups of Metazoa toward true Articulata.

Some groups of the Vendian Bilateria are not yet assigned to any classes or even phyla of the Metazoa. These forms have either a body plan unknown among later Metazoa, or one that is characterized by a combination of features each typical of another phylum.

The most abundant and typical elements of the Vendo-Ediacarian metazoan fauna became extinct at the end of the Vendian. This was the fate of the main groups of Radiata, in particular those with concentric body plan and with symmetry of infinitely high order; many Bilateria, especially those with imperfect metamerism and the symmetry of glide reflection; and also some other peculiar groups such as the Petalonamae. This mass extinction may have been partially due to competition for food and other resources with more progressive lineages that continued into the Phanerozoic. However, the very fact that those characteristically Vendian Metazoa are essentially different from all later phyla, virtually resolves the problem of their belonging to the problematica (see Chapter 1 by Bengtson, this volume).

The high level of differentiation of the animal

world at the beginning of the Vendian and the abundance of invertebrates at highly variable levels of organization may have been a result of pre-Vendian evolution of the Metazoa. Indirect and rare paleoichnologic data support this conjecture. We have, then, every reason to search for metazoan fossils in older divisions of the Precambrian.

REFERENCES

Anderson, M. M. 1976. Fossil Metazoa of the Late Precambrian Avalon fauna, Southeastern Newfoundland. *Geol. Soc. Am. Abstr. Progr.* **8**, 754.

———. 1978. Ediacaran fauna. *In* Lapedes, D. N. (ed.). *Yearbook of Science and Technology.* McGraw-Hill, New York, pp. 146–149.

——— and Conway Morris, S. 1982. A review, with description of four unusual forms, of the soft-bodied fauna of the Conception and St. John's Groups (Late Precambrian), Avalon Peninsula, Newfoundland. *Proc. 3rd North Am. Paleont. Conv.* **1**, 1–8.

——— and Misra, S. B. 1968. Fossils found in the Precambrian Conception Group in Southeastern Newfoundland. *Nature* **200**, 680–681.

Beklemishev, V. N. 1964. *Principles of Comparative Anatomy of Invertebrates.* Nauka, Moscow [in Russian].

Clark, R. W. 1964. *Dynamics of Metazoan Evolution: The Origin of the Coelom and Segments.* Clarendon, Oxford.

Debrenne, F. and Naud, G. 1981. Méduses et traces fossiles supposées précambriennes dans la formation de San Vito, Sarrabus, Sub-Est de la Sardaigne. *Bull. Soc. Géol. Fr.* **23**, 23–31.

Fedonkin, M. A. 1981a. *White Sea Biota of the Vendian. Precambrian Non-skeletal Fauna of the Russian Platform North.* Nauka, Moscow [in Russian].

———. 1981b. Precambrian fauna of the Russian Platform. *U.S. Geol. Surv. Open-File Rep.* **81-743**, 91–92.

———. 1982. Precambrian soft-bodied fauna and the earliest radiation of invertebrates. *Proc. 3rd North Am. Paleont. Conv.* **1**, 165–167.

———. 1983a. Organic world of the Vendian. Stratigraphy and paleontology. *Itogi Nauki Tekhniki. VINITI AN SSSR* **12**, 127 pp. [in Russian].

———. 1983b. Non-skeletal fauna of Podolia (Dniester River Valley). *In* Velikanov, V. A., Asseeva, E. A., and Fedonkin, M. A. (eds.). *The Vendian of The Ukraine.* Naukova Dumka, Kiev, pp. 128–139 [in Russian].

———. 1983c. Promorphology of Vendian Radialia as a key to understanding the early evolution of Coelenterata. *Abstr. 4th Int. Symp. Fossil Cnidaria, Washington, D.C.*, p. 6.

———. 1984. Promorphology of the Vendian Radialia. *In Stratigraphy and Paleontology of*

the Earliest Phanerozoic. Nauka, Moscow, pp. 30–58 [in Russian].

Glaessner, M. F. 1972. Precambrian paleozoology. *In* Jones, J. B. and McGowran, B. (eds.). *Stratigraphic Problems of the Later Precambrian and Early Cambrian.* Univ. of Adelaide, Centre for Precambrian Research, Special Paper **1**, pp. 43–52.

———. 1979. Precambrian. *In* Robison, R. A. and Teichert, C. (eds.). *Treatise on Invertebrate Paleontology,* part A. Geological Society of America and Univ. of Kansas Press, Lawrence, pp. A79–A118.

———. 1980. *Parvancorina*—an arthropod from the Late Precambrian (Ediacarian) of South Australia. *Ann. Naturhist. Mus. Wien* **83**, 83–90.

———. 1984. *The Dawn of Animal Life. A Biohistorical Study.* Cambridge Univ. Press, Cambridge, 244 pp.

——— and Wade, M. 1966. The Late Precambrian fossils from Ediacara, South Australia. *Palaeontology* **9**, 599–628.

Hadži, J. 1944. Turbellariyska teoriya Knidaryev. *Razpr. Mat. Prir. Akad. Ljubljana* **3**(1), 1–239 [in Slovenian].

———. 1963. *The Evolution of Metazoa.* Pergamon, London.

Hahn, G. and Pflug, H. D. 1980. Ein neuer Medusen-Fund aus dem Jung-Präkambrium von Zentral-Iran. *Senckenb. Leth.* **60**, 449–461.

Harrington, H. J. and Moore, R. C. 1956. Dipleurozoa. *In* Moore, R. C. (ed.). *Treatise on Invertebrate Paleontology,* part F, *Coelenterata.* Geological Society of America and Univ. of Kansas Press, Lawrence, pp. F24–F27.

Hyman, L. H. 1940. *The Invertebrates: Protozoa through Ctenophora.* McGraw-Hill, New York.

Ivanov, A. V. 1968. *Origin of Multicellular Animals. Phylogenetic Essays.* Nauka, Leningrad [in Russian].

———. 1976. On the monophyletic nature of Metazoa. *Zool. Zh.* **46**(10), 1446–1455 [in Russian].

Jägersten, G. 1955. On the early phylogeny of the Metazoa. The bilatero-gastrea theory. *Zool. Bidr. Uppsala* **30**, 321–354.

———. 1959. Further remarks on the early phylogeny of the Metazoa. *Zool. Bidr. Uppsala* **33**, 79–108.

———. 1972. *Evolution of the Metazoan Life Cycle.* Academic Press, London.

Keller, B. M. 1969. Imprint of unknown animal from the Valdai Series of Russian Platform. *Geol. Inst. Trans.* **206**, 175 [in Russian].

——— and Fedonkin, M. A. 1977. New organic fossil finds in the Precambrian Valdai Series along the Syuzma River. *Int. Geol. Rev.* **19**, 924–930.

Lemche, H. 1959. Protostomian relationships in

the light of *Neopilina. Proc. 15th Int. Congr. Zool.,* pp. 381–389.

————. 1960. A possible central place for *Stenothecoides* Resser 1939 and *Cambridium* Horny 1957 (Mollusca, Monoplacophora) in invertebrate phylogeny. *Proc. 21st Int. Geol. Congr.*

Masterman, A. T. 1898. On the theory of archimeric segmentation and its bearing upon the phyletic classification of the Coelomata. *Proc. R. Soc. Edinb.* **22,** 1–270.

Melnikov, O. A. 1971. On the primary heteronomy of body segments in the Articulata. *Zh. Obshch. Biol.* **32**(5), 597–611 [in Russian].

————. 1977. To the promorphology of articulated animals. *Zh. Obshch. Biol.* **38**(3), 393–408 [in Russian].

Missarzhevsky, V. V. 1974. New data on the earliest fossils of Early Cambrian of Siberian Platform. *In Biostratigraphy and Paleontology of the Lower Cambrian of Europe and Asia.* Nauka, Moscow, pp. 179–189 [in Russian].

Moore, R. C. and Harrington, H. J. 1956. Scyphomedusae. *In* Moore, R. C. (ed.). *Treatise on Invertebrate Paleontology,* part F, *Coelenterata.* Geological Society of America and Univ. of Kansas Press, Lawrence, p. F38.

Palij, V. M., Posti, E., and Fedonkin, M. A. 1979. Soft-bodied Metazoa and trace fossils of Vendian and Lower Cambrian. *In Upper Precambrian and Cambrian Paleontology of East European Platform.* AN ASSR, Moscow, pp. 49–82 [in Russian].

Remane, A. 1950. Die Entstehung der Metamerie der Wirbellosen. *Verh. Dtsch. Zool. Ges.,* pp. 16–23.

————. 1952. *Die Grundlagen des Natürlichen Systems, der Vergleichendend Anatomie und der Phylogenetik.* Akad. Verlagsgesellsch., Leipzig.

Runnegar, B. 1982. Oxygen requirements, biology and phylogenetic significance of the Late Precambrian worm *Dickinsonia,* and the evolution of the burrowing habit. *Alcheringa* **6,** 223–239.

Sedgwick, A. 1884. On the origin of metameric segmentation and some other morphological questions. *Q. J. Microscop. Sci.* **24,** 1–43.

Sokolov, B. S. 1972. Vendian and Early Cambrian Sabelliditida (Pogonophora) of the USSR. *Proc. 23rd Int. Geol. Congr.,* pp. 79–86.

———— and Fedonkin, M. A. 1984. The Vendian as the terminal system of the Precambrian. *Episodes* **7,** 12–19.

Valkov, A. K. 1982. *Biostratigraphy of the Lower Cambrian of the East Siberian Platform (Uchur-Maya Region).* Nauka, Moscow [in Russian].

Valkov, A. K. and Sysoev, V. A. 1970. Cambrian angustiochreids of Siberia. *In Stratigraphy and Paleontology of Proterozoic and Cambrian of the East Siberian Platform.* Izdatelstvo AN SSSR, Yakutsk, pp. 94–100 [in Russian].

Wade, M. 1969. Medusae from the uppermost Precambrian or Cambrian sandstones, Central Australia. *Palaeontology* **12,** 351–365.

————. 1971. Bilateral Precambrian chondrophores from the Ediacara fauna, South Australia. *Proc. R. Soc. Vict.* **84,** 183–188.

————. 1972a. Hydrozoa and Scyphozoa and other medusoids from the Ediacara fauna, South Australia. *Palaeontology* **15,** 197–225.

————. 1972b. *Dickinsonia:* polychaete worms from the Late Precambrian Ediacara fauna, South Australia. *Mem. Queensland Mus.* **16,** 171–190.

7. CHONDROPHORINE HYDROZOANS AS PROBLEMATIC FOSSILS

GEORGE D. STANLEY, JR.

Because of our instinctive curiosity about unusual organisms as well as our fascination with oceanic voyages and open-sea navigation, the imagination is easily captured by *Velella,* the living, by-the-wind sailor. These floating, medusoid cnidarians are wind propelled, by virtue of a large erect sail that rises from their disclike tentaculated tissues. Thousands of these organisms are frequently driven onto shore by prevailing winds and currents, and their remains litter some beaches in great abundance (Fig. 1).

While these hydrozoans are unusual in their mode of wind-directed travel, they are also unique in representing a small, morphologically distinct group of organisms that have somehow survived unchanged over a vast amount of geologic time. Today the entire group, the suborder Chondrophorina, consists of two families each containing a single genus, and in each genus only a few species. Because they are composed of predominantly soft tissues and have no calcified parts, it seems intuitively obvious that they would afford only a remote chance, if any chance at all, for preservation. This is not the case, however, because chondrophorines secrete within their coenosarcal tissues an internal chitinoid, gas-filled disc or float. Fossils attributable to such a float were first described by Hall (1847), but initially they were confused with unrelated fossils such as corals.

Ruedemann (1916) ascertained the true affinities of a flattened, enigmatic specimen from the Upper Devonian rocks of New York State, as a chondrophorine. He named his disclike fossil *Plectodiscus,* and showed it to be closely related to living *Velella.* Later, Caster (1942) studied other Late Devonian specimens of *Plectodiscus* and properly assessed their true nature.

Exquisitely preserved, but rare Devonian examples from the Hunsrück Slate have recently revealed soft tissues and hard parts in the genus *Plectodiscus* (Yochelson et al. 1983). These show remarkable morphologic similarity with living *Velella. Plectodiscus* has also been described from Carboniferous rocks (Chamberlain 1971; Stanley and Yancey 1986), and almost identical, probably congeneric *Silurovelella* comes from the Later Silurian (Fisher 1957). The antiquity and slow evolution of chondrophorines is further shown in Late Precambrian Ediacaran chondrophorines (Wade 1971; 1972), which are little differentiated from Carboniferous and Holocene examples. Chondrophorines are thus a very ancient group whose roots most likely extend back to that critical interval of metazoan evolution in the Late Precambrian, when the initial body plans of cnidarians were becoming established (Scrutton, 1979; see also Chapter 6 by Fedonkin, this volume).

A few living species that populate our present oceans appear so little different from their distant ancestors that one can regard them as "living fossils." There is, however, the apparent unexplained absence of chondrophorines between Late Carboniferous and Holocene. I am highly dubious of this presumed absence during roughly 300 Ma. A recent discovery from the Late Cretaceous (Stanley and Kanie 1985) suggests that such a hiatus is artificial, and due to the almost complete lack of proper criteria with which to assess chondrophorine fossils.

The purpose of this paper is to clarify the problematic nature of chondrophorines through review of the general morphology and anatomic structure of living and fossil forms, and to present some new findings relevant to their taphonomy and geologic distributions.

THE CHONDROPHORINE ORGANISM

The biology of extant chondrophorines is based on study of extant *Velella* and *Porpita.* The higher systematic position of chondrophorines is still a subject of debate. Some workers placed these genera in the order Siphonophorida (Hyman 1940; Harrington and Moore 1956), while others regard them as hydroids under the order Hydroida. Another question is whether chrondrophorines are specialized, solitary individuals (Mackie 1959) or polymorphic colonies. In this paper I take the traditional view of chondrophorines as floating hydroid colonies, composed of various specialized individuals.

Review of morphology and structure of the Chondrophorina is provided by Hyman (1940), Caster (1942), and Harrington and Moore (1956). The general biology of velellid chondrophorines was more recently reviewed by Yochelson et al. (1983).

Fig. 1. Dead and dying *Velella* along the coast of California near San Francisco. Hundreds of individuals each about 8 to 10 cm long, probably represent a whole flotilla that has been washed up on the sandy beach. Such occurrences are common today and may also have been a frequent phenomenon in the geologic record.

Velella consists of an explanate, disclike coenosarc, slightly convex on the aboral side, from which rises a crest or sail (Fig. 2). The oral surface is tentaculated and bears a number of fundamentally different types of specialized appendages (Fig. 2C): (1) gonozooids, (2) long marginal fringes of dactylozooids that serve a sensory and protective function, (3) short, globular reproductive organs called gonophores that eventually detach to become free-swimming adult medusae, and (4) a single, centrally located gastrozooid that serves as a feeding polyp. Directly above the gastrozooid, within a centradenial cone, is a complex system of gastrodermal canals. This region also houses a massive hepatic or "liver" organ. The anatomy of a generalized velellid in floating position is given in Fig. 3.

All these organs and tissues make up the soft-bodied component of the chondrophorine organism, but from the paleontologic point of view, the internal float or pneumatophore is of greatest interest. It is disc shaped, secreted entirely within the coenosarcal tissues, and consists of a series of concentrically enveloping, tightly adpressed air chambers or pneumatocysts. The composition of this float is here termed "chitinoid," because there is no confirmation of a true chitin composition (Rudall 1955). Nevertheless, it is very thin and flexible, and is composed of relatively durable

material (Fig. 2A). Unlike the surrounding tissues, this material possesses considerably more potential for preservation. Upon death, the pneumatophore is liberated and can float for a long time before becoming waterlogged and sinking. F. M. Bayer (personal communication) noted goose barnacles attached to the floating pneumatophore of *Velella*, confirming that after death it can continue floating for a considerable length of time.

Not surprisingly, it is the float of chondrophorines that constitutes the majority of fossil discoveries. Only in one exceptional situation was preservation of tissues possible, and the complete, tentaculated organism has been fossilized (Yochelson et al. 1983). The discovery of such fossils allowed close comparisons with living chondrophorines and revealed striking anatomic similarities.

The genus *Porpita* (Fig. 4) is radically different from *Velella*. Unlike *Velella*, it is radially symmetric, has a low, caplike shape in profile, and lacks a sail. Like *Velella*, the internal air chamber of the pneumatophore is concentrically developed, but unlike *Velella*, it is divided radially into numerous partitions (gastrovascular canals) that are strongly superimposed upon the more weakly developed, concentric symmetry (Fig. 4). While the asymmetric, sailed *Velella* utilizes prevailing winds for unidirectional movement, *Porpita* is

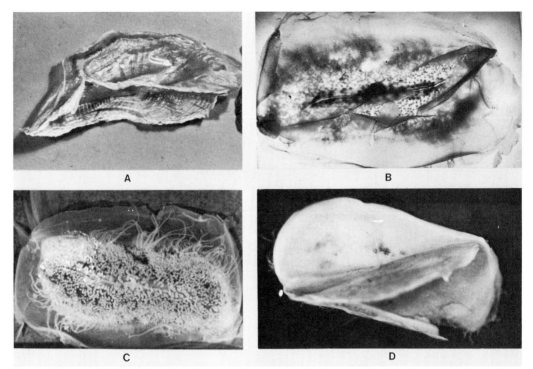

Fig. 2. Specimen of *Velella* about 6 cm long at various stages of desiccation. A. Dried pneumatophore showing much wrinkling. The sail is folded over the disc. B. Aboral view of the soft tissues after the pneumatophore was removed through the aboral surface. Note the remaining flaps of sail tissue that compare well with the fossil example (see Fig. 10). C. Oral view showing nature of the specialized tentacles. D. Intact specimen (aboral) with high sail and covering of coenosarcal tissues. Pneumatophore and furrow are visible faintly through tissues.

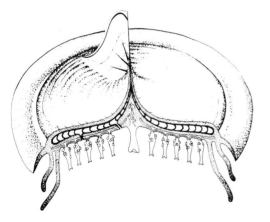

Fig. 3. Reconstruction of a velellid chondrophorine shown in floating position with a cutaway view of the interior, revealing the internal pneumatophore (black) and enveloping soft tissues. The tentacular appendages include a marginal group of dactylozooids, inner circles of reproductive gonozooids, and the single, centrally located gastrozooid containing the mouth. Directly below the sail is the hepatic or "liver" organ. Reconstruction based on living *Velella* and Devonian fossils. Natural size.

oriented more toward multidirectional movement and utilizes tentacles as well as currents and waves for propulsion. Like velellids, fossils of porpitids are also known from the internal float.

SIMILARITIES WITH OTHER GROUPS

Fossil chondrophorines could be confused with scales of bony fish, discinoid inarticulate brachiopods, corals, the scutella of certain crustaceans, the aptychi of cephalopods, trace fossils, jellyfish, and the shells of some mollusks. Ruedemann (1916) discussed a wide array of misinterpretations for some fossil velellids. Caster (1942) easily dismissed many of these possibilities when he considered the morphologic details of living velellids. Other fossil groups, like the large inarticulate brachiopods *Lindstroemella* and *Oehlertella* of the Devonian, may have some apparent similarities with velellids, but the presence of muscle scars, the pedicle foramen, and other characteristics serve to distinguish such brachiopods from Devonian velellids.

Despite the detailed work of Rauff (1939) and Caster (1942) on the Devonian chondrophorine *Plectodiscus,* confusion continues over its affinities (Yochelson et al. 1983). For example, Seilacher (1961, table 11, fig. 4) illustrated it as a pos-

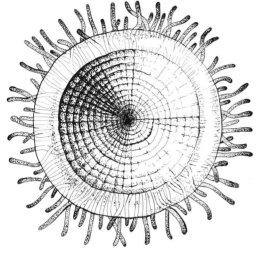

Fig. 4. *Porpita,* a living chondrophorine lacking a sail and displaying distinct radial, gastrodermal canals and the concentric pneumatocysts. Approximately natural size.

sible holothurian, and Fuchs (1915, pl. 1, tables 3, 3a) misidentified it as the underside (epitheca) of the coral *Pleurodictyum hunsruckianum.*

Fish scales

Paleozoic bony fish possess an armor consisting of overlapping rows of heavy scales (Romer 1964), which may be disclike, circular to elliptical, or rhomboidal in shape. They have a thin, enamel-like covering and concentric growth lines (Fig. 5). In general they are in the same size range as some chondrophorine fossils, and therefore possibilities for misidentification exist. They differ, however, in composition, surface sculpturing, and microstructure.

Trace fossils

Trace fossils also can be confused with chondrophorines. Caster (1942) described a problematic fossil *Palaeoscia floweri,* as a large porpitid chondrophorine. It is corrugated with ringlike markings and a vague, centrally located radial structure. This fossil, however, is clearly not a chondrophorine but a trace fossil (Osgood 1970). The concentrically coiled trace fossil *Spirorhaphe* Fuch, consisting of circular, concentric grazing or

feeding surface trails, seems to resemble the concentric penumatocysts of small *Plectodiscus* (Stanley and Yancey 1986). These fossils, however, lack the specific details of true chondrophorine discs (Fig. 8F).

Cephalopod aptychi

Aptychi are fossils usually interpreted either as the apertural covers or the lower part of the jaw of cephalopods, but there is still much debate about their exact function. Although I know of no cases where these fossils have been confused with chondrophorines, a Devonian aptychus illustrated by Yochelson et al. (1983) shows that striking similarities exist.

Jellyfish

Harrington and Moore (1956) listed under *incertae sedis* a number of vaguely chondrophorinelike forms. Most of these resemble porpitids rather than velellids, but they lack the clear characteristics of true porpitids, particularly the clear radial

Fig. 5. Bony fish scales. Note the superficial similarities with chondrophorines such as those on Fig. 8. Length across each photograph is 1.8 cm. Mississippian Heath Formation, Bear Gulch Limestone, Montana.

Fig. 6. *Plectodiscus discoideus* (Rauff). Note the crumbled, deformed nature of the once-convex disc, which has been severely flattened. The radial furrow is present, but the sail is not visible because it most likely lies on the reverse side; the specimen is therefore upside down. Diameter of disc is 12 cm. Royal Scottish Museum 1905.62.11 (Yochelson et al. 1983, fig. 2). Reproduced by permission of E. Schweizerbart'sche Verlagsbuchhandlung.

or concentric ornament. Many of these may be jellyfish casts, but others clearly show relationships to the Chondrophorina. The Lower Cambrian *Velumbrella czarnockii*, a flat form with strong radial canals, described as a medusoid jellyfish (Stasińska 1960), is most likely a porpitid.

Jellyfish fossils (scyphozoans), unlike chondrophorines, are entirely soft bodied and lack the internal pneumatophore and its gas chambers. They are usually preserved as impressions or casts and exhibit features uncharacteristic of chondrophorines: branching radial canals, pouches, lobes, marginal fringes, symmetric "gonads," and others. Classic examples of jellyfish occur in the Solenhofen Limestone (Maas 1902). The morphology of jellyfish is usually less distinctive than chondrophorines as a result of lack of hard parts and the mold and cast preservation; but superficial similarity with chondrophorine porpitids may exist, as in the case of Precambrian forms (Wade 1972; Hahn and Pflug 1980).

Mollusks

Yochelson and Stanley (1981) and Yochelson et al. (1983) pointed out the similarities of chondrophorines with univalved mollusks, particularly limpets and monoplacophorans. The flat to low, cap- or disc-like shape, coupled with the fine, evenly spaced corrugations of the concentric pneumatocysts, can produce a fossil with close morphologic parallels to univalved mollusks. A large and distinctive Devonian velellid *Plectodiscus*, illustrated by Yochelson et al. (1983, fig. 2),

has flattened, evenly spaced, concentric pneumatocysts that resemble molluscan growth lines (Fig. 6). The radial sinus of the chondrophorine disc may be misidentified as the selenizone of a patelliform gastropod (Fig. 8D, J). This mistake has been made at least on two occasions (Donaldson 1962; Kanie 1975), and other examples of misidentification may also exist (Yochelson and Stanley 1981; Stanley and Kanie 1985). Some chondrophorines also resemble certain monoplacophorans (Fig. 8C, E, I); for example, Cambrian *Parmorphella* Matthew or *Scenella* Billings. *Palaeacmaea* Hall and Whitfield, a supposed Cambrian monoplacophoran (Fig. 8D, I), has a low, conical shell without muscle scars (Yochelson and Stanley, 1981). Yochelson and Gil Cid (1984) have assigned *Scenella* to the Chondrophorina on the basis of specimens from the Lower Cambrian of Spain. They further propose that all Lower Cambrian *Scenella* are chondrophorines, and therefore they cannot be ancestral to mollusks. If some, or all, of these early molluscan fossils were found to be coelenterates, it would drastically change our views of the origin and evolution of the mollusks. My point here is that careful study needs to be done on any fossils resembling fossil chondrophorines.

DEATH AND PRESERVATION: FACTORS OF TAPHONOMY

Occasionally chondrophorine remains are locally abundant. Huckriede (1967), for example, reported > 100 specimens on one bedding surface, and Stanley and Yancey (1986) found numerous specimens stacked one upon another (Fig. 8F). Such concentrations most likely represent flotillas that either sank together in deep water or were driven onto shallow beaches or intertidal areas.

Yochelson et al. (1983) have shown that the sail may be flattened upon the disc so that it becomes difficult, if not impossible, to recognize it (Fig. 8F). Also, it may be compressed so tightly into the corrugated surface of the disc that it may become almost indistinguishable from the disc itself (Figs. 7 and 8F). Furthermore, if a specimen is preserved upside down with oral surface in view, as frequently occurs in shales or argillites, the sail in the matrix below will not be visible on a parallel bedding surface. This is demonstrated in Devo-

nian examples (Fig. 9) in which radiographic study reveals the presence of the sail.

A straight ridge or low furrow in the disc of fossil velellids appears as a radial line traversing the surface of the pneumatophore (Figs. 6 and 7). Where it encounters the edge of the disc, an emargination is produced. This is the principal vascular sinus, which in living *Velella* accommodates a large vascular vessel of the soft, coenosarcal tissues. This distinctive feature of velellids can also be altered during taphonomy by flattening and compression of the disc, producing either a ridge or a sulcus. It can also be plastically offset (Fig. 8D).

Fossil chondrophorines are found in a variety of rock types ranging from sandstone and silt-

Fig. 7. *Plectodiscus cortlandensis* Caster. Holotype whitened with magnesium oxide. The concentric corrugations of the numerous pneumatocysts and the radial furrow in the lower part (light arrow) are clearly visible. The possible trace of a bilobed sail (dark arrow) can be faintly discerned. Short diameter of the disc is 10 cm. Middle Devonian, New York.

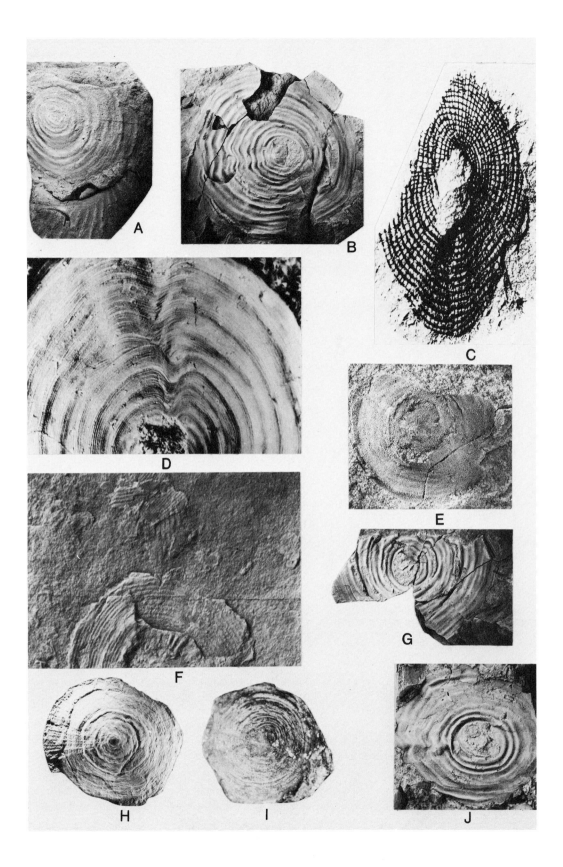

74

stone to shales and limestone and dolomite. Caster (1942) and Fisher (1957) reported them from basinal, dark shale deposits; other examples include Late Paleozoic deep-water deposits in Malaysia (Stanley and Yancey, 1986) (Fig. 8F), Pennsylvanian flysch deposits (Chamberlain 1971), the Ediacara rocks (Glaessner and Wade 1966; Glaessner 1971), and coarser-grained, terrigenous clastic rocks of shallow-water origin (Stanley and Kanie 1985). Because these organisms were planktonic, living in the warm surface waters, their remains occur without regard to depth or rock type. One example is even found in carbonate, tidal flat environments where it probably became beached by winds and tides (Yochelson and Stanley 1981).

The taphonomy of Devonian velellid *Plectodiscus* was given by Yochelson et al. (1983). They described a wide array of fossils found in various stages of preservation, ranging from intact, pyrite-replaced soft tissues, through partially decayed specimens bearing some tissues, to compressed remains of the organic pneumatophore disc. The fine preservation of such unusual fossil material allows close analogies to be made with modern *Velella*.

After death, chondrophorines can sink intact, together with both coenosarcal tissues and the internal disc, but generally the tissues quickly separate from the pneumatophore (Fig. 2). A fossil example of this is given in Fig. 10. Predation is also a strong factor in the preservational potential, and evidence of nibbling may exist (Yochelson and Stanley 1981). After the onset of decay, the disc usually separates from the tissues through a rupture along the edge of the coenosarc. Liberation of the disc could be accelerated by wave action or predation, in which case the disc may become damaged. Beached specimens lose all tissues by desiccation, decomposition, or land predation, and pneumatophores thus freed may

be wind-blown on the beach for days before becoming buried (Fig. 1). Desiccation may cause a high degree of wrinkling of the disc—an unlikely phenomenon in specimens buried in a water medium.

When the pneumatophore becomes waterlogged and sinks to the sediment surface, it may come to rest with either the aboral (sail) or oral (mouth) side up (Fig. 6). The case of stacking of discs is interpreted as transportation by weak bottom currents (Stanley and Yancey, 1986).

After burial, the sail may become flattened directly to the more resistant corrugated surface of the disc, or sediment may accumulate between sail and disc. Soon after burial, compaction begins to flatten the water-filled gas chambers, and it may also alter the original convexity of the disc (Fig. 6). In rare cases, sediment may enter the pneumatocysts and result in their three-dimensional preservation (Huckriede 1967). The degree of compaction and water loss depends, of course, on the grain size of the sediments. Specimens from fine-grained rocks show the maximum degree of flattening of the disc, while those in coarse sediments show some preservation of the original disc convexity (Fig. 8A). Flattening can sometimes produce a shinglelike overlapping pattern (Fig. 11) in the original even-spaced concentric gas chambers. Because the walls of a gas-inflated pneumatophore are exceedingly thin, complete compaction can produce a disc of only filmlike thickness (Fig. 12). In slates of the Devonian Hunsrück, an additional tectonic stress component of horizontal shear has resulted in slight distortion of the concentric pneumatocysts across the disc surface (Yochelson et al. 1983).

Since the walls of the sail are even thinner, they would be very difficult to detect; this has been confirmed on numerous fossil velellids studied so far. Thus preserved, the compressed pneumatophore indeed strikingly resembles the growth lines

---

Fig. 8. Problematic fossils probably attributable to chondrophorines. A. *Palaelophacmaea annulata* (Yokoyama). Cast of the pneumatophore and the bivalve *Pterotrigonia*. This chondrophorine was initially classified as a large limpet gastropod and was reevaluated by Stanley and Kanie (1985). Lower Cretaceous, Japan, ×.66. B. *P. annulata* (Yokoyama). Cast of another flattened specimen. Note wrinkles and irregularities of the surface and the radial feature interpreted as the impression of the vascular canal. Lower Cretaceous, Japan, ×.66. C. *Scenella morenensis* Yochelson and Gil Cid. Initially confused with a monoplacophoran. First referred to as "Forma A" by Gil Cid (1972). Lower Cambrian, Spain. ×.33. D. *Palaelophacmaea criola* (Donaldson). Showing irregularities of the surface including offsetting of the radial feature. First classified as a patelliform gastropod and reevaluated by Yochelson and Stanley (1981). Lower Ordovician, Pennsylvania. ×4.6. E. *Palaeacmaea irvingi* (Whitfield). An early monoplacophoran most likely relatable to the Chondrophorina. Upper Cambrian, Wisconsin. USNM 27512. ×.8. F. *Plectodiscus* sp. The flattened pneumatophore of a sailed velellid. One specimen stacked upon another. Sail is flattened tightly against the disc of the upper specimen but the radiating support struts of the sail are clearly visible on the surface. Note the close spacing of the individual pneumatocysts. Debris, partly sail fragments, is scattered above in the matrix. Early Carboniferous, Malay Peninsula. ×2.66. G. *Palaelophacmaea annulata* (Yokoyama). Cast of specimen in B. ×.66. H. *Scurriopsis (Scurriopsis) cycloides* (Tichy). Described as a patelliform gastropod (Tichy 1980, tables 1, 1a-b). Note distinct radial elements, concentric corrugations, an apical "pore," and irregularities of the disc surface. Upper Triassic, Italy. ×1.33. I. *Palaelophacmaea criola* (Donaldson). Circular, conical fossil traditionally interpreted as a limpet by Donaldson (1962) and later reevaluated by Yochelson and Stanley (1981). J. *P. annulata* (Yokoyama). Note the irregularities of the concentric pneumatocysts and other features of the disc surface. Lower Cretaceous, Japan. ×.66.

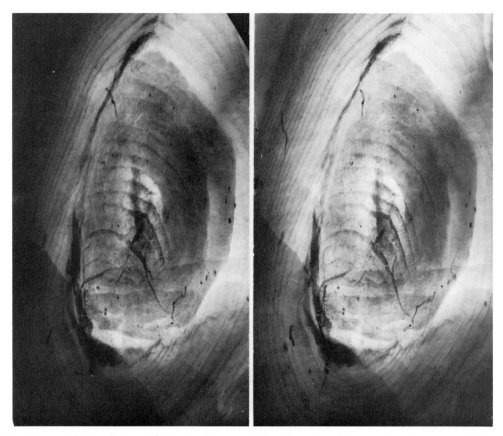

Fig. 9. Stereo-pair X-radiograph of the central region of *Plectodiscus discoideus* revealing the remnant of the sail (darker portion). Distance across each radiograph is 6.5 cm. WS 4415, Lower Devonian, West Germany. WS numbering system in this and all subsequent figures denotes specimens in the collection of Wilhelm Stürmer (Erlangen, West Germany).

of univalved mollusk shells. In the porpitid forms the additional strong radial partitions of the pneumatophore may look remarkably like the radial ornament of patelliform gastropods (Fig. 8A, B, G, H, J). The extinct *Palaelophacmaea* Donaldson, from Ordovician and Cretaceous (Yochelson and Stanley 1981; Stanley and Kanie 1985), is a fundamentally different type of chondrophorine. These forms are distinctly bilaterally symmetric and they share certain characteristics with velellids. They lack a sail, however, but have a radial ridge or groove somewhat like *Plectodiscus* but much more pronounced (Fig. 8D). This resemblance of radial ridge to a selenizone of limpets is even more striking in *Palaelophacmaea,* which was originally described as a limpet.

Such specimens as those illustrated by Ruedemann (1916) and Caster (1942) show a variety of irregular radial lineations or pleatlike folds of the disc surface. These appear to be artifacts of wrinkling of the pneumatophore during desiccation and compression (Fig. 2A). Since the pneumatophore is composed of thin, flexible material, once

compressed, it can be greatly altered by objects lying below or coming to rest above. Such objects, including coarser sediment grains, can leave various impressions on the disc surface. Total collapsing and compression of the sightly convex disc produce a great deal of wrinkling, especially toward the center, and disruption of the concentric corrugations of the pneumatocysts (Fig. 6).

The pneumatophore is composed of an organic substance akin to chitin. However, in fossils this material is altered diagenetically, and sometimes only a thin organic film is preserved. In carbonate rocks, calcite replacement has occurred (Yochelson and Stanley 1981). The Devonian *Plectodiscus* of the Hunsrück has a high concentration of organics, and conversion to iron sulfides, especially pyrite, has produced exquisite fossils (Yochelson et al. 1983). In most other specimens the organic material is missing, and mold and cast preservation is predominant. In the Ediacara fauna, this is the only mode of preservation. In pneumatocysts infilled by sediment, the corrugation appears three-dimensional and is less likely to confuse the pneumatophore identification with a mollusk shell. However, because of the tight sealing of the resilient pneumatocysts, infilling occurs rarely.

STRATIGRAPHIC OCCURRENCES AND DISTRIBUTIONS

Table 1 is a list of valid fossil chondrophorine genera. Considering their total time span—Late Precambrian to Holocene—the number of genera is unusually small. The majority are clustered in the Paleozoic with only one Mesozoic genus (Cre-

taceous) and none from the Tertiary or younger rocks.

Precambrian forms

The oldest chondrophorines are Ediacaran (Glaessner and Wade 1966; Wade 1971); they include the porpitid *Eoporpita,* and a unique group of bilaterally symmetric chondrophorines somewhat similar to velellids but so different that a new family, Chondroplidae, has been erected for them (Wade 1971). Although these chondrophorines are preserved as molds and casts, a remark-

Fig. 10. Remains of the soft tissues of *Plectodiscus discoideus* on the dorsal surface. The ragged trace of the tissues that once covered the sail is visible near the center (arrow). X-radiograph has confirmed the absence of a pneumatophore and the presence of tentacles beneath the specimen. This fossil is similar to recent *Velella* specimens, where the whole pneumatophore was removed. Note irregular nature of the concentric features. Distance across bottom of photograph is 5.8 cm. WS 807, Lower Devonian, West Germany.

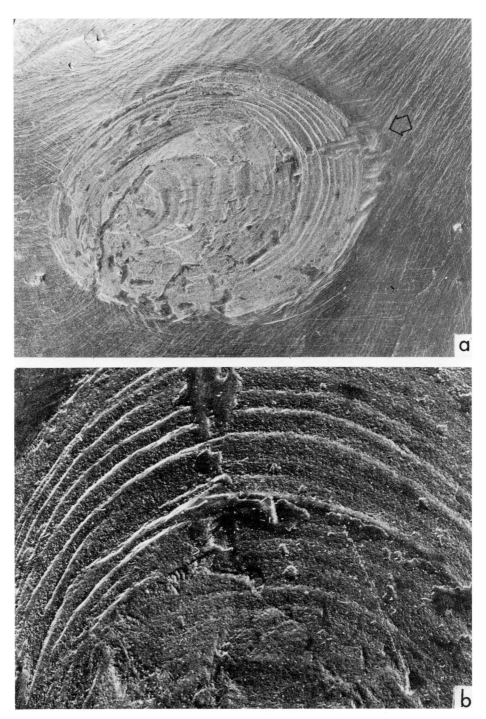

Fig. 11. Deformation in *Plectodiscus discoideus*. Compression and directional shear have accentuated the oval shape and produced much crumpling and deformation, particularly at the upper left. The trace of the radial furrow is still visible (arrow). Specimen has been badly scratched by mechanical steel brush preparation. Long direction of disc is 9 cm. Details of surface enlarged approximately × 3. WS 11917, Lower Devonian, West Germany.

Fig. 12. A portion of the disc of *Plectodiscus discoideus* resting against the carapace of the large phyllocariid crustacean *Nahecaris.* This specimen illustrates the thin, filmlike nature of the compressed pneumatophore. Note how the originally pliable disc has conformed to the shape of the underlying crustacean. Distance across photo is 6.5 cm. WS 4415, Lower Devonian, West Germany.

able amount of detail reveals, for example, the resistant internal chambers of the pneumatophore and their three-dimensional shapes. This distinguishes them from superficially similar but fundamentally different medusoids such as *Cyclomedusa* and *Brachina,* which are most likely scyphozoans (Wade 1971). *Eoporpita,* clearly a porpitid, has features closely comparable to living *Porpita. Eoporpita* has a low, flat pneumatophore characterized by closely adpressed pneumatocysts and fine, radiating striae, representing the gastrodermal canals of *Porpita.* One specimen shows details of tentacles that appear differentiated into three specialized types (Wade 1971, pl. 40, fig. 5).

Ovatoscutum concentricum shows remarkable similarities to Paleozoic velellid *Plectodiscus* (see Glaessner and Wade 1966). It has an oval pneumatophore with strong concentric pneumatocysts. Like *Plectodiscus,* it is bilaterally symmetric with a notch at the distal end, but the narrow axial furrow of *O. concentricum* passes across the disc, dividing it into two symmetric parts, while in *Plectodiscus* only a radial furrow is present. The

pneumatophore of *Ovatoscutum* resembles the juvenile stage of *Velella,* which has bilateral symmetry and a furrow that transverses the entire surface of the disc. In the adult *Velella* the pneumatophore becomes twisted, producing sigmoidal asymmetry and the loss of earlier bilateral symmetry (Caster 1942). Glaessner and Wade (1966) and Wade (1971) state that *Ovatoscutum* lacks a sail and illustrate two complete, sailless individuals. However, the taphonomic factors, previously discussed, sometimes make preservation and recognition of the sail in Paleozoic forms nearly impossible (Fig. 6).

The other Ediacaran bilateral chondrophorine, *Chondroplon bilobatum* Wade, is radically different from velellids. It consists of a bilobed pneumatophore with strong notches or emarginations at either side. It is slightly convex and marked by a distinct, blunt, keel-like ridge that divides the pneumatophore into two equal parts. The blunt medial ridge never developed into a sail. There are good reasons for erecting a new family for *Chondroplon,* but I question whether *Ovatoscutum* should be included.

Cambrian forms

There is surprisingly little known about Cambrian chondrophorines. *Velumbrella czarnockii* was described as a medusoid from the Lower Cambrian (Stasińska 1960) and is preserved as molds and casts of a circular, platelike fossil with distinct radial symmetry. There is a strong suggestion of porpitids, possibly related to *Eoporpita.*

No definite velellids come from Cambrian rocks, but a number of Cambrian fossils referred to *incertae sedis* or monoplacophorans may in fact be chondrophorines. Such problematic monoplacophorans were illustrated by Gil Cid (1972) from the Lower Cambrian of Spain. Yochelson and Gil Cid (1984) have assigned these fossils to *Scenella* but recognize them as bilateral chondrophorines, morphologically similar to *Vellela* but lacking a sail (Fig. 8C). Yochelson and Stanley (1981) have already pointed out that on a morphologic basis, a number of other Cambrian monoplacophorans could be chondrophorines, directly related to well-established Ediacaran groups. A relation to *Chondroplon* is very possible for *Scenella.*

Ordovician and Silurian forms

Middle Ordovician strata of New York have yielded an elliptic but otherwise very modern porpitid, *Discophyllum peltatum* Hall (Harrington and Moore 1956). This may be related to the Cambrian *Velumbrella,* but it differs from it in the fine spacing and detailed structure of the radial canals.

Another univalved mollusk, originally referred

TABLE 1. *Age and preservation of chondrophorine taxa*

Taxa	Description	Age	Preservation Type
Archaeonectris Huckriede	Convex, semiglobular successively enveloped chambers; sail lacking	Late Ordovician or Early Silurian	Steinkerns
Chondroplon Wade	Bilateral, chambered disc with two strong lobes and an axial furrow producing an emargination at either end; sail lacking	Late Precambrian	Mold and cast
Discophyllum Hall	Discoidal, elliptical with radial bands crossed by concentric lines; sail lacking	Middle Ordovician	? Mold and cast
Eoporpita Wade	Convex, circular porpitid disc with radial and concentric canals, raised central region; sail lacking	Late Precambrian	Mold and cast
Ovatoscutum Glaessner and Wade	Bilaterally symmetrical, concentrically chambered, oval disc with axial furrow; ? sail lacking	Late Precambrian	Mold and cast
Scenella Billings (as redefined by Yochelson and Gil Cid 1984)	Bilateral conical disc, concentric chambers uniformly spaced with superimposed radial canals, producing a reticulate pattern; sail lacking	Early Cambrian	Mold and cast; ? organic film
Palaelophacmaea Donaldson	Bilaterally symmetric, convex, chambered disc with distinct radial ridge; sail lacking	Middle Ordovician– Early Cretaceous	Mold and cast
Paropsonema Clarke	Convex to discoidal disc with closely spaced radial canals in cycles; sail lacking	Devonian–Silurian	? Mold and cast
Plectodiscus Ruedemann	Circular to elliptical, bilaterally symmetric disc with concentric chambers; sail present	Early Devonian– Pennsylvanian	Mold and cast; organic film; pyrite replacement and soft-tissue preservation
Porpita Lamarck	Circular, disc or cone; concentric chambers with strong radial canals; sail lacking	Holocene	—
Silurovelella Fisher	Circular to elliptical, bilaterally symmetric disc; sail present	Later Silurian	Mold and cast
Velella Lamarck	Asymmetric, sigmoidal disc with concentric chambers; sail present	Holocene	—
Velumbrella Stasińska	Circular, discoidal plates with radial canals and central depressed region; sail lacking	Lower Cambrian	? Mold and cast

to a patelliform gastropod (*Palaelophacmaea criola* Donaldson), was reinterpreted as a chondrophorine by Yochelson and Stanley (1981). These fossils are characterized by concentric corrugations of the pneumatocysts, bilateral symmetry, strong radial ridge, and the lack of a sail (Fig. 8D). Stanley and Kanie (1985) described a Cretaceous species of *Palaelophacmaea* and assigned it and the Ordovician example to the extinct family Chondroplidae.

Another somewhat unique chondrophrine, *Ar-chaeonectris benderi* Huckriede, is either latest Ordovician or earliest Silurian in age. It lacks a sail and consists of concentrically enveloping pneumatocysts producing a globular, biscuit shape. Because of its unique pneumatocyst arrangement, it may also represent a new family-level group. No other examples, living or fossil, are known.

Paropsonema mirabile Chapman, clearly a porpitid, comes from the Silurian of Australia and is quite similar to the Devonian *P. cryptophya*

Clarke (Harrington and Moore 1956). Both are very modern-appearing porpitids and relate well to the Ordovician *Discophyllum,* to which presumably they are related. In addition, a Late Silurian sailed velellid, *Silurovelella casteri* Fisher, demonstrates the presence of both velellid and porpitid families in this period. Yochelson et al. (1983) referred *Silurovelella* to the Devonian *Plectodiscus,* discussed below.

Devonian forms

The best preserved fossils are from the Devonian. These are from Lower and Middle Devonian dark shales of West Germany and New York State. *Plectodiscus discoideus* (Rauff), with soft-part preservation, is from West Germany (Yochelson et al. 1983), and *P. cortlandensis* Caster (Fig. 7) and *P. molestus* Ruedemann from New York. These velellids are circular to ellipsoidal, bilaterally symmetric, and possess sails. Morphologically, they form with *Silurovelella* a closely related if not congeneric group. The detailed anatomic reconstruction of the soft parts of the Devonian velellids from West Germany has revealed great similarities with living *Velella* (Fig. 13). As mentioned above, a Devonian porpitid, *Paropsonema*

cryptophya Clarke, is also known and presumed to be related to the Silurian forms.

Late Paleozoic forms

Plectodiscus is also known from the Late Carboniferous (*P. circus* Chamberlain); a new Carboniferous species comes from Malaysia (Stanley and Yancey 1986). These Late Paleozoic species demonstrate the long persistence of the Silurian and Devonian *Plectodiscus* group.

Post-Paleozoic forms

It is surprising that no post-Paleozoic chondrophorines have been found. The problem is clearly one of recognizing the problematic nature of these fossils. The rediscovery of *Palaelophacmaea annulata* (Yokoyama) as the first Mesozoic chondrophorine from Lower Cretaceous rocks of Japan (Stanley and Kanie 1985) is a case in point (Fig. 8A, B, G, J). For over 100 years, this fossil was regarded as a somewhat problematic patelliform gastropod, and only recently was it recognized as a chondrophorine. This illustrates the often difficult problem of correctly recognizing and identifying chondrophorine fossils.

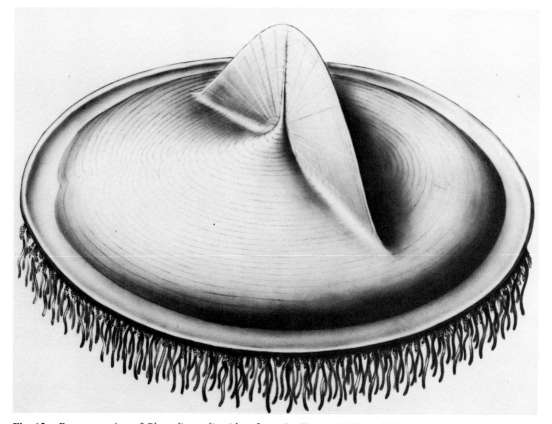

Fig. 13. Reconstruction of *Plectodiscus discoideus* from the Hunsrück Slate of West Germany (Yochelson et al. 1983, fig. 1). Reproduced with permission of E. Schweizerbart'sche Verlagsbuchhandlung.

Clearly they have at least moderate chances for preservation, and their remains must exist in post-Cretaceous rocks. Another rather unusual shieldlike patelliform gastropod was described as *Scurriopsis* (S.) *cycloides* from the Late Triassic (Tichy 1980). The wrinkled, irregular "growth lines" of this supposed gastropod and the apparent thinness of the shell are suggestive of porpitids (Fig. 8H), and future study might prove it to be the first Triassic example.

We thus find chondrophorine fossils classified as holothurians, rugose corals, trace fossils, *incertae sedis,* jellyfish, monoplacophorans, and gastropods. Clearly, if these fossils are to be properly identified, all possible variations of morphology must be recognized.

EVOLUTION

Chondrophorines are slowly evolving lineages whose roots begin in the Precambrian. Although their geologic record is spotty, some generalizations about their evolution can be made. Figure 14 shows a possible outline of chondrophorine phylogeny.

Three families, the Porpitidae, the Chondroplidae, and possibly the Velellidae, were established in the Precambrian. Of these, only the Chondroplidae are now extinct. Based on fossil evidence and comparisons with living taxa, it is suggested that at the generic level chondrophorines were never diverse but throughout their long existence evolved at very slow rates. Such slow rates of evolution and low diversity are indeed characteristic of living fossils. This extreme conservatism is typified within the velellid *Plectodiscus* group, which seems to have maintained almost the same, identical pneumatophore plan from at least the Late Silurian until Late Carboniferous—a total of > 100 Ma in duration. No sail is visible, but if the Precambrian *Ovatoscutum* is also a velellid, closely allied with *Plectodiscus,* then this family will extend back to > 600 Ma ago. Living *Velella* differs from *Plectodiscus* in having a sigmoidal twist of the disc (Caster 1942), which imparts asymmetry to the bilateral pneumatophore (Fig. 14). It is apparent that *Velella,* in earlier juvenile stages, has clear bilateral symmetry and the development of sigmoidal asymmetry occurs secondarily during ontogeny.

The sail of velellids was a significant feature, setting this distinctive group of by-the-wind sailors apart from other chondrophorines. Sailed propulsion once evolved, was retained and only slightly elaborated upon during the long velellid existence. Judging from Devonian fossils, the sail seems to have increased in size, until, as in living *Velella,* it occupied the whole diameter of the disc. The sail does not appear to have been developed in the porpitid group or in the extinct Chondroplidae.

Like the velellids, the porpitids were established in the Precambrian, and have undergone little change. Fossil porpitids, even more than the velellids, appear similar to their living descendant *Porpita.* The internal, chambered pneumatophore is similar in porpitids and in velellids, and differences between them are primarily in the presence and density of the radial, gastrovascular canals and the lack of a sail. The porpitid *Discophyllum* may have developed an oval rather than circular shape, but this trend appears not to have been significant.

The Ordovician or Silurian *Archaeonectris,* while clearly a chondrophorine, is unusual. It was described as a possible porpitid but is so different from all known forms that it could represent a new family. It may possibly possess the type of tentacles, which would demonstrate its chondrophorine affinities, but the globose, enveloping, involute mode of pneumatocyst construction sets it apart from all other chondrophorines. Wade (1972) suggested that *Archaeonectris* could have been derived from the modification of either velellid or porpitid type pneumatophores. As Wade pointed out, the problem seems to be one of increasing the volume of a float chamber without loss of strength or increase in weight. Since no other representatives are known, little can be said about the evolutionary position of *Archaeonectris.* The possibility exists that these fossils represent a dead-end evolutionary lineage and thus were only an experiment in pneumatophore construction, as inferred in Fig. 14.

Finally, the Chondroplidae are a fundamentally different group, also with roots in the Precambrian. The discovery of *Palaelophacmaea* in Ordovician and Cretaceous reveals the continuity from the Paleozoic into the Mesozoic. The strongly convex shape, concentric pneumatocysts, bilateral symmetry, and the lack of a sail are characteristics that make *Palaelophacmaea* unique. Morphology of *Ovatoscutum* more closely allies it with the Cambrian *Scenella* and the Silurian–Carboniferous *Plectodiscus* group. In fact, as Wade (1971) suggested, this Precambrian genus may belong to Velellidae rather than Chondroplidae. Detailed work on Precambrian forms will be required to establish this relationship. The lack of fossil chondrophorines from the Early Cretaceous to the present still needs to be investigated. During this interval the chondroplids became extinct, but the other two groups have survived.

The persistence of the fundamental body plans of the Chondrophorina allows some generalizations to be made regarding their extraordinarily extensive geologic history and slow rates of evolution. Both the Velellidae and Porpitidae are very ancient. The Velellidae range from the Late Silurian—a duration of some 415 Ma. If the Precambrian *Ovatoscutum* is also a member of this family, it extends it to > 600 Ma. The Porpitidae,

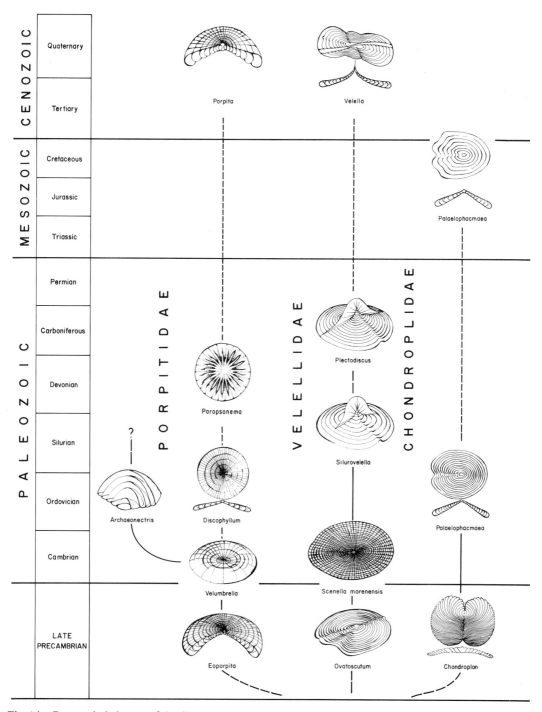

Fig. 14. Proposed phylogeny of the Chondrophorina and possible relationships among the various genera within the three families. This figure illustrates the simple, conservative nature of the group, the general forms of the various genera, and the absence of taxa in the late Paleozoic, Mesozoic, and Cenozoic. Refer to Table 1 for details.

originating with the Precambrian *Eoporpita*, also are a long-lived taxon. This long time span may be unparalleled by any other family. Individual velellid genera such as *Plectodiscus* show a duration of at least 100 Ma, and this seems typical of the group. The small number of taxa and their extensive time ranges, coupled with the exceedingly low rates of origination, are typical of so-called living fossils.

All evidence from fossils indicates that chondrophorines have followed consistent life-styles during their lengthy existence, as planktotrophic, wind- and current-dispersed open-ocean mariners. The key to their unparalleled low rates of evolution may be linked to their broad dispersal patterns and the unspecialized nature of their open-ocean planktonic life-styles.

As Wade (1972) points out, the earliest chondrophorines of the Ediacara fauna had already developed a chitinoid approach to biomineralization of their pneumatophore. The lightweight, tough, pliable nature of the internal pneumatophore float appears to have been the key to their success as wind-dispersed planktonic organisms. It also accounts for the preservational record of these organisms. Elaboration in the arrangement and disposition of the concentric pneumatocysts appears to have been relatively stable in the Paleozoic with the possible exception of *Archaeonectris*.

Both bilateral and radially symmetric forms were also established at the onset of their evolutionary history. Bilateral symmetry appears, at least in the Paleozoic, to have been associated with the development of a sail, which is simply a structure produced by the closed-off, noninflated upward extensions of the gas-filled pneumatocysts. The sail may have been a feature of the Precambrian *Ovatoscutum*, but this cannot be demonstrated. The bilateral chondrophorine *Scenella morenensis* Yochelson and Gil Cid from the Lower Cambrian resembles both *Ovatoscutum* and *Velella*, but like the former, shows no evidence of a sail. The evolving Ediacaran stock may have not developed a sail until sometime prior to the Silurian. In any event, a well-developed sail in the Silurian *Silurovelella* attests to the importance of this distinctive feature in wind propulsion. This was a structure of paramount significance in the wide dispersal of the group. Although the sail appears to have increased in size, reaching the total length of the disc in modern *Velella*, its fundamental organization has changed very little.

CRITERIA FOR RECOGNITION

Considering what is now known regarding chondrophorine morphology, taphonomy, and evolution, certain criteria may allow these fossils to be distinguished from unrelated groups that they may resemble. Of all the groups considered earlier, two appear to cause the most confusion. These are trace fossils and mollusks.

Many trace fossils, like the concentrically coiled surface-feeding burrows of *Spirorhaphe*, could be confused with such Paleozoic velellids as *Plectodiscus*. Other radial and concentric trace fossils may be confused with porpitids, as was *Palaeoscia floweri* Caster. Trace fossils lack the detailed pneumatophore morphology of chondrophorines. All chondrophorines possess in their pneumatophores a single, centrally located, circular embryonic air chamber and a central pore. Furthermore, chondrophorines, but not trace fossils, usually display fine surface detail related to the chitinoid disc. The presence of the predictably arranged, superimposed radial disc can also serve to distinguish chondrophorines from morphologically similar trace fossils. The confusion between chondrophorines and the shells of univalved mollusks is more difficult to resolve, and the problem is compounded by diagenesis, which may alter the original mineralogic composition. Indeed, molds and casts are the usual style of preservation and chondrophorines are of the same size range as many mollusks. They also may occur in the same environments as mollusks. The circular to elliptical shells of patellid gastropods and monoplacophorans present real potential for confusion, as evidenced by the recurring examples of fossil chondrophorines found mislabeled in museums. The multiple criteria listed below offer the most operational methods of distinguishing chondrophorines from univalved mollusks: (1) the film-like organic nature of the compressed chondrophorine pneumatophore, (2) evidence of extensive plastic deformation and irregularities of the concentric corrugations of the "shell" without evidence of fracturing, (3) presence of an apical pore in the central pneumatocysts, (4) presence of a sail, (5) distinctive radial ridge and groove features of the disc, (6) absence of muscle scars, (7) occurrence of fossils that appear out of place with respect to the environment of particular mollusk groups, and (8) evidence of predation such as nibbling marks on the organic float.

Such criteria were used to distinguish nonsailed chondroplid Cretaceous chondrophorines from true limpet gastropods, with which they had been erroneously classified (Stanley and Kanie 1985). In this case the radial ridge or groove of the principal vascular vessel bore a striking resemblance to the selenizone. However, the thin organic film-like nature of the presumed shell, coupled with the lack of muscle scars and presence of an apical pore, served to distinguish these fossils from limpets occurring in the same environment. The plastic deformation of the pneumatophore including wrinkling is characteristic, and it produces the irregularities of the pneumatophore disc. Moreover, possible predation marks are present. Impressions of the vascularized tissues may also appear on the disc surface. All these characteristics are incompatible with the shell morphology and type of deformation usually associated with a true calcareous shell (Yochelson and Stanley 1981).

Presence or absence of a sail may not in itself always be diagnostic, because this thin, delicate structure can be made almost invisible by compression upon the disc. Also, if the disc is upside down, the sail will not be seen except perhaps in X-radiography. In some specimens of *Plectodiscus,* the outline of the sail was apparent even though the sail itself was not (Fig. 7). Other criteria, such as absence of muscle scars, are also useful in distinguishing univalved mollusks from chondrophorine fossils.

It appears to me that the single most diagnostic characteristic of chondrophorine fossils is the exceedingly thin nature of the compressed disc. In the case of Devonian fossils, thickness was not measured but estimated to be filmlike and only fractions of a millimeter thick (Fig. 12). The organic, chitinoid disc is frequently diagenetically altered, but geochemical studies might detect the original composition. In the case of rare specimens where infilling with sediment occurred prior to fossilization, the original pneumatocyst relief of 2 to 3 mm is still present.

Clearly these ancient by-the-wind sailors have populated oceans and seaways since before the dawn of the Phanerozoic. Their remains clearly have potential for preservation and can be expected in marine rocks of all ages and many environments. The extensive and unexplained stratigraphic gaps in chondrophorine occurrences will eventually be filled as more critical and objective attention is directed to these problematic fossils.

Acknowledgments

My curiosity about chondrophorines and their potentially problematic nature was first stimulated by Ellis L. Yochelson who introduced me to some strange fossils in the mollusk collections of the U.S. National Museum. I thank Raylene Beier, who assisted in organizing and editing the manuscript during various stages of its production. Roxanne Jumer skillfully drew the artistic reconstructions in Figs. 3 and 4, as well as the individual sketches in Fig. 14. The Department of Geology, University of Montana, arranged for typing the manuscript.

REFERENCES

Caster, K. E. 1942. Two new siphonophores from the Paleozoic. *Palaeontogr. Am.* **3,** 56–90.

Chamberlain, C. K. 1971. A "by-the-wind-sailor" (Velellidae) from the Pennsylvanian flysch of Oklahoma. *J. Paleont.* **45,** 724–728.

Donaldson, A. C. 1962. A patelliform gastropod, *Palaelophacmaea criola:* a new genus and species from the Lower Ordovician of central Pennsylvania. *Proc. West Virginia Acad. Sci.* **34,** 143–149.

Fisher, D. W. 1957. Lithology, paleoecology and paleontology of the Vernon Shale (Late Silurian) in the type area. *N.Y. State Mus. Sci. Serv. Bull.* **364,** 1–31.

Fuchs, A. 1915. Der Hunsrückschiefer und die Unterkoblenzschichten am Mittelrhein (Loreleigegend). Teil 1. Beitrag zur Kenntnis der Hunsrückschiefer und Unterkoblenzfauna der Loreleigegend. *Abh. K. Preuss. Geol. Landesanst., N.F.* **79,** 1–80.

Gil Cid, M. D. 1972. Aportacion al conocimiento del Cambrico inferior de Sierra Morena. *Bol. R. Soc. Española Hist. Nat. (Geol.)* **70,** 215–222.

Glaessner, M. F. 1971. The genus *Conomedusites* Glaessner and Wade and the diversification of the Cnidaria. *Paläont. Z.* **45,** 7–17.

—— and Wade, M. 1966. The Late Precambrian fossils from Ediacara, South Australia. *Palaeontology* **9,** 599–628.

Hahn, G. and Pflug, H. D. 1980. Ein neuer Medusen-Fund aus dem Jung-Präkambrium von Zentral-Iran. *Senckenb. Leth.* **60,** 449–461.

Hall, J. 1847. *Paleontology of New York,* vol. 1., *Nat. Hist. N.Y.* **6,** 380 pp.

Harrington, H. J. and Moore, R. C. 1956. Siphonophorida. *In* Moore, R. C. (ed.). *Treatise on Invertebrate Paleontology,* part F, *Coelenterata,* Geological Society of America and Univ. of Kansas Press, Lawrence, pp. F145–F152.

Huckriede, R. 1967. *Archaeonectris benderi* n. gen. n. sp. (Hydrozoa), eine Chondrophore von der Wende Ordovicium/Silurium aus Jordanien. *Geol. Palaeont.* **1,** 101–109.

Hyman, L. H. 1940. *The Invertebrates: Protozoa through Ctenophora.* McGraw-Hill, New York.

Kanie, Y. 1975. Some Cretaceous patelliform gastropods from the northern Pacific region. *Sci. Rep. Yokosuka City Mus.* **21,** 1–44.

Maas, O. 1902. Über Medusen aus dem Solenhofer Schiefer und der unteren Kreide der Karpathen. *Palaeontographica* **48,** 297–322.

Mackie, G. O. 1959. The evolution of the Chondrophora (Siphonophora-Disconathe): new evidence from behavioral studies. *Trans. R. Soc. Canada* **53**(5), 7–20.

Osgood, R. G. 1970. Trace fossils of the Cincinnati area. *Palaeontogr. Am.* **6,** 281–444.

Rauff, H. 1939. *Palaeonectris discoidea* Rauff, eine Siphonophoride Meduse aus dem rheinischen Unterdevon nebst Bemerkungen zur umstrittenen *Brooksella rhenana* Kinkelin. *Paläont. Z.* **21,** 194–213.

Romer, A. S. 1964. *Vertebrate Paleontology.* Univ. of Chicago Press, Chicago.

Rudall, K. M. 1955. The distribution of collagen and chitin. *Symp. Soc. Exp. Biol.* **9,** 49–70.

Ruedemann, R. 1916. Paleontologic contributions from the New York State Museum. *N.Y. State Mus. Bull* **189**, 1–225.

Scrutton, C. T. 1979. Early fossil Cnidarians. *In* House, M. R. (ed). *The Origin of Major Invertebrate Groups,* Academic Press, London, pp. 161–207.

Seilacher, A. 1961. Holothurien im Hunsrückschiefer (Unter-Devon). *Notizbl. Hess. Landesamt Bodenforsch.* **89**, 66–72.

Stanley, G. D., Jr. and Kanie, Y. 1985. The first Mesozoic chondrophorine (medusoid hydrozoan) from the Lower Cretaceous of Japan. *Palaeontology* **28**, 101–109.

———— and Yancey, T. 1986. A new Late Paleozoic chondrophorine (Hydrozoa. Velellidae) By-the-Wind-Sailor from Malaysia. *J. Paleont.* **60**, 76–83.

Stasińska, A. 1960. *Velumbrella czarnockii* n. gen., n. sp.—Méduse du Cambrien Inférieur des Monts de Sainte-Croix. *Acta Palaeont. Polon.* **5**, 337–348.

Tichy, V. G. 1980. Gastropoden und Scaphopoden aus der Raibler Gruppe (Karn) von Raibl (Cave del Predil), Italien, *Verh. Geol. B.-A.* **3**, 443–461.

Wade, M. 1971. Bilateral Precambrian chondrophores from the Ediacara fauna, South Australia. *Proc. R. Soc. Victoria* **84**, 183–188.

————. 1972. Hydrozoa and Scyphozoa and other medusoids from the Precambrian Ediacara fauna, South Australia, *Palaeontology* **15**, 197–225.

Yochelson, E. L. 1981. A Devonian aptychus (Cephalopoda) from Alabama. *J. Paleont.* **55**, 124–127.

———— and Stanley, G. D., Jr. 1981. An Early Ordovician patelliform gastropod, *Palaelophacmaea,* reinterpreted as a coelenterate. *Lethaia* **15**, 322–330.

————, Stürmer, W., and Stanley, G. D., Jr. 1983. *Plectodiscus discoideus* (Rauff): a redescription of a chondrophorine from the Early Devonian Hunsrück Slate, West Germany. *Paläont. Z.* **57**, 39–68.

———— and Gil Cid, M. D. 1984. Reevaluation of the systematic position of *Scenella. Lethaia* **17**, 331–340.

8. PROBLEMATICA OF THE EARLY CAMBRIAN

A. YU. ROZANOV

Intensive studies during the years since 1965, especially those stimulated by Project 29 of the International Geological Correlation Program (IGCP), strongly substantiated the view that a unique phenomenon occurred at the boundary of the Precambrian and Cambrian: the mass appearance of various skeletal fossils. It has also been established that practically all invertebrate groups already existed in the Early Cambrian. Early Cambrian fossils include in fact foraminifers, radiolarians, sponges, coelenterates, mollusks, arthropods, brachiopods, echinoderms, and others. Some of these fossils are abundant in terms of both biomass and taxonomic diversity; others are insignificant in Cambrian communities.

There now exists quite strong evidence that the first geologic appearance of these skeletal fossils does not coincide in time with the actual origination of the taxa they represent, but only with the acquisition of the ability to develop a fossilizable skeleton. Many Late Precambrian and Early Cambrian fossils represent only partially scleritized body parts. These are most commonly outer structures (as in the Wiwaxiidae, Siphogonuchitidae, Chancelloriidae, etc.) or polychaetelike tubular lorica (hyolithelminths), but "inner" structures may also occur (Anabaritidae, etc.). With the timing of the first appearance of certain sclerites taken into account, the process of "scleritization" of the animal kingdom appears to have begun already in the Late Precambrian (not later than in Redkino time of the Vendian, as sabelliditids occur in those strata).

We still do not fully understand the reasons for the widespread acquisition of the ability to form a skeleton at the beginning of the Cambrian. However, a number of points should be noted.

First, a remarkable change took place in the proportions of magnesium and calcium in carbonate storage from the Riphean to the Early Cambrian; the diagram in Fig. 1 is based on a rough calculation of dolomite and limestone volumes on the greatest platforms. Limestone sedimentation gains much more importance at the beginning of the Cambrian, especially in the Botomian. At the same time, salt sedimentation reached its maximum and resulted in the greatest industrial deposits.

Second, a sudden increase in phosphate storage in the Tommotian (Fig. 1) led to formation of the huge phosphorite deposits in Asia (Kazakhstan, Mongolia, China, and possibly also India and Iran). The amount of P_2O_5 in other sedimentary rocks considerably increased in the Tommotian (Rozanov 1979). This reflects an increase in P_2O_5 contents in the ocean waters, which apparently allowed for the metabolic use of phosphates by many animal groups. The thermodynamic inefficiency of this process, as compared to the use of carbonates, entailed rapid extinction of a number of groups.

At the beginning of the Cambrian, primarily red-colored carbonate sediments were predominant in many geographic areas. Stromatolites, so characteristic of the Riphean and Vendian, underwent a sharp decline. Numerous calcareous algae appeared. Phytoplankton diversity increased considerably, although its total biomass may have remained unchanged. Finally, the nature of biotic penetration into the sea bottom changed (Fedonkin 1981) as mudeaters began actively to rework the sediment.

In the recent literature on Early Cambrian paleontology and stratigraphy, small shelly fossils (SSF) are often recognized as a distinct organic group. A large and sometimes even overwhelming proportion of these forms cannot be identified with other Phanerozoic groups of organisms. Practically all these fossils are very small (< 1 to at most a few millimeters). The notion of Problematica applies to them precisely.

It is impossible to discuss the systematic position of all these groups within the limits of a single article. There is a voluminous literature on these fossils, and the interpretations range from most fantastic (Fig. 2) to quite plausible ones (Bengtson and Missarzhevsky 1981). In this paper I briefly characterize the best known Early Cambrian problematic fossils and attribute them to a number of as yet informal groups.

Classification of the Early Cambrian SSF varies among individual paleontologists. For example, Missarzhevsky (in Rozanov et al. 1969) distinguished hyolithelminths, fossils wtih a calcareous tubular shell, camenids (the name replaced later by tommotiids), and other fossils. Missarzhevsky and Mambetov (1981) recognized hyolithelminths, tommotiids, "conodonts," chalkieriids, cambroclavitids, anabaritids, and others. Grigorieva (Yesakova) (in Sokolov and Zhuravleva 1983) subdivided all SSF into hyolithelminths, tubular problematica (including Anabaritidae and

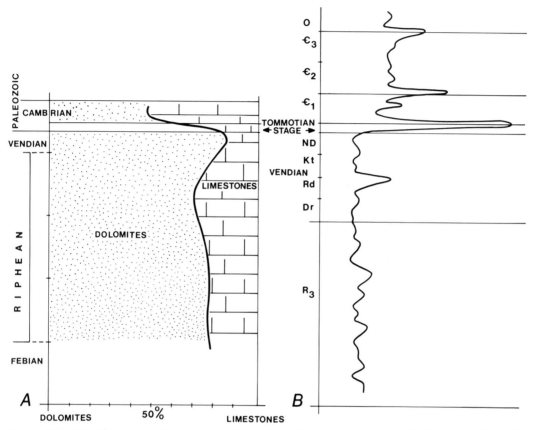

Fig. 1. Diagrammatic representation of storage changes in late Precambrian and early Cambrian. A. Proportion of magnesium and calcium carbonate storage. B. Increase in content of P_2O_5 in the ocean water in the Tommotina Stage. Letter symbols denote consecutive stages of the Late Precambrian.

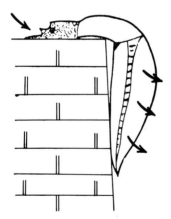

Fig. 2. Reconstruction of wiwaxiid animal according to Luo Huilin et al. (1982, fig. 51).

Colleolidae), tommotiids, and other skeletal problematica. Similar groups are recognized by Voronin et al. (1982).

Thus, the following groups can be identified among SSF: hyolithelminths, anabaritids, "cono-

donts," tommotiids, chalkieriids (= siphogonuchitids), cambroclavitids, and other fossils. All these problematica have one feature in common: they all can be chemically extracted from the rock because they generally have a phosphate "shell." In contrast, SSF with carbonate skeletons often remain unnoticed or are studied in thin sections together with algae and archaeocyathids with which they are usually associated. In this paper I briefly review all these groups except for "conodonts" or, to be more precise, conodont-like fossils, which have been thoroughly discussed by Bengtson (1983).

HYOLITHELMINTHS

By convention, hyolithelminths include curved cylindrical or narrow conical tubes. They are one of the most prevalent groups of tubular problematica in the Early Cambrian. Their best known representatives are *Hyolithellus* and *Torellella*. At present, the great majority of paleontologists regard the hyolithelminths as an order with two well-defined families: Hyolithellidae and Torellellidae (Fig. 3). The former family includes *Hyolithellus* Billings and *Pseudorthotheca* Cobbold; the latter includes *Torellella* Holm and *Koksuja* Missarzhevsky.

Other tubular forms are, as a rule, described as "other fossils." Among tubular problematica with phosphate tubes, three genera with very unique morphology have been described recently: *Mongolitubulus* Missarzhevsky (Fig. 4), *Tommototubulus* Fedorov (Fig. 5), and *Karatubulus* Missarzhevsky (Fig. 6). *Mongolitubulus* has a laminate phosphate wall with very peculiar scalelike sculpture at the outer surface. *Tommototubulus* has a similar but less orderly arranged sculpture. By analogy to hyolithelminths, a new family, Mongolitubulidae, might be erected for these forms. *Karatubulus* is a tube with egg shaped bulges stretched along the axis. No suggestions have ever

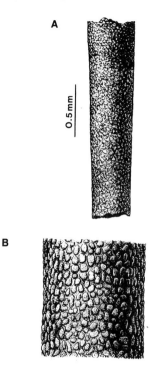

Fig. 5. *Tommototubulus savitzkii* Fedorov.

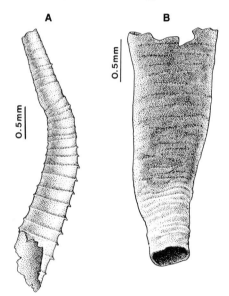

Fig. 3. Hyolithelminths. A. *Hyolithellus vladimirovae* Missarzhevsky. B. *Torellella lentiformis* (Sysoev).

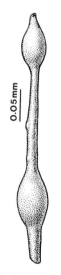

Fig. 6. *Karatubulus nodosus* Missarzhevsky.

been made concerning the systematic position of these fossils, and it is indeed very difficult to tell anything, except to note that the tube of *Karatubulus* somewhat resembles the foraminifer test. Hyolithelminths are widely recorded in many Early Cambrian rocks all over the world; in turn, *Mongolitubulus* has thus far been found only in Mongolia, Karatau, and perhaps China, *Tommototubulus* only in southern Siberia, and *Karatubulus* only in Karatau.

Fig. 4. *Mongolitubulus squamifer* Missarzhevsky.

Among tubular fossils with carbonate tubes, *Colleoloides* Walcott and anabaritids (see below) are best known. *Colleoloides* is a narrow cylindrical tube with longitudinal sculpture slightly inclined toward the axis. It has been described from many localities in North America, Europe, and Siberia, most recently by Bengtson and Fletcher (1983).

ANABARITIDS

Anabaritids include a vast group of tubular problematica with triradiate symmetry (Fig. 7). Their nomenclature was discussed by Glaessner (1979). Many authors assign also *Cambrotubulus* and *Tiksitheca* to anabaritids, but it is entirely possible that these two genera are not related to this group.

Although anabaritid tubes are outer skeletons (Abaimova 1978), it is now clear that they are inner structures in the sense that they resided inside the organism. They are widely believed to consist of a carbonate substance, but no firm evidence to this effect exists.

PLATYSOLENITES AND *VOLBORTHELLA*

The genera *Platysolenites* and *Volborthella* are unique among tubular problematica of the Early Cambrian in that their tubes are silicate in composition. The structure and systematic position of these fossils were analyzed by Glaessner (1978) and Rozanov (1979, 1983a).

Platysolenites is very simple in structure, up to a few centimeters in length and a few millimeters in diameter. Complete specimens occur rather rarely. The tube is usually covered with more or less evenly arranged thin grooves (Fig. 8). As a rule, grooves are absent from the oldest parts of the tube; such fragments were commonly described as *Serpulites* (?). *Platysolenites* is widely distributed in the Tommotian of the East Euro-

Fig. 8. *Platysolenites antiquissimus* Eichwald.

pean Platform and northern Europe. It has also been reported from the western part of the United States and northern Siberia.

Volborthella has a more complex structure (Fig. 9). It is a conical tube with cylindrical canal along the axis of the cone. The space between the central canal and the outer surface is divided into subhorizontal zones. These zones were previously regarded as cephalopod septa. However, they more probably reflect developmental phases of a worm tube. *Volborthella* attains up to a few millimeters in size. It is best known from the Atdabanian of the western part of the East European Platform, but it has been recorded also in younger rocks.

TOMMOTIIDS

Tommotiids are phosphatic fossils particularly characteristic of the Early Cambrian. They are generally recognized as a separate order Tommotiida, including, somewhat conventionally, at least three families: Tommotiidae, Lapworthellidae, and Kelanellidae (see Chapter 10 by Dzik, this volume).

They probably were first described by Cobbold (1921), who discussed *Lapworthella* and attributed it to the hyolithids. Later on, *Lapworthella* was described by Poulsen (1942), Lochman (1956), and others. But tommotiids have been more thoroughly studied only since the late 1960s, after abundant *Lapworthella, Tommotia, Camenella,* and *Kelanella* were found in Siberia

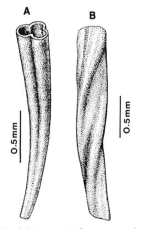

Fig. 7. Anabaritids. A. *Anabarites trisulcatus* Missarzhevsky. B. *Anabarites ternarius* Missarzhevsky.

Fig. 9. *Volborthella tenuis* Schmidt.

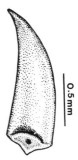

Fig. 11. *Sachites trianguliformis* (Mambetov).

(Rozanov and Missarzhevsky 1966; Rozanov et al. 1969).

Tommotiids are pyramidal, asymmetric sclerites with laminated walls (Fig. 10). Sometimes they are flattened-conoid, horn- or shell-like in shape (Missarzhevsky and Grigorieva 1981). Tommotiids are widely variable morphologically, often with sinistral and dextral forms. This group is known from the lower part of the Middle Cambrian on all the continents. It is very important stratigraphically and has been successfully used for zonation.

COELOSCLERITOPHORA

Bengtson and Missarzhevsky (1981) assign to this group the majority of hollow sclerites with empty inner cavity and external opening. The Coeloscleritophora include three families: Wiwaxiidae, Siphogonuchitidae, and Chancelloriidae. Their typical Early Cambrian representatives are the wiwaxiid *Sachites* Meshkova (Fig. 11), *Siphogonu-*

chites Qian (Fig. 12A), and *Chancelloria* Walcott (Fig. 13).

Some of these fossils were previously assigned to sponges, bryozoans, and others, but I essentially agree with Bengtson's and Missarzhevsky's interpretation (but see Chapter 10 by Dzik, this volume). All the Coeloscleritophora are widely distributed geographically and stratigraphically (probably throughout the entire Cambrian). The group is extremely abundant and diverse. The sclerite morphology shows much intrapopulation variability, as is clearly exemplified by *Wiwaxia corrugata* from the Middle Cambrian Burgess

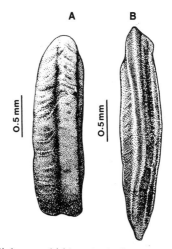

Fig. 12. Siphogonuchitids. A. *Siphogonuchites triangulatus* Qian. B. *Palaeosulcachites biformis* Qian.

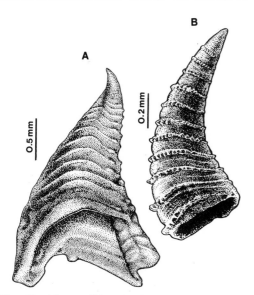

Fig. 10. Tommotiids. A. *Tommotia zonata* Missarzhevsky. B. *Lapworthella bella* Missarzhevsky.

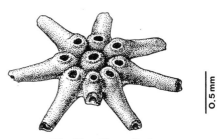

Fig. 13. *Chancelloria* sp.

Shale (see Chapter 13 by Briggs and Conway Morris, this volume).

CAMBROCLAVITIDS

Cambroclavitids were discovered at about the same time in Kazakhstan and southern China. These are small, bilaterally symmetric, hollow sclerites consisting of a more-or-less flat plate and a perpendicular, slightly inclined spiculum. The wall is definitely phosphatic. *Cambroclavus* Mambetov (= *Zhijinites* Qian) (Fig. 14) is the most typical representative of this group. Several other genera have also been described.

Cambroclavitids are widespread in Asia. They seem to have first appeared in the Atdabanian. They may be related to the earliest conodonts, or they may represent a separate high-level taxon.

OTHER PROBLEMATICA

I will first consider a number or forms that somewhat resemble mollusk shells in shape.

Maikhanella was described by Zhegallo (in Voronin et al. 1982) from Mongolia; it probably occurs also in China. It is low and caplike, with its apex slightly displaced toward the edge of the shell (Fig. 15). Its very peculiar outer surface is covered with tubercles arranged in parallel rows. The inner surface is smooth. In the microstructure, Zhegallo noticed three peculiar layers. This microstructure and the external ornamentation suggest that *Maikhanella* is not a mollusk.

Rozanoviella was described by Missarzhevsky (1981) and Voronin et al. (1982) as a bellerophontid archaeogastropod, but I consider it problematic. It has a chain of openings on the carina (Fig. 16). In many cases, however, the openings are displaced laterally or even absent.

Tannuolina was described in detail by Fonin and Smirnova (1967). It is cap shaped, and variable in height, with a slightly angular base. The outer surface is covered with growth lines, but one of the lateral sides bears pseudopores (Fig. 17). True *Tannuolina* is known only from the Botomian of the Altai-Sayan area and Mongolia, although there are also some reports on *Tannuolina* from China.

The forms described as *Sunnaginia* Missarzhevsky are very enigmatic. They were first found at the very base of the Tommotian on the Siberian Platform, but they are also known to occur in

Fig. 16. *Rozanoviella atypica* Missarzhevsky.

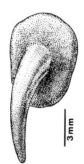

Fig. 14. *Cambroclavus* (= *Zhijinites*) *undulatus* Mambetov.

Fig. 15. *Maikhanella multa* H. Zhegallo.

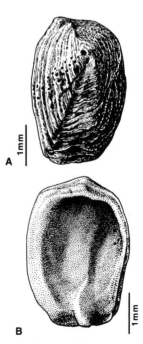

Fig. 17. *Tannuolina multifora* Fonin and Smirnova. A. Outer surface.   B. Inner surface.

England and Mongolia; they may also range higher up in stratigraphic sections. *Sunnaginia* has the form of a pyramid with a wrinkled surface; the wrinkles are parallel to the aperture (Fig. 18). The outer carinae show up as grooves at the inner surface. The shell is generally thought to be phosphatic, but its mineral composition has not been studied in detail.

Tumulduria Missarzhevsky is rather unique among the other so-called problematica. It was first described from the base of the Tommotian on the Aldan River. The original description by Missarzhevsky (in Rozanov et al. 1969) follows:

Shells minute (in millimeters), phosphatic, multilayered, bilaterally symmetric, flattened, oval in outline. Rounded fold, covering a little more than a half of the entire shell width, runs longitudinally. Shell edges lie practically in one plane. Sharp transversal folds with uneven, curved surface vary in width and amplitude; they are slightly inclined toward one edge of the shell. All the folds are discernible, although slightly flattened, on the inner shell surface.

Tumulduria has a peculiar morphology (Fig. 19), and its fragments may resemble trilobite fragments. It is such fragments that were described as the most ancient trilobites from the base of the Tommotian in Siberia (Fedorov et al. 1979).

A number of problematic disc and globe shaped fossils were described from the Early Cambrian. *Gaparella* Missarzhevsky, *Archaeooides* Qian, *Olivoides* Jiang, *Lenargyrion* Bengtson, and *Microdictyon* Bengtson et al. are perhaps most conspicuous in this group.

All these forms are phosphatic, but their diagenetic alteration cannot be ruled out. Some of these names may be synonyms. Thus, it is difficult to distinguish between *Gaparella* and *Archaeooides*. *Archaeooides* is rounded, hollow, pierced, probably by pores on numerous tubercles (Fig. 20). Sometimes it also bears various depressions. Missarzhevsky points out a remote morphologic resemblance of *Gaparella* to some sacklike archaeocyathids; however, possible analogs can also be sought among foraminifers. These globose problematica are widespread in the Tommotian and Atdabanian of Kazakhstan, Mongolia, and China. Despite their abundance, they have not yet been studied thoroughly.

Lenargyrion and *Microdictyon* have been investigated in much greater detail (Bengtson 1977; Bengtson and Missarzhevsky 1981; Missarzhevsky and Mambetov 1981). *Lenargyrion* (Fig. 21) was described in detail by Bengtson (1977). These are small, button shaped forms with one side smooth and slightly salient; the other side is wide and conical, with a flat top covered by conical tubercles. The morphology and wall structure of these fossils are well known, and their skeleton is phosphatic. Bengtson suggests that they may be related to vertebrates (see Chapter 16 by Dzik, this volume). *Lenargyrion* occurs in the upper part of the Atdabanian in Siberia. Similar or perhaps identical fossils were described from the Botomian of Turkey and Spitsbergen as *Hadimopanella* Gedik (Gedik 1977; Wrona 1982). Wrona regards *Lenargyrion* as a junior synonym of *Hadimopanella*.

Microdictyon is even more enigmatic (see Chapter 9 by Bengtson et al., this volume). It is dome shaped and pierced with dense apertures,

1mm

Fig. 18. *Sunnaginia imbricata* Missarzhevsky.

0·5mm

Fig. 20. *Archaeooides* sp.

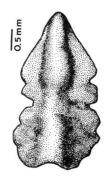

0.5mm

0.02mm

Fig. 21. *Lenargyrion* (= *Hadimopanella*) *knappologicum* Bengtson.

Fig. 19. *Tumulduria incomperta* Missarzhevsky.

each surrounded by six smaller openings that give the impression of a hexahedral outline (Fig. 22). The skeleton is phosphatic. These sclerites have much stratigraphic importance because they occur at the same level, near the Atdabanian–Botomian boundary, in North America, Europe, Siberia, and Kazakhstan.

Two more groups of "other problematica" from the Early Cambrian are also worth mentioning. One group includes *Salanacus* Grigorieva (Fig. 23) and *Koksodus* Missarzhevsky (Fig. 24). These fossils are aggregates of conical teeth on a plate. They may resemble the gastropod radula. Their association into one group, however, is only tentative because *Salanacus* appears to be composed of a siliceous substance, whereas *Koksodus* may be phosphatic in composition. These are fairly common sclerites, both in Tommotian and Atdabanian strata.

The second group includes *Paracarinachites* Qian and Jiang (Fig. 25A) and *Brushenodus* Jiang (Fig. 25B). They were first discovered in China only a few years ago. Qian and Jiang (personal communication) believe that these fossils are related to conodonts.

CONCLUSIONS

I would like to emphasize three points. First, this review of Early Cambrian problematic fossils

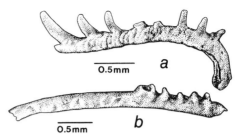

Fig. 25. Conodontlike fossils. A. *Paracarinachites sinensis* Qian and Jiang. B. *Brushenodus prionnodes* Jiang.

does not cover the groups discussed by Bengtson (1983). I provisionally retain *Lapworthella* among tommotiids. Two other fossils, *Sabellidites* and *Mobergella,* are so well known that they need not be redescribed here. I also note that the assignment of *Rhombocorniculum* (Fig. 26) to conodonts remains questionable, and that the nature of *Fomitchella* Missarzhevsky (Fig. 27) still is ambiguous (see Chapter 16 by Dzik, this volume).

Second, I do not consider in this paper numerous problematica with carbonate skeletons, which are, as a rule, studied in thin sections. This group includes cribricyathids, hydroconozoans (Fig. 28), and radiocyathids (see Chapter 4 by Zhuravlev, this volume), but the diversity of carbonate prob-

Fig. 22. *Microdictyon effusum* Bengtson et al.

Fig. 23. *Salanacus voronini* N. Grigorieva.

Fig. 24. *Koksodus serratus* Missarzhevsky.

Fig. 26. *Rhombocorniculum cancellatum* (Cobbold).

Fig. 27. *Fomitchella infundibuliformis* Missarzhevsky.

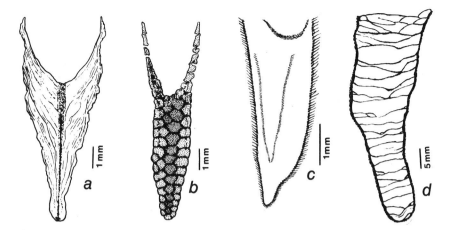

Fig. 28. Longitudinal thin sections. A. *Hydroconus mirabilis* Korde. B. *Cambroporella granulosa* (Vologdin). C. *Cribricyathus* sp. D. *Tabulaconus kordeae* Handfield.

lematica is almost as great as of those with phosphatic or siliceous skeletons.

Third, I would like to emphasize that Early Cambrian rocks contain numerous remains of very peculiar organisms, both animals and plants, most of which are unknown after the Cambrian. I tend to think that numerous high-level taxa developed in the Early Cambrian and rapidly became extinct (Rozanov 1983b).

Acknowledgments ·

I am grateful to N. I. Krasnova and A. Yu. Zhuravlev for their assistance, and to V. D. Kalganov for preparation of the drawing.

REFERENCES

Abaimova, G. P. 1978. Anabaritids—the oldest fossils with calcareous skeleton. *In* Sukhov, S. V. (ed.). *New Materials on Stratigraphy and Paleontology of Siberia.* SNIIGGIMS, Novosibirsk, pp. 77–83 [in Russian].

Bengtson, S. 1977. Early Cambrian button-shaped phosphatic microfossils from the Siberian Platform. *Palaeontology* **20**, 751–762.

———. 1983. The early history of Conodonta. *Fossils and Strata* **15**, 5–19.

——— and Fletcher, T. P. 1983. The oldest sequence of skeletal fossils in the Lower Cambrian of southwestern Newfoundland. *Can. J. Earth Sci.* **20**, 525–536.

——— and Missarzhevsky, V. V. 1981. Coeloscleritophora, a major group of enigmatic Cambrian metazoans. *U.S. Geol. Surv. Open-File Rep.* **81–743**, 19–21.

Cobbold, E. S. 1921. The Cambrian horizons of Comley (Shropshire) and their Brachiopoda,

Pteropoda, Gastropoda, etc. *Q. J. Geol. Soc. Lond.* **76**, 305–410.

Fedonkin, M. A. 1981. White sea biota of Vendian (Precambrian non-skeletal fauna of the Russian Platform North). *Trans. Geol. Inst. Acad. Sci.* USSR **342**, 3–100 [in Russian].

Fedorov, A. B., Egorova, L. I., and Savitsky, V. E. 1979. The first finding of oldest trilobites in the lower part of the stratotype of the Tommotian stage of the Lower Cambrian (R. Aldan) *Doklady AN SSSR Geol.* **249**, 1188–1190 [in Russian].

Fonin, V. D., and Smirnova, T. N. 1967. New group of problematical Early Cambrian microorganisms and some methods of their preparation. *Paleont. Zh.* **1967** (2), 15–27.

Gedik, I. 1977. Conodont biostratigraphy in the Middle Taurus. *Bull. Geol. Soc. Turkey* **20**, 35–48.

Glaessner, M. F. 1978. The oldest Foraminifera. *B.M.R. Bull.* **192**, 61–65.

———. 1979. Precambrian. *In* Robison, R. A. and Teichert, C. (eds). *Treatise on Invertebrate Paleontology,* part A. Geological Society of America and Univ. of Kansas Press, Lawrence, pp. A79–A135.

Lochman, C. 1956. Stratigraphy, paleontology and paleogeography of the *Ellipsocephala asaphoides* strata in Cambridge and Hoosick quadrangles. *N.Y. Bull. Geol. Soc. Am.* **67**, 1313–1397.

Luo Huilin, Jiang Zhiwen, Wu Ziche, Song Xueliang, and Onyang Lin. 1982.*The Sinian–Cambrian Boundary in Eastern Yunnan, China.* People's Publishing Co., Yunnan, 300 pp. [in Chinese].

Missarzhevsky, V. V. 1981. Early Cambrian hyolithes and gastropods of Mongolia. *Paleont. Zh.* **1981** (1), 21–28 [in Russian].

———— and Grigorieva, N. V. 1981. About new representatives of order Tommotiida. *Paleont. Zh.* **1981** (4), 91–97 [in Russian].

———— and Mambetov, A. M. 1981. Stratigraphy and fauna of Cambrian and Precambrian boundary beds of Maly Karatau. *Trans. Geol. Inst. Acad. Sci. USSR* **326**, 3–92.

Poulsen, C. 1942. Some hitherto unknown fossils from the Exsulans limestones of Bornholm. *Medd. Dansk Geol. Foren.* **10**(2), 212–235.

Rozanov, A. Yu. 1979. Some problems of studying of ancient skeletal organisms. *Biulleten' Moskovskogo Obshchestva Ispytatelej Prirody Otd. Geol.* **54**(3), 62–69 [in Russian].

————. 1983a. *Platysolenites. In* Urbanek, A. and Rozanov, A. (eds). *Upper Precambrian and Cambrian palaeontology of the East-European Platform.* Wyd. Geol., Warszawa, pp. 94–100.

————. 1983b. The main directives of ancient Phanerozoic organisms researched. *In* Tatarinov, L. P. (ed). *The Main Problems of Paleontological Research in the U.S.S.R.* Nauka, Moscow, pp. 90–96 [in Russian].

———— and Missarzhevsky, V. V. 1966. Biostratigraphy and fauna of the low horizons of the Cambrian. *Trudy Geol. Inst. Acad. Sci. USSR* **148**, 3–126 [in Russian].

————, Missarzhevsky, V. V., Volkova, N. A., Voronova, L. G., Krylov, I. N., Keller, B. M., Korolyuk, I. K., Lendzion, K., Michniak, R., Pykhova, N. G., and Sidorov, A. D. 1969. The Tommotian Stage and the Cambrian Lower Boundary Problem. *Trans. Geol. Inst. Acad. Sci. USSR* **206**, 3–380 [in Russian]. Eng. trans. 1981, Raaben, M. E. (ed.). Amerind Publ. Co. Pvt. Ltd., New Delhi. 359 pp.

Sokolov, B. S. and Zhuravleva, I. T. (eds). 1983. *Lower Cambrian Stage Subdivisions of Siberia. Atlas of Fossils.* Nauka, Moscow, 216 pp. [in Russian].

Voronin, Yu. I., Voronova, L. G., Grigorieva, N. V., Drozdova, N. A., Zhegallo, H. A., Zhuravlev, A. Yu., Ragozina, A. L., Rozanov, A. Yu., Sayutina, T. A., Sysoiev, B. A., and Fonin, V. D. 1982. The Precambrian/Cambrian boundary in the geosynclinal areas (the reference section of Salany-Gol, MPR). *Trans. Soviet-Mongolian Paleont. Exped.* **18**, 5–152 [in Russian].

Wrona, R. 1982. Early Cambrian phosphatic microfossils from southern Spitsbergen (Nornsund Region). *Palaeont. Polon.* **43**, 9–16.

9. THE CAMBRIAN NETLIKE FOSSIL *MICRODICTYON*

STEFAN BENGTSON, S. CROSBIE MATTHEWS, AND VLADIMIR V. MISSARZHEVSKY

The recent surge of work on the skeletalized faunas of the earliest Cambrian has helped to create a picture of a marine biota dominated by short-lived groups very different from what is known from later deposits and from present-day seas. Fossils of such organisms are by their very nature bound to be of problematic affinity. A considerable proportion of these early skeletalized organisms had phosphatic hard parts, and large quantities of their remains have been extracted by chemical means from suitable rocks all over the world.

In the course of such investigations there were a number of encounters with a reticulate phosphatic fossil of obscure affinity. Colleagues in several countries have subsequently drawn our attention to occurrences of similar forms over a wide geographic area. These fossils deserve comment for several reasons. They are widespread, have some stratigraphic significance, are truly enigmatic in nature, and—not least important—are exceptionally beautiful.

The reticulate fossil has figured in earlier publications (e.g., Matthews and Missarzhevsky 1975, pl. 4:2, 5, 8; Missarzhevsky and Mambetov 1981, pl. 13:3, 5). The generic name was formally introduced as *Microdictyon* Bengtson, Matthews et Missarzhevsky 1981 by Missarzhevsky and Mambetov (1981), who recognized a type species, *M. effusum*, and described another form, *?M. inceptor,* which they assigned, with a clear expression of doubt, to *Microdictyon.* Further examination of this species confirms that it is not congeneric with or even closely related to *Microdictyon effusum.*

We present here fuller documentation and closer analysis of the genus *Microdictyon.*

PROVENANCE OF MATERIAL

The following list gives the areas known to us at which *Microdictyon* has been found.

Asia

In Asia, *Microdictyon* is known in the Lower Cambrian both on the Siberian Platform and in regions where rocks of equivalent age have been relatively severely deformed (Karatau, Bateny Hills, Altaj, Tamdytau). At some localities the beds containing *Microdictyon* are within continuous fossiliferous sequences and the exact stratigraphic position is obvious. Elsewhere, the stratigraphic relationships of the occurrence of *Microdictyon* must be judged on the evidence of associated fossils. But in all such cases the data available justify reference to a range close to the boundary (as defined by Rozanov 1973) between the Atdabanian and the Botomian stages of the Lower Cambrian. [The Botomian Stage, as defined by Spizharsky et al. (1983), corresponds to the lower part of the Lenian Stage of previous usage.] The specimen described here as *M.? tenuiporatum* is derived from the older Tommotian Stage.

In what follows, the words "horizon," "formation," and "member" represent the terms *gorizont, svita,* and *pachka* as used by Soviet geologists.

1. Isit', River Lena, Yakutia, Tommotian Stage, coll. V. V. Missarzhevsky, sample No. M45-75. (Ref.: Rozanov et al. 1969, pp. 29–31). *Microdictyon? tenuiporatum.*

2. Achchagy-Kyyry-Taas, River Lena, Yakutia, Perekhodnaya Formation, 3rd Member, upper Atdabanian or lower Botomian Stage, *Fansycyathus lermontovae* Zone, coll. S. Bengtson, sample No. Sib73-1-SB. (Ref.: Keller et al. 1973, pp. 30–34, 64, 99–102; Bengtson 1977, pp. 751–752). *Microdictyon* sp.

3. Exposure north of Bograd, Bateny Hills, eastern part of Kuznetskij Alatau Range, Bed 3 or 4 of Zadorozhnaya et al. (1973, fig. 2), upper Atdabanian or lower Botomian Stage, coll. V. V. Missarzhevsky, sample No. M69-8/2. (Ref.: Rozanov et al. 1969, fig. 20, third column; Zadorozhnaya 1973, fig. 2). *Microdictyon rhomboidale.*

4. River Isha, near mouth of Shilovka, Gornyj Altaj, Kameshki Horizon, upper Atdabanian or lower Botomian Stage, coll. V. V. Missarzhevsky, sample No. M69-4/11. (Ref.: Repina et al. 1964, p. 36, unit 2). *Microdictyon* sp.

5. River Uchbas, Malyj Karatau, Geress Member, Shabakty Formation, upper Atdabanian Stage, *Rhombocorniculum cancellatum* Zone, coll. V. V. Missarzhevsky, sample No. M617/1, coll. N. P. Meshkova, sample No. 264. (Ref.: Mambetov and Missarzhevsky 1972; Missarzhevsky and Mambetov 1981, pp. 11–14). *Microdictyon effusum.*

6. 200 km north of Bukhara, northern Tamdytau (Kyzyl-Kum), upper Atdabanian Stage(?),

coll. B. V. Yaskovich, coll. T. I. Khajrulina, samples No. 474d and 444/1. (Ref.: Zhuravleva et al. 1970). *Microdictyon rhomboidale.*

Northwestern Europe

Most of the *Microdictyon* specimens so far found in Europe are fragmentary. Moreover, the stratigraphic relationships of the Scandinavian occurrences are in no case immediately clear, since the fossils have been found in condensed and possibly reworked sections (Scania and Bornholm) or have been found only in glacial drift material (South Bothnian area). The occurrence in Shropshire, England, is stratigraphically a clearer case.

7. Comley, Shropshire, England, *Strenuella* Limestone (Ac$_4$ of Cobbold 1921), mid-Lower Cambrian [probably correlatable with the Atdabanian–Botomian transition of the Siberian Platform; cf. Matthews (1973) and Bengtson (1977)]. (Ref.: Cobbold 1921; Matthews 1973). *Microdictyon* cf. *effusum.*

8. Kalby, Læså Rivulet, Bornholm, Kalby Marl, Middle Cambrian, *Ptychagnostus gibbus* Zone (possibly reworked Lower Cambrian material), coll. V. Berg-Madsen. (Ref.: Grönwall 1902; C. Poulsen 1942; V. Poulsen 1963; Berg-Madsen 1981). *Microdictyon* sp.

9. Two localities, 1 km south of Brantevik and 600 m north of Gislövshammar, southeastern Scania, "Fragment Limestone" (bed F in Bergström and Ahlberg 1981, fig. 6, and a bed overlying bed G in their fig. 8), Middle Cambrian, *Ptychagnostus gibbus* Zone or immediately below (possibly reworked Lower Cambrian material), coll. S. Bengtson, samples No. S70-2-SB and S71-35-SB. (Ref.: Bergström and Ahlberg 1981). *Microdictyon* sp.

10. South Bothnian area, glacial drift, Lower Cambrian, coll. C. Wiman, sample Eggegrund No. 3. (Ref.: Wiman 1903; Thorslund and Axberg 1979). *Microdictyon effusum.*

North and Central America

With the exception of the Middle Cambrian *Microdictyon* from Utah (see item 15, below), the American occurrences, according to current stratigraphic interpretation (e.g., Bergström 1981), all seem to correspond to approximately the same broad stratigraphic interval as the Asian ones—that is, the Atdabanian–Botomian transition—but it should be added that the biostratigraphic correlation between the different faunal provinces in the Lower Cambrian is by no means secure.

11. Rensselaer Polytechnic Institute, New York State, Schodack Limestone, Lower Cambrian, *Elliptocephala asaphoides* beds, sample provided by Dr. D. W. Fisher. (Ref.: Lochman

1956; Bird and Rasetti 1968, p. 26; Landing 1984). *Microdictyon* sp.

12. Eastern Massachusetts, Station No. 2 of Shaler and Foerste (1888), Hoppin Formation, Lower Cambrian, *Callavia* Zone, coll. E. Landing. (Ref.: Shaler and Foerste 1888; Landing and Brett 1982). *Microdictyon* sp. (= "*Milaculum*-like fragments" of Landing and Brett 1982; specimens seen by S. B.).

13. South end of Sekwi Range, Mackenzie Mountains, British Columbia, Sekwi Formation, Lower Cambrian, lower part of *Nevadella* Zone, coll. W. H. Fritz, GSC localities 73026, 73027, 73036, 73039, and 73073. (Ref.: Fritz 1972, 1979, and references therein). *Microdictyon* cf. *rhomboidale.*

14. Cassiar Mountains, British Columbia (section 8 of Fritz 1980, fig. 25.2), Rosella Formation, Lower Cambrian, lower part of *Nevadella* Zone, coll. W. H. Fritz. (Ref.: Fritz 1980; Conway Morris and Fritz 1984). *Microdictyon* sp.

15. Topaz Mountain, Drum Mountains, Millard County, Utah (about 15 m east of mine road in SE1/4SE1/4 sec. 1, T. 15 S., R. 11 W.; 15-minute quadrangle map, U.S. Geol. Surv., 1953), Swasey Limestone, uppermost bed, Middle Cambrian, *Ptychagnostus gibbus* Zone or immediately below, coll. R. A. Robison, sample No. 123. *Microdictyon robisoni.*

16. White Mountains, California, upper Campito and lower Poleta Formation, Lower Cambrian, *Nevadella* Zone, M. C. Tynan, personal communication (1981). (Ref.: Nelson 1978; Tynan 1983). *Microdictyon* cf. *rhomboidale.*

17. About 42 km south of Goldfield, Esmeralda County, Nevada, Campito Formation, Montenegro Member, Lower Cambrian, lower part of *Nevadella* Zone, coll. J. P. Albers and J. H. Stewart, sample 4131-CO. (Ref.: Albers and Stewart 1972, p. 11). *Microdictyon* n.sp. 1.

18. Sierra el Rajón, Caborca region, Mexico, Unit 2, Puerto Blanco Formation (correlatable with the Poleta Formation of California), Lower Cambrian, M. A. S. McMenamin, personal communication (1983). (Ref.: McMenamin 1984). *Microdictyon* sp.

South Australia

The correlation of the South Australian sequences in terms of Siberian stages is not entirely clear, as the taxonomy of the rich assemblages of skeletal fossils has been worked out only to a minor extent. *Microdictyon* occurs together with a fauna (Faunal Assemblage 2 of Daily 1956) that seems to be of approximately the same age as the Asian occurrences.

19. Bunyeroo Creek, Flinders Ranges, Wilkawillina Limestone, mid-Lower Cambrian, coll. B. Daily. (Ref.: Daily 1956, 1972; Thomson et al. 1976). *Microdictyon* n.sp.

20. Quarry about 1 km south-southwest of Curramulka, Yorke Peninsula, Parara Limestone, about 43 m above base, mid-Lower Cambrian, coll. B. Runnegar et al., samples No. L1763b and L1846. (Ref.: Daily 1956, pp. 109–110). *Microdictyon* n.sp.

GENUS *MICRODICTYON* BENGTSON, MATTHEWS AND MISSARZHEVSKY 1981

Derivation of Name.   Greek *mikros,* small, and *diktyon,* net.

Type Species.   *Microdictyon effusum* Bengtson, Matthews and Missarzhevsky 1981.

Species Included.   M. *effusum,* M. *rhomboidale* n.sp., M. *robisoni* n.sp., and, provisionally, *M.? tenuiporatum* n.sp. Note that ?*M. inceptor* (Missarzhevsky and Mambetov 1981) does not belong to *Microdictyon.*

Diagnosis. Phosphatic netlike plates with a crudely hexagonal meshwork. Outline varies from rounded oblong through rhombic to more complex shapes. Plate length ranges from about 0.5 to about 2.5 mm. Holes in meshwork round, from about 10 to about 130 μm in diameter, penetrating the plate or terminating with a thin, usually half-spherical bottom. Mushroom shaped or spiky nodes protrude at the junctions of the wall between the holes. Holes usually decrease in size toward the plate periphery; edge of plate formed by peripheral girdle. Plates constructed of two main layers: a *framework* forming the walls and the bottoms (if present) of the holes, and a *capping* making up the nodes and the crests of the walls.

External and Internal Structure. As the general structure does not appear to vary significantly between the species of *Microdictyon,* it is described here jointly for all members of the genus.

The outline of the plates is variable between, but fairly constant within what is here interpreted as species (see below). The structure of the plates, however, is very characteristic and makes it possible to identify with certainty even very small fragments as pieces of *Microdictyon.* There is an upper side, usually convex, on which the nodes are expressed. These nodes are regularly situated at the junctions of the walls between the holes. They are commonly mushroom shaped, with a distinct brim (e.g., Fig. 4F), but may in some forms be flattened and tablelike (Figs. 4D and 12E) or produced into thorns (Figs. 8, 9, and 12D). The holes are round, usually larger toward the center of the plate. The opening of the holes is marked by a conspicuous sharp edge (Figs. 4F, 8C, 9F, I, 10D, E, G, and 13A, arrow). The holes may continue through the plate or terminate by a thin wall on the lower side of the plate (Figs. 2 and

3). This wall is continuous with the side walls of the holes (Fig. 3).

The network is basically hexagonal, and the nodes are then situated at three-wall junctions. Dislocations in this lattice may occur, particularly in plates of more irregular outline. Some holes may then be surrounded by less than or more than six other holes. In those instances where such dislocations force two wall junctions toward each other the nodes may merge, the end result being a four-wall junction with an enlarged node. The arrows in Fig. 4A point at some such junctions (a) as well as one case of two normal junctions being so close to each other as to make the nodes almost merge (b).

Sections and fractures through the plates reveal a basically twofold internal structure (Figs. 3, 9F, 10D–G, and 13). The lower parts of the walls between the holes and the thin bottoms that may terminate the holes are made up of a continuous tissue here referred to as the *framework.* The crests of the walls and the nodes are made up of another tissue, here termed *capping.*

The framework consists of clear apatite that shows only faint birefringence, indicating a weakly preferred orientation of C-axes parallel to the surfaces of the framework. Occasionally there are traces of lamination present, with laminae about 1 μm thick. The laminae may be interlayered with more porous substance and may even form tabulate partitions underneath the nodes, as in Fig. 3 (the regions beneath nodes 3 and 4).

The capping appears to be somewhat more complex in structure. The color in transmitted plane-polarized light is yellowish, darker than the framework. There is usually a thin (about 2 to 3 μm) outermost layer (Figs. 3 and 13), which in the section in Fig. 3 is characterized by strong birefringence and C-axis orientation perpendicular to the surface. The layer is continuous around the edges of the nodes (the "brims" of the "mushrooms") and terminates at the sharp edges marking the orifices of the holes (Figs. 10D, E and 13A, arrow). In the nodes, this layer is underlaid by thicker layers that have the same yellowish color but usually no pronounced birefringence. The substance here appears more porous and opaque (e.g., Fig. 13).

In the high parts of the plates (i.e., in the region of the nodes) there is, in addition to the porous parts of capping and framework tissue, a considerable amount of porous substance between these two tissues and in the cavities below the framework. The space between the walls formed by the framework is usually seen to be empty when viewed from the lower side (e.g., Figs. 6 and 9M).

In some instances there is a basal layer of granular apatite (Figs. 7D, E, 10A, C, and 11E) underlying the framework. Although the sectioned specimen in Fig. 3 shows such a layer as a probably secondary incrustation of phosphorite, the struc-

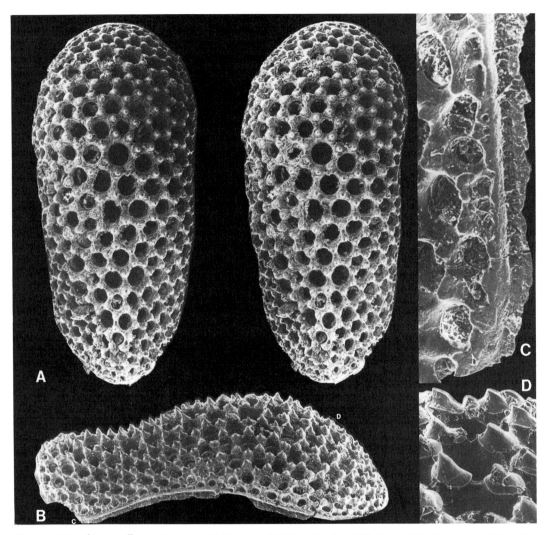

Fig. 1. *Microdictyon effusum* Bengtson, Matthews and Missarzhevsky 1981. Geress Member, base of Shabakty Formation, River Uchbas, Malyj Karatau, upper Atdabanian, coll. N. P. Meshkova(?), sample 264. Swedish Museum of Natural History, Stockholm (SMNH), No. X 2109.   A. Stereo-pair. ×50.   B. ×50. Positions of C and D indicated.   C. Detail of B. ×250.   D. Detail of B. ×200.

ture in most other cases seems to reflect a primary, nonperforated tissue.

Microdictyon effusum Bengtson, Matthews and Missarzhevsky 1981.

(Figs. 1–3)

Synonymy. • 1975—Gen. et sp. indet. (Matthews and Missarzhevsky 1975, p. 300, pl. 4:2, 5, 8). • 1979—*Microdiction* [sic] *effusum* Bengts., Matt., Miss. (Mambetov and Repina 1979, pp. 100–113). • 1981—*Microdictyon effusum* (in one position spelled *effussum;* see below) Bengtson, Matthews and Missarzhevsky (Missarzhev-

sky and Mambetov 1981, p. 78, pl. 13:3, 5). • *non* 1983—*Microdictyon effussum* Bengtson, Matthews et Missarzhevsky 1981 (Grigor'eva in Sokolov and Zhuravleva 1983, p. 169, pl. 64:9).

Derivation of Name. Latin *effusus*, widespread, referring to the wide geographic distribution of this species. [The spelling *effussum* in the heading of p. 78 in Missarzhevsky and Mambetov (1981) is a printing error; in all other positions in this publication the name is correctly spelled.]

Holotype (by Monotypy). Missarzhevsky and Mambetov (1981, pl. 13:3, 5). Geological Institute of the USSR Academy of Science, Moscow, No. 4296/30.

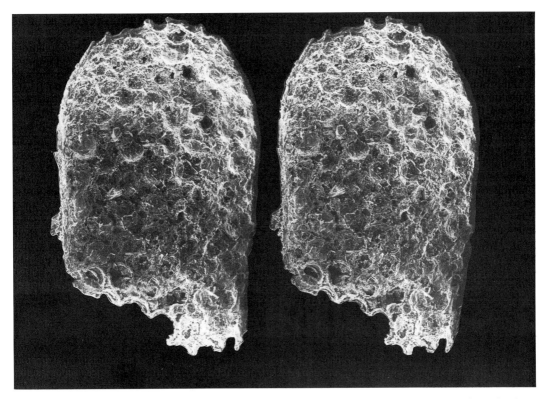

Fig. 2. *Microdictyon effusum* Bengtson, Matthews and Missarzhevsky 1981. Same sample as specimen in Fig. 1. SMNH No. X 2110. Stereo-pair, view from the lower side. ×60.

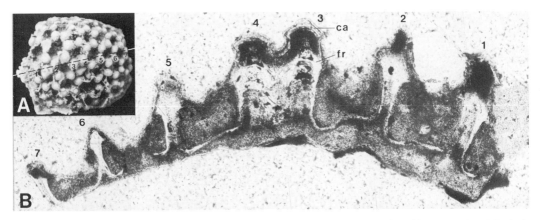

Fig. 3. *Microdictyon effusum* Bengtson, Matthews and Missarzhevsky 1981. Glacial drift boulder, South Bothnian area, Lower Cambrian, coll. C. Wiman, sample Eggegrund No. 3. Palaeontological Museum, Uppsala, No. B 577. A. Unsectioned specimen, line indicates plane of sectioning in B, nodes numbered as in B. ×32. B. Thin section of the same specimen, plane-polarized light, nodes numbered as in A, capping (ca) and framework (fr) indicated. ×160.

Type stratum. River Uchbas, Malyj Karatau, Geress Member, base of Shabakty Formation (cf. Mambetov and Missarzhevsky 1972; Missarzhevsky and Mambetov 1981).

Material. Seven specimens.

Occurrence. Malyj Karatau, South Bothnian area; Lower Cambrian.

Diagnosis. Microdictyon species with strongly convex plates, egg shaped in outline. Nodes mushroom shaped. Holes often closed from the lower side.

Description

The plate outline is typically egg shaped, one end being more blunt than the other (Fig. 1). Maximum observed length dimension is 2.30 mm, and the length–width ratio varies from 1.8:1 to 2.3:1. The hexagonal pattern of holes and nodes is mostly regular, but deviations may occur. Mesh size decreases toward the periphery; hole diameters may vary within one specimen from 130 μm in the center to about 10 μm at the girdle.

The nodes have a distinct brim and usually a subcentrally placed apex, the displacement of the apices being toward the narrower end of the plate (Fig. 1B, D). Node diameter is about 50 μm; larger nodes occur at occasional quadruple wall junctions (e.g., Fig. 1, lower central part).

The holes are frequently closed basally by a thin hemispherical part of the framework (Figs. 2 and 3). In the sectioned specimen from Eggegrund the bottom of the holes is not hemispherical but rather flattened with a slight central elevation (Fig. 3).

Microdictyon rhomboidale n.sp.

(Figs. 4–6)

Derivation of Name. Latin *rhomboidalis,* rhombic, referring to the prevalent plate outline.

Holotype. See Fig. 4A–D. Swedish Museum of Natural History, Stockholm, No. X 2111.

Type Stratum. Bateny Hills, Tamdytau, Lower Cambrian [bed 3 or 4 of Zadorozhnaya et al. (1973); cf. also Zhuravleva et al. (1970)].

Material. Eleven specimens.

Occurrence. Only known from the type stratum.

Diagnosis. Microdictyon species with weakly convex to flattened, thin plates; outline rhombic with more or less pronounced reentrant at one of the obtuse angles. Nodes mushroom shaped. Holes not closed from below.

Description

The plate outline is roughly rhombic (Figs. 4A, E, 5A, and 6A), with a maximum observed dimen-

sion of 1.1 mm. In the plane of the longest axis the girdle protrudes into distinct points. The outline of the girdle on one side of this axis (oriented upward in Figs. 4A, E and 5A) is approximately semicircular, but commonly there are two weaker angular protrusions about midway between the lateral points and the midline of the plate (e.g., Fig. 5A). On the opposite side of the long axis (oriented downward in Figs. 4A, E and 5A) the outline may also be semicircular, but in most cases there is a more or less marked medial reentrant (Figs. 4A, E, and 6A).

As in *M. effusum,* the mesh size tends to decrease toward the periphery. The maximum observed hole diameter is 70 μm, whereas at the girdle it may be about 10 μm.

The nodes are similar in shape and size to those of *M. effusum.*

The plates are only slightly domed. They are typically thin and have a delicate appearance. Holes with basal closure as in *M. effusum* (Figs. 2 and 3) have not been observed. The sides of the holes beneath the escarpment formed at the orifice are approximately cylindrical (although they may in some instances start to taper in the very lowermost part, which suggests that also in this species the holes may have been occasionally sealed), and have a distinct pattern of wrinkles parallel to the cylinder axes (Figs. 4F and 6).

Microdictyon cf. *rhomboidale*

(Fig. 7)

Material. Nine incomplete specimens and a few smaller fragments.

Occurrence. Mackenzie Mountains, British Columbia, Canada, Lower Cambrian, Sekwi Formation, lower part of *Nevadella* Zone (cf. Fritz 1972, 1979).

Description

The specimens from the Mackenzie Mountains are very similar to *M. rhomboidale* described herein with regard to shape of nodes and thickness of plates. There are also basic similarities in the plate outline: the specimen in Fig. 7I, for example, shows the same kind of lateral points and medial reentrant as *M. rhomboidale,* but the side opposite the reentrant is drawn out into a very acute point (broken in the specimen) that does not recall any similar feature in *M. rhomboidale.* Larger Mackenzie specimens (Fig. 7A, D) suggest a more complicated outline than has been observed in *M. rhomboidale,* but the material is too fragmentary to allow determination of the exact plate shape. Although some of these differences may be size related (no such large specimens are known from the type area of *M. rhomboidale*), the available material suggests that the species represented in the Mackenzie Mountains is distinct

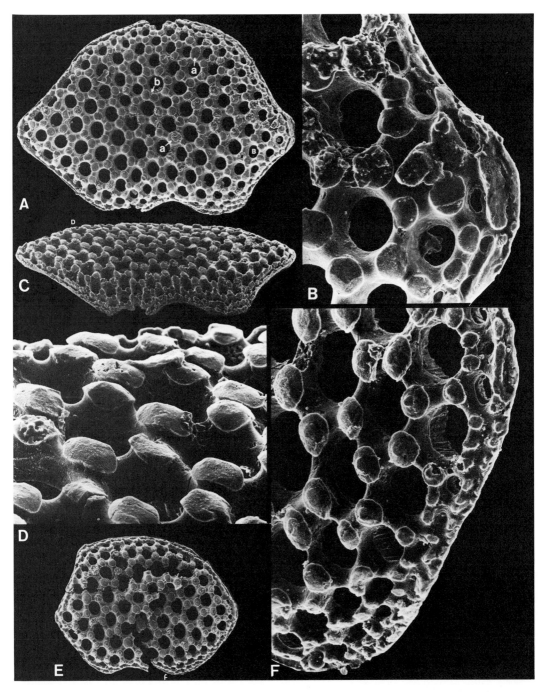

Fig. 4. *Microdictyon rhomboidale* n.sp. Tamdytau (Kyzyl-Kum), upper Atdabanian, coll. T. I. Khajrulina, sample No. 444/1. A–D. Holotype, SMNH No. X 2111. A. ×75. Two four-wall junctions (a) and one case of two three-wall junctions approaching each other (b) indicated with arrows (cf. text). Position of B indicated. B. Detail of A. ×375. C. ×75. Position of D indicated. D. Detail of C. ×375. E–F. *M. rhomboidale* n.sp. Same region and collector as A–D, sample No. 7. SMNH No. X 2112. E. ×75. Position of F indicated. F. Detail of girdle, different tilt. ×375.

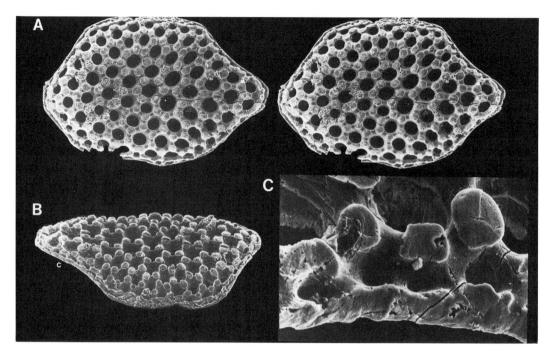

Fig. 5. *Microdictyon rhomboidale* n.sp. Tamdytau (Kyzyl-Kum), upper Atdabanian, coll. T. I. Khajrulina, sample No. 474d. SMNH No. X 2113.   A. Stereo-pair. ×90.   B.   ×90.   C.   Detail from upper right margin in A, different tilt. ×750.

from *M. rhomboidale.* Because of the incomplete nature of the available material, the Canadian species is referred to under open nomenclature.

One of the specimens (Fig. 7C–E) shows what appears to be a primary basal plate that closes the holes from below.

Microdictyon robisoni n.sp.

(Figs. 8 and 9)

Derivation of Name.   In the honor of Dr. Richard A. Robison, who discovered the species and provided the material for this investigation.

Holotype.   See Fig. 8. Swedish Museum of Natural History, Stockholm, No. X 2115.

Type Stratum.   Topaz Mountain, Drum Mountains, Millard County, Utah, Middle Cambrian, Swasey Limestone, uppermost bed, *Ptychagnostus gibbus* Zone or just below.

Material.   Twenty incomplete specimens and numerous fragments.

Occurrence.   The species is only known from its type stratum.

Diagnosis.   *Microdictyon* species with strongly convex plates and tall, narrow, spike shaped nodes with flattened ends. Holes not closed from below.

Description

The available specimens are not complete enough to show the exact outline of the plates. Most larger fragments suggest a round to oval outline and strongly convex plates, but fragments of straight or even concave girdles (Fig. 8) show that the outline may be somewhat more complicated. The maximum observed dimension is 1.8 mm (broken specimen), and the holes in this specimen range from 20 to 80 μm in diameter, but fragments of plates with larger holes may indicate the occurrence of larger specimens. The largest observed hole diameter is 100 μm.

The nodes are tall and narrow, ending in slight expansions (Figs. 8C and 9C, F, I). The upper surfaces of these expansions may form an angle with the plate surface (Fig. 8C) and may also be elaborated into platforms bearing small tubercles (Fig. 9I). As in *M. effusum* and *M. rhomboidale,* the upper surfaces of the nodes may be inclined in a certain direction (Figs. 8C and 9C, I).

The plates are fairly thin and have a delicate appearance. No basal closure of holes has been observed. The surface of the walls beneath the edge of the capping shows parallel wrinkles similar to those of *M. rhomboidale* (Fig. 9F).

Remarks.   The fragment from the Middle(?) Cambrian of Scania illustrated in Fig. 12C, D may possibly belong to this species.

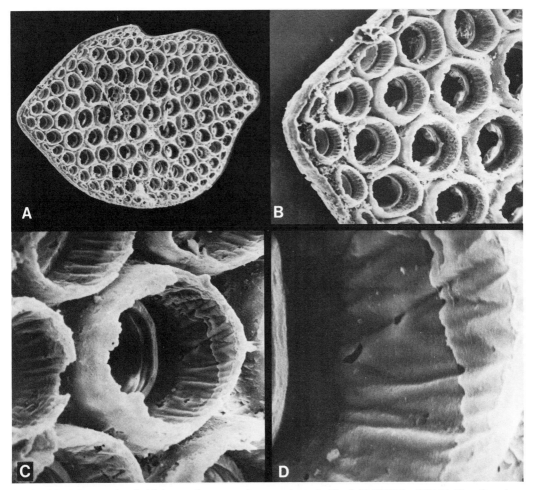

Fig. 6. *Microdictyon rhomboidale* n.sp. Same sample as specimen in Fig. 5, SMNH No. X 2114. View from lower side. A. ×75. B. ×225. C. ×750. D. ×2250.

Microdictyon n.sp. 1

(Fig. 10)

Material. Four fragments.

Occurrence. 42 km south of Goldfield, Esmeralda County, Nevada, Campito Formation, Montenegro Member, Lower Cambrian, lower part of *Nevadella* Zone (cf. Albers and Stewart 1972).

Description

The available material is insufficient to determine the exact outline of the plates; fragments indicate a simple rounded outline and highly convex plates.

The nodes are mushroom shaped, and the whole surface pattern does not differ significantly from that of *M. effusum* or *M. rhomboidale*. However, the specimens are conspicuous for their thick and robust nature (see Fig. 10F, G), which seems to differentiate them so markedly from the former that they are likely to belong to a new species.

The holes are generally open downward although there is a tendency toward the formation of rounded bottoms (Fig. 10F, G). The specimen in Fig. 10A–C has a lower plate of granular apatite similar to that seen in *M.* cf. *rhomboidale* (Fig. 7C, D).

Microdictyon? tenuiporatum n.sp.

(Fig. 11)

Derivation of Name. Latin *tenuis,* small, and *poratus,* having pores; referring to the small dimensions of the pores.

Holotype. See Fig. 11. Swedish Museum of Natural History, Stockholm, No. X 2122.

Type Stratum. Isit', River Lena, Yakutia, Tommotian Stage (cf. Rozanov et al. 1969).

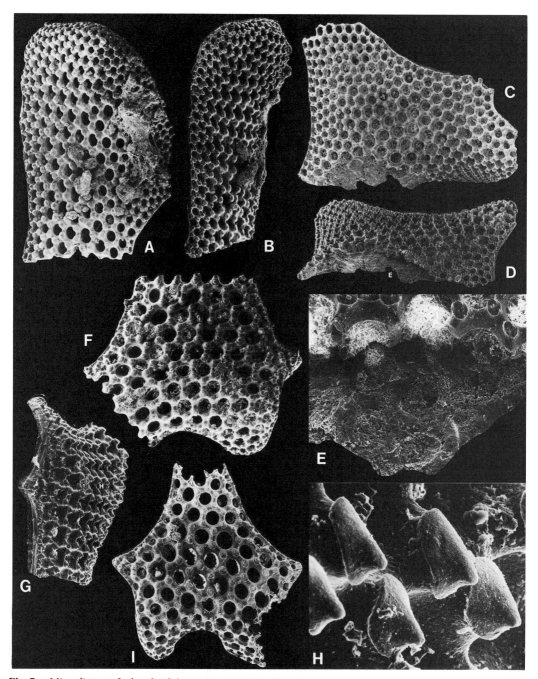

Fig. 7. *Microdictyon* cf. *rhomboidale*. South end of Sekwi Range, Mackenzie Mountains, British Columbia, Canada, lower part of Sekwi Formation, *Nevadella* Zone, GSC localities 73026 (A, B) and 73073 (C–H), coll. W. H. Fritz. A, B. Geological Survey of Canada, Ottawa, No. GSC 79515. ×25. C–E. No. GSC 79516. C, D. ×25. Position of E indicated in D. E. Detail of D. ×150. F–H. No. GSC 79517. F, G. ×50. Position of H indicated in G. H. Detail of G. ×375. I. No. GSC 79518. ×50.

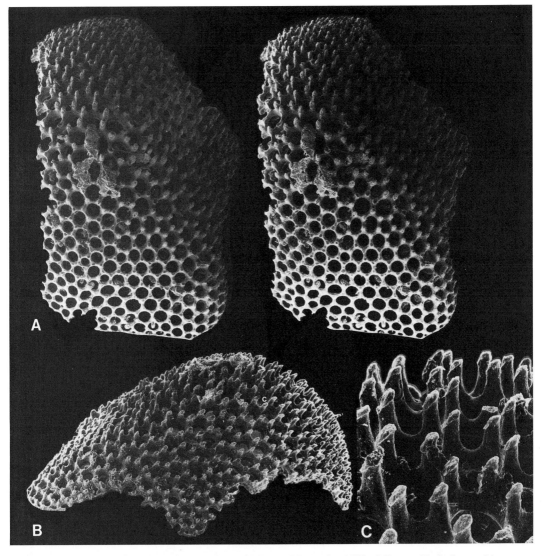

Fig. 8. *Microdictyon robisoni* n.sp. Topaz Mountain, Drum Mountains, Millard County, Utah, Swasey Limestone, Middle Cambrian, *Ptychagnostus gibbus* Zone or just below, coll. R. A. Robison, sample No. 123. Holotype, SMNH No. X 2115. A. Stereo-pair. × 60. B. × 60. Position of C indicated. C. Detail of B. ×240.

Material. One specimen (the holotype).

Diagnosis. Round, domed plate with irregularly or tetragonally arranged holes, about 20 μm in diameter. Nodes short and pillarlike, with distinct flat tops. Inner layer finely granular, without perforations, but with occasional reflections of the tetragonal pattern.

Description

The single specimen (Fig. 11) has an almost perfect round outline with a diameter of 1.4 mm. It is dome shaped with the highest point near the center of the plate.

As in the species referred without question

mark to *Microdictyon,* the structure is shaped of pores and intervening nodes. However, the pores of *M.? tenuiporatum* are considerably smaller than in the other species. The diameter of normal pores is about 20 μm; as in the unquestioned *Microdictyon,* some pores may be larger or smaller depending on whether there are dislocations in the normal lattice (Fig. 11D, E), and pores near the periphery tend to be smaller. Also, the arrangement of the pore lattice is different in *M.? tenuiporatum,* in that it is tetragonal rather than hexagonal. Dislocations within the lattice are common, however, and there is a tendency toward concentrically arranged pore rows around the center of the plate (Fig. 11A).

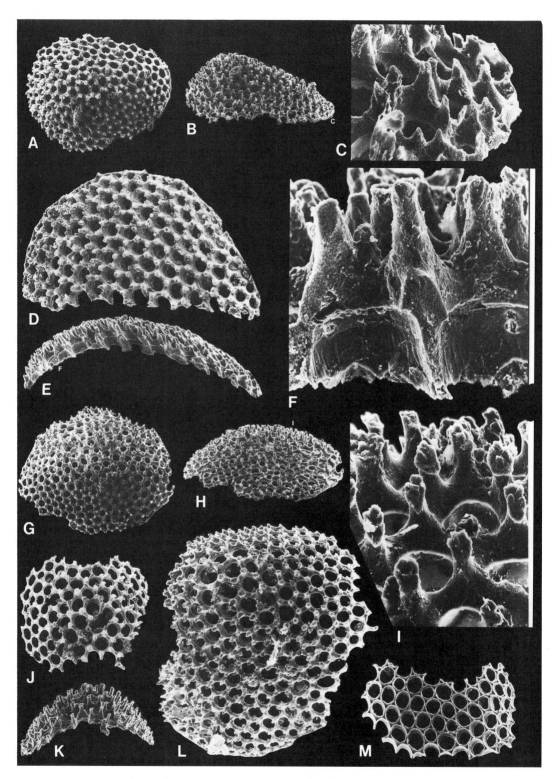

Fig. 9. *Microdictyon robisoni* n.sp. Same sample as specimen in Fig. 8. A–C. SMNH No. X 2116. A, B. ×75. Position of C indicated in B. C. Detail of B. ×600. D–F. SMNH No. X 2117. D, E. ×75. Position of F indicated in E. F. Detail of E. ×600. G–I. SMNH No. X 2118. G, H. ×75. Position of I indicated in H. I. Detail of H. ×600. J, K. SMNH No. X 2119. ×75. L. SMNH No. X 2120. ×75. M. SMNH No. X 2121. ×75.

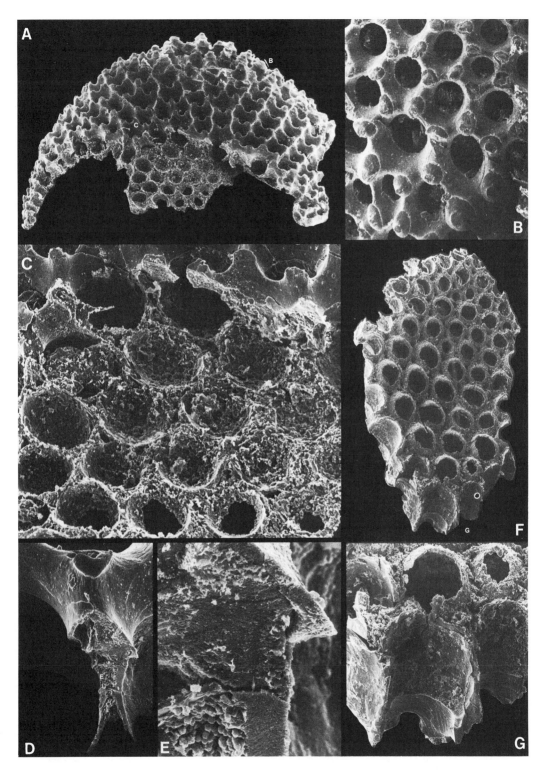

Fig. 10. *Microdictyon* n.sp. 1. About 42 km south of Goldfield, Esmeralda County, Nevada, Campito Formation, Montenegro Member, Lower Cambrian, lower part of *Nevadella* Zone, coll. J. P. Albers and J. H. Stewart, sample No. 4131-CO. A–C. U.S. National Museum of Natural History, Washington, D. C., No. USNM 381038. A. ×50. Positions of B and C indicated. B. Detail of A, different tilt. ×100. C. Detail of A. ×200. D, E. No. USNM 381039. Broken wall near node. D. ×250. Position of E indicated. E. Detail of D. ×1250. F, G. No. USNM 38140. F. View from below. ×50. Position of G indicated. G. Detail of F. ×125.

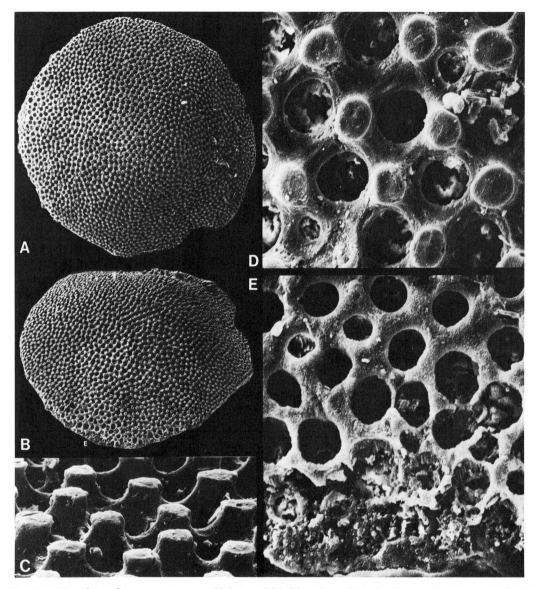

Fig. 11. *Microdictyon? tenuiporatum* n.sp. Holotype. Isit', River Lena, Yakutia, Tommotian Stage, coll. V. V. Missarzhevsky, sample No. M45-75. SMNH No. X 2122.   A, B.   ×50. Position of E indicated in B.   C. Detail of B. ×750.   D. Detail of A. ×750.   E. Detail of B. ×500.

The fine structure of the plate has not been studied in detail because only one specimen is available. A division into framework and capping seems to be present, and longitudinal plications on the side walls of the pores below the edges of the capping (Fig. 11E) seem to correspond well with those of unquestioned *Microdictyon* (Figs. 4F, 6, and 9F). In this plate, however, there is an additional inner layer, a few tens of micrometers in thickness, that is not perforated but appears to be primary (Fig. 11E, bottom). It is finely granulated and has a fairly coarse surface. The lower side of this layer is mostly without distinct fea-

tures, but in some places there is a reflection of the outer tetragonal pattern of pores.

Remarks

The general agreement in gross and fine structure (as far as it has been observed) between this species and the other species described in this paper leaves little doubt that they are closely related. A question mark is here applied to the generic name because the other species form a closely defined group from which *M.? tenuiporatum* differs in some important respects. It should most probably be referred to a new genus, but the available single specimen of a single species is considered to be insufficient material to establish this genus formally.

FORMATION AND FUNCTION OF THE *MICRODICTYON* PLATES

Formation

It is clear from the morphology and structure of the *Microdictyon* plates that they were not formed by accretionary growth of the mineralized tissue. The distribution of the nodes and holes and the presence of a definite edge, the girdle, in all complete specimens preclude the incorporation of earlier growth stages in larger specimens. Thus the evidence suggests that they were formed in one single phase of deposition, although the presence of internal laminations (e.g., Fig. 3) may be taken to indicate that there were steps involved in this

formation. These were, however, not related to ontogenetic growth.

The capping always seems to be present and to cover all of the upper surface of the framework; it is not possible to determine if either of the two tissues was formed before the other one. The capping, which forms the sometimes very protrusive nodes and shows a more regular crystallographic structure than the framework, was by all the evidence originally mineralized. It thus formed a hard, possibly protective structure.

Whether or not the framework was originally mineralized is less evident. It formed a comparatively resilient structure, by the evidence of its regular preservation, but the varying appearance of the lower parts of the framework suggests that it may have been pliable. The bottoms (where present) of the pores are commonly hemispherical (e.g., Fig. 2), but the sectioned Eggegrund specimen of *M. effusum* (Fig. 3) shows bottoms that

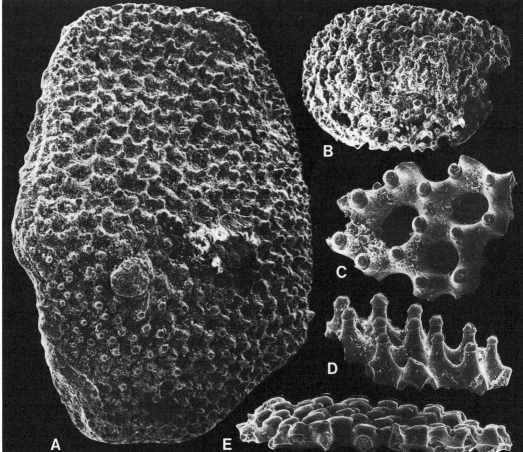

Fig. 12. *Microdictyon* spp. A,B. Kalbygård, Laeså, Bornholm, Kalby Marl, coll. V. Berg-Madsen. A. SMNH No. X 2123. × 50. B. SMNH No. X 2124. ×100. C,D. SMNH No. X 2125, 1 km south of Brantevik, Scania, "Fragment Limestone," coll. S. Bengtson, sample No. S70-2-SB. ×100. E. SMNH No. X 2126. 600 m north of Gislövshammar, Scania, "Fragment Limestone," coll. S. Bengtson, sample No. S71-35-SB.

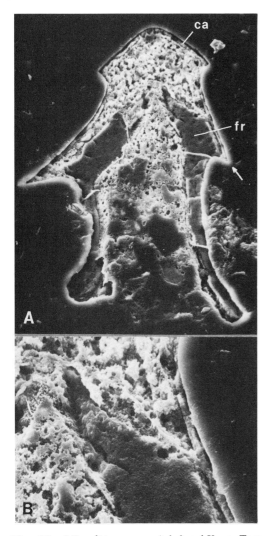

Fig. 13. *Microdictyon* sp. Achchagyj-Kyyry-Taas, River Lena, Yakutia, Perekhodnaya Formation, 3rd Member, upper Atdabanian, coll. S. Bengtson, sample No. Sib73-1-SB. SMNH No. X 2127. A. Polished and etched section through node, showing relationship between capping (ca) and framework (fr). ×400. B. Detail of A. ×1200.

appear to have been buckled without evidence of breakage. (The possibility cannot be excluded, however, that this was their original shape.) Because the framework is so consistently preserved in all specimens, it is likely that it was impregnated with apatite crystallites, but it probably lacked an interlocking crystalline fabric.

The relationship between downward open and closed holes is particularly intriguing. The cylindrical profile and regular lower rims of the open holes in specimens such as that in Figs. 4F, 6, and 9M make it very unlikely that the bottoms have

been secondarily lost by abrasion or solution. On the other hand, the difference in basal closure between the holes seen in Fig. 10F and G is most probably a feature of preservation. The presence of primarily open or closed holes seems to be a species character: rounded, closed bottoms are found in *M. effusum* and *M.* n.sp. 1, whereas cylindrical open holes occur in *M. rhomboidale, M.* cf. *rhomboidale, M. robisoni,* and *M.? tenuiporatum.* It should be noted that both types of holes may abut against a basal, probably primary, nonperforated sheet of apatite (Figs. 7D, E, 10A, C, and 11E). Thus we may conclude that the presence of open holes does not reflect a particular developmental stage and that it was not crucial for the function of the plates.

Function

We consider here two possible alternative interpretations of the function of *Microdictyon* plates: (1) basal supports of an encrusting organism, colony, or egg deposit, and (2) dermal sclerites serving to toughen the skin and increase surface friction.

The first hypothesis is inspired by the superficial resemblance of *Microdictyon* plates to the basal skeletons of organisms such as bryozoans, tabulates, and sclerosponges. The protruding, sometimes spiniform nodes could then be seen as a means of protecting the soft tissues between and below them.

There are two important arguments, however, against a close analogy with any of these organisms: (1) the lack of a constant unimodal or polymodal size distribution of the holes would be highly atypical of a colonial organism, and (2) all these organisms of colonies grow or grew from an initial founding individual, whereas *Microdictyon* plates, as argued above, were not growing structures.

To interpret the plates as substrates for egg deposits would remove the objection regarding growth, but the size distribution of the holes would be equally difficult to explain with this model. To account for the great size range of holes in some plates and the sometimes great difference in hole size between plates of the same species, we would have to assume that the eggs were very small (10–20 μm in diameter), the size of the holes reflecting the number of eggs deposited therein. Whichever model is applied, it is difficult to interpret the holes in terms of direct relationship to soft tissues.

The relative constancy of plate shape within a species may in an encrusting organism be seen either as an intrinsically derived character or a reflection of the shape of a selected substrate. Unfortunately, because of the sparsity of *Microdictyon* plates in the sediments we have not been able to identify individual plates in rock sections. If this could be done it might give information on whether the plates were encrusting a carbonate substrate.

The second hypothesis, that the plates were dermal sclerites in a larger animal, also found its inspiration in possibly analogous structures in living animals, namely the sieve plates of, for example, holothurians. There is obviously no homology involved with holothurian sclerites, and the dense capping and protruding nodes of *Microdictyon* plates would suggest them to be exposed to the external medium (for protection and/or increase of surface friction of the body wall) rather than being wholly embedded in the skin. The high convexity of plates of *M. effusum* and *M. robisoni* would perhaps not be expected in sclerites covering a considerably larger animal, but this is not a prohibitive argument against the hypothesis. Thus it seems difficult to refute it at present, but in the absence of known homologous sclerites, evidence of articulating edges or facets, or specimens showing natural associations of several sclerites, the dermal sclerite hypothesis is without corroboration and must be considered highly speculative.

Other hypotheses that seem at first consideration possible are in fact weak. For example, if the *Microdictyon* plates were analogous or homologous to the cuticular parts of arthropod composite eyes, the presence of a probably primary basal sheet (Figs. 7D, E, 10A, C, and 11E) would not be readily explainable. If they were parts of organs for grasping or holding, one might expect the taller spines in, for example, *M. robisoni* to be frequently broken and the plate shape to be less convex than in the latter and in *M. effusum*.

AFFINITIES

The phosphatic plates of *Microdictyon* show distinct similarities to representatives of the problematic genus *Milaculum* Müller 1973. The first report of this fossil was by Ethington and Clark (1965), who found its small nodulose sclerites in etching residues from the Lower Ordovician Columbia Ice Fields section in Alberta, Canada (Ethington and Clark 1965, p. 204, pl. 1:20). *Milaculum* is now known to contain a substantial diversity of forms ranging in age from the Late Cambrian to the Middle Ordovician and having a broad geographic distribution (e.g., Müller 1973; Nitecki et al. 1975; Ethington and Clark 1981).

Milaculum is characterized by small (about 250 to 500 μm), phosphatic, usually elongated and arched plates carrying a more-or-less regular surface ornament of nodes and/or ridges. The plates may be perforated by regularly spaced pores, but are usually without perforations. The lower, concave side may show poorly defined nodes or pits,

or (in the case of perforated specimens) pores, but is commonly without conspicuous structures.

The outline of *Milaculum* plates varies from egg shaped to rod shaped, with the lower surface defining a more-or-less perfect semicylindrical space. The intermediate forms include various less regular shapes; a common variety has flaring ends that tend to produce an incomplete (e.g., *M. muelleri*—Ethington and Clark 1981, pl. 14:7, 9) or complete (e.g., *M. ethinclarki*—Müller 1973, pl. 34:5, 6, 8) "dog biscuit" shape.

The fine structure of *Milaculum* plates is in need of further investigation. Müller studied thin sections and SEM mounts, but the preservation of the material did not allow a proper understanding of the histologic and mineralogic structure. There appear to be needle shaped crystallites forming the upper surface of *M. perforatum* plates, oriented in a radial pattern around the perforations and meeting in a midline on the walls between perforations (Müller 1973, p. 218, pl. 32:1c). There is also some evidence of laminated structures at the lower openings of the perforations (Müller 1973, p. 218, pl. 32:6d).

The plates of *Microdictyon* and *Milaculum* could have been formed in a homologous manner by related organisms. They share a phosphatic composition, a lack of growth structures, a convex shape with clearly defined upper and lower sides, a generally nodular upper surface, and tendencies toward perforations. The most striking differences are that *Microdictyon* plates are larger by almost one order of magnitude, have prominent round pores that more often than not open to the underside, and have a generally hexagonal (as opposed to quadragonal or irregular in *Milaculum*) disposition of pores and nodes. In view of the patterns of variation in both genera, none of these differences appear to preclude close affinity; the species referred to here as *Microdictyon? tenuiporatum*, in addition to having nodes and pores very similar to those of *Microdictyon*, has a quadragonal disposition of nodes and pores and is more fine-meshed than other *Microdictyon*.

If the plates of *Microdictyon* and *Milaculum* are indeed homologous, it would have implications on the interpretation of their functional morphology. The foregoing discussion on formation and function of *Microdictyon* plates is in all matters of significance applicable to *Milaculum*. Thus it was concluded that the conspicuous holes in the former are not a vital feature for the function of the plates. The presence of holes is then no obstacle to affinity with the nonperforated *Milaculum*. If microstructural investigations of *Milaculum* plates are to bring out structures similar to those of *Microdictyon*, a close affinity may be postulated. In the absence of such evidence, however, we prefer not to formalize this idea in a taxonomic decision.

The further affinities of *Microdictyon* (and *Milaculum*) are—as befits fossils under discussion in the present volume—problematic.

Acknowledgments
We are grateful to the following persons for bringing occurrences of *Microdictyon* to our attention and/or for giving us access to material: Vivianne Berg-Madsen, Stig M. Bergström, Simon Conway Morris, Brian Daily, Richard A. Davis, Donald W. Fisher, William H. Fritz, Ed Landing, Mark A. S. McMenamin, Nina P. Meshkova, John E. Repetski, Richard A. Robison, Bruce Runnegar, and Mark C. Tynan. Meit Lindell, Monica Siewertz, and Tommy Westberg gave technical assistance, and Simon Conway Morris read the manuscript. Stefan Bengtson's work has been supported by grants from the Swedish Natural Science Research Council.

REFERENCES

Albers, J. P. and Stewart, J. H. 1972. Geology and mineral deposits of Esmeralda County, Nevada. *Nev. Bur. Mines Geol. Bull.* **78**, 1–80.

Bengtson, S. 1977. Early Cambrian button-shaped phosphatic microfossils from the Siberian Platform. *Palaeontology* **20**, 751–762.

Berg-Madsen, V. 1981. The Middle Cambrian Kalby Clay and Borregård Members of Bornholm, Denmark. *Geol. Fören. Stockh. Förhandl.* **103**, 215–231.

Bergström, J. 1981. Lower Cambrian shelly faunas and biostratigraphy in Scandinavia. *U. S. Geol. Surv. Open-File Rep.* **81-743**, 22–25.

——— and Ahlberg, P. 1981. Uppermost Lower Cambrian biostratigraphy in Scania, Sweden. *Geol. Fören. Stockh. Förhandl.* **103**, 193–214.

Bird, J. M. and Rasetti, F. 1968. Lower, Middle, and Upper Cambrian faunas in the Taconic sequence of eastern New York: stratigraphic and biostratigraphic significance. *Geol. Soc. Am. Spec. Pap.* **113**, 66 pp.

Cobbold, E. S. 1921. The Cambrian horizons of Comley (Shropshire) and their Brachiopoda, Pteropoda, Gasteropoda, etc. *Q. J. Geol. Soc. Lond.* **76**, 325–386.

Conway Morris, S. and Fritz, W. H. 1984. *Lapworthella filigrana* n. sp. from the Lower Cambrian of the Cassiar Mountains, northern British Columbia, Canada. *Paläontol. Z.* **58**, 197–209.

Daily, B. 1956. The Cambrian in South Australia. *In El sistema cambrica, su paleogeografia y el problema de su base. Part II: Australia, America.* 20th Int. Geol. Congr., Mexico, pp. 91–147.

———. 1972. The base of the Cambrian and the first Cambrian faunas. *In* Jones, J. B. and McGowran, B. (eds.). *Stratigraphic Problems of the Later Precambrian and Early Cambrian.* Univ. of Adelaide, Centre for Precambrian Research, Special Paper **1**, pp. 13–41.

Ethington, R. L. and Clark, D. L. 1965. Lower Ordovician conodonts and other microfossils from the Columbia Ice Fields section, Alberta, Canada. *Brigham Young Univ. Geol. Stud.* **12**, 185–205.

——— and ———. 1981. Lower and Middle Ordovician conodonts from the Ibex area, western Millard County, Utah. *Brigham Young Univ. Geol. Stud.* **28**(2), 1–160.

Fritz, W. H. 1972, Lower Cambrian trilobites from the Sekwi Formation type section, Mackenzie Mountains, northwestern Canada. *Geol. Surv. Can. Bull.* **212**, 1–58.

———. 1979. Eleven stratigraphic sections from the Lower Cambrian of the Mackenzie Mountains, northwestern Canada. *Geol. Surv. Can. Pap.* **78-23**, 1–19.

———. 1980. Two new formations in the Lower Cambrian Atan Group, north-central British Columbia. *Curr. Res., Part B, Geol. Surv. Can. Pap.* **80-18**, 217–225.

Grönwall, K. A. 1902. Bornholms Paradoxideslag og deres fauna. *Danmarks Geol. Undersøg.* **2**(13), 250 pp.

Keller, B. M., Rozanov, A. Yu., Missarzhevsky, V. V., Repina, L. N., Shabanov, Yu. Ya., and Egorova, L. I. 1973. *Putevoditel' ekskursii po rekam Aldanu i Lene. Mezhdunarodnaya ekskursiya po probleme granitsy kembriya i dokembriya* [Excursion guide to the Aldan and Lena rivers. International excursion on the problem of the Precambrian-Cambrian boundary]. AN SSSR, Moscow-Yakutsk, 118 pp. (in Russian and English).

Landing, E. 1984. Skeleton of lapworthellids and the suprageneric classification of tommotiids (Early and Middle Cambrian phosphatic problematica). *J. Paleont.* **58**, 1380–1398.

——— and Brett, C. E. 1982. Lower Cambrian of eastern Massachusetts: microfaunal sequence and the oldest known borings. *Geol. Soc. Am. Abstr. Progr.* **14**, 33.

Lochman, C. 1956. Stratigraphy, paleontology, and paleogeography of the *Elliptocephala asaphoides* strata in Cambridge and Hoosick quadrangles, New York. *Geol. Soc. Am. Bull.* **67**, 1331–1396.

Mambetov, A. M. and Missarzhevsky, V. V. 1972. Novye dannye ob okamenelostyakh iz fosforitonosnykh tolshch Malogo Karatau [New data on fossils from the phosphorite-bearing beds of Malyj Karatau]. In *Stratigrafiya dokembriya Kazakhstana i Tyan'-Shanya.* Izdatel'stvo MGU, Moscow, pp. 217–221.

——— and Repina, L. N. 1979. Nizhnij kembrij Talasskogo Ala-Too i ego korrelyatsiya s razrezami Malogo Karatau i Sibirskoj platformy. [The Lower Cambrian of the Talassky Ala-Too and its correlation with the sections of

Malyj Karatau and the Siberian Platform.] *In* Zhuravleva, I. T. and Meshkova, N. P. (eds.). *Biostratigrafiya i paleontologiya nizhnego kembriya Sibiri.* Nauka, Novosibirsk, pp. 98–138.

Matthews, S. C. 1973. Lapworthellids from the Lower Cambrian *Strenuella* Limestone at Comley, Shropshire. *Palaeontology* **16**, 139–148.

——— and Missarzhevsky, V. V. 1975. Small shelly fossils of late Precambrian and early Cambrian age: a review of recent work. *J. Geol. Soc.* **131**, 289–304.

McMenamin, M. A. S. 1984. Paleontology and stratigraphy of Lower Cambrian and Upper Proterozoic sediments, Caborca Region, Northwestern Sonora, Mexico. Unpublished Ph.D. dissert., Univ. of California at Santa Barbara, 218 pp.

Missarzhevsky, V. V. and Mambetov, A. M. 1981. Stratigrafiya i fauna pogranichnykh sloev kembriya i dokembriya Malogo Karatau [Stratigraphy and fauna of the Precambrian–Cambrian boundary beds of Malyj Karatau]. *Trudy Geol. Inst. AN SSSR* **326**, 90 pp.

Müller, K. J. 1973. *Milaculum* n.g., ein phosphatisches Mikrofossil aus dem Altpaläozoikum. *Paläont. Z.* **473/4**, 217–228.

Nelson, C. A. 1978. Late Precambrian-Early Cambrian stratigraphic and faunal succession of eastern California and the Precambrian-Cambrian boundary. *Geol. Mag.* **115**, 121–126.

Nitecki, M. H., Gutschick, R. C., and Repetski, J. E. 1975. Phosphatic microfossils from the Ordovician of the United States. *Fieldiana: Geol.* **35**(1), 1–9.

Palmer, A. R. 1964. An unusual Lower Cambrian trilobite fauna from Nevada. *U.S. Geol. Surv. Prof. Pap.* **483-F**, F1–F12.

Poulsen, C. 1942. Nogle hidtil ukendte fossiler fra Bornholms Exulanskalk. *Medd. Dansk Geol. Foren.* **10**(2), 212–235.

Poulsen, V. 1963. The Lower Middle Cambrian Kalby-ler (Kalby Clay) on the Island of Bornholm. *Medd. Danske Vidensk. Selskab* **23**(14), 14 pp.

Repina, L. N., Khomentovsky, V. V., Zhuravleva, I. T., and Rozanov, A. Yu. 1964. *Biostratigrafiya nizhnego kembriya Sayano-Altajskoj skladchatoj oblasti* [Lower Cambrian biostratigraphy of the Altaj-Sayan folded area]. Nauka, Moscow, 364 pp.

Rozanov, A. Yu. 1973. Zakonomernosti morfologicheskoj evolyutsii arkheotsiat i voprosy yarusnogo raschleneniya nizhnego kembriya [Regularities in the morphological evolution of Archaeocyathans and questions regarding the

stage division of the Lower Cambrian]. *Trudy Geol. Inst. AN SSSR* **241**, 164 pp.

———, Missarzhevsky, V. V., Volkova, N. A., Voronova, L. G., Krylov, I. N., Keller, B. M., Korolyuk, I. K., Lendzion, K., Michniak, R., Pykhova, N. G., and Sidorov, A. D. 1969. Tommotskij yarus i problema nizhnej granitsy kembriya. [The Tommotian Stage and the Problem of the Lower Boundary of the Cambrian]. *Trudy Geol. Inst. AN SSSR* **206**, 380 pp.

Shaler, N. S. and Foerste, A. F. 1888. Preliminary description of North Attleborough fossils. *Bull. Mus. Comp. Zool.* **16**, 27–41.

Sokolov, B. S. and Zhuravleva, I. T. (eds.). 1983. *Yarusnoe raschlenenie nizhnego kembriya Sibiri. Atlas okamenelostej* [Lower Cambrian Stage Subdivision in Siberia. Atlas of fossils]. Nauka, Moscow, 216 pp.

Spizharsky, T. N., Ergaliev, G. Kh., Zhuravleva, I. T., Repina, L. N., Rozanov, A. Yu., and Chernysheva, N. E. 1983. Yarusnaya shkala kembrijskoj sistemy [A stage subdivision of the Cambrian System]. *Sovetskaya geologiya* **1983**(8), 57–72.

Thomson, B. P., Daily, B., Coats, R. P., and Forbes, B. G. 1976. Late Precambrian and Cambrian geology of the Adelaide "Geosyncline" and Stuart Shelf, South Australia. *25th Int. Geol. Congr. Excursion Guide No. 33A,* 53 pp.

Thorslund, P. and Axberg, S. 1979. Geology of the southern Bothnian Sea. Part I. *Bull. Geol. Inst. Univ. Uppsala, N. S.* **8**, 35–62.

Tynan, M. C. 1983. Coral-like microfossils from the Lower Cambrian of California. *J. Paleont.* **57**, 1188–1211.

Wiman, C. 1903. Studien über das Nordbaltische Silurgebiet. I. Olenellussandstein, Obolussandstein und Ceratopygeschiefer. *Bull. Geol. Inst. Univ. Uppsala* **6**, 12–76.

Zadorozhnaya, N. M., Osadchaya, D. V., and Repina, L. N. 1973. Novye dannye po biostratigrafii nizhnego kembriya okrestnostej pos. Bograd (Batenyovskij kryazh) [New data on the Lower Cambrian biostratigraphy of the area around the village Bograd (Bateny Hills)]. *In* Zhuravleva, I. T., (ed.). *Problemy paleontologii i biostratigrafii nizhnego kembriya Sibiri i Dal'nego vostoka.* Nauka, Novosibirsk, pp. 119–151.

Zhuravleva, I. T., Repina, L. N., Yaskovich, B. V., Khajrulina, T. I., Poniklenko, I. A., and Luchinina, V. A. 1970. *K poznaniyu rannego kembriya yuzhnogo Tyan'-Shanya* [To the knowledge of the Early Cambrian of Tien-Shan]. Fan, Tashkent, 53 pp.

10. TURRILEPADIDA AND OTHER MACHAERIDIA

JERZY DZIK

The name Machaeridia was originally proposed (Withers 1926) for Early Paleozoic animals with a bilaterally symmetric body covered with calcitic sclerites arranged in longitudinal rows. Their armors (or strobili; Pope 1975) are only rarely found articulated, and their isolated sclerites are not easily identifiable. Therefore, they do not attract much attention. In calcitic residues from Ordovician limestones treated with dilute acetic acid, however, their remnants frequently dominate in the identifiable skeletal debris, which demonstrates their paleoecologic significance.

The Ordovician–Carboniferous Machaeridia with well-calcified sclerites (genera *Plumulites, Mojczalepas* gen. nov., *Turrilepas, Deltacoleus, Clarkeolepis, Plicacoleus* gen. nov., *Lepidocoleus, Aulakolepos,* and *Carnicoleus* gen. nov.; see systematic paleontology section below) are here assigned to the order Turrilepadida Pilsbry 1916. There are also many other fossils—in the Cambrian (Tommotiida, Sachitida) and in the Middle Paleozoic (Hercolepadida ord. nov.)—that in several aspects resemble the Ordovician turrilepadids. It is the aim of the present paper to discuss phylogenetic relationships of the Turrilepadida to those groups as well as to other taxa.

BODY PLAN

The genus *Plumulites* is the most primitive known representative of the Turrilepadida. Its body was bilaterally symmetric and dorsoventrally depressed, with four rows of sclerites. The sclerites were thin walled and probably weakly calcified. They developed by mineralization of dorsal covers of leaflike protrusions of the body. Because of their analogy (perhaps also homology; see below) to the annelid elytra, I propose to call these protrusions elytra. Their protective function is self-evident, and hence, I consider as dorsal that side of the body that is covered with elytra.

As judged from the mode of preservation of the articulated strobili of *Plumulites,* its elytra had little mobility in the transverse plane, but the animal may have been able to roll its body in the sagittal plane. Jell (1979) interpreted a specimen of *Plumulites* lacking the anterior part as a heteromorph and reconstructed peculiar palps at its anterior end. The anterior part of that specimen, however, may be tucked under its body, as in the Cambrian *Wiwaxia* (Conway Morris 1982: W).

The "palps" would, then, be posterior tips of elytra.

Mobility in the sagittal plane was strongly restricted in *Turrilepas.* Its body was approximately isometric in cross section, with elytra of the outer (or lateral) rows oriented vertically, as in most Machaeridia. The elytra were mobile in the transverse plane, perhaps up to bringing the lateral elytra in contact along the venter (Fig. 1B). In *Aulakolepos,* in turn, the lateral elytra underwent reduction, and the body was covered nearly exclusively with dorsal elytra (Fig. 1C).

Elytra are metamerically arranged in all machaeridian strobili. The segments of *Plumulites* and other turrilepadids include four elytra each. Two first segments of *Plumulites* (Jell 1979) and *Turrilepas* (Withers 1926), however, bear only two dorsal elytra each; furthermore, the elytra of the first segment are much smaller and more medially located than the others in *Turrilepas* (Withers 1926). The tagma composed of those first segments is here called the head. The subsequent segments bear already four elytra each, but their lateral elytra differ in shape from the more posterior ones. This is the thoracic tagma. In *Plumulites,* the lateral elytra have concentric apical rugae (Figs. 1 and 2; see also Jell 1979). Beginning with the sixth segment, the lateral elytra attain their regular shape and, except for some decrease in size and a slight increase in elongation in the caudal part of the body, the abdominal sclerites are largely uniform (Figs. 2 and 3).

SCLERITE GROWTH AND MORPHOLOGY

In most Turrilepadida, the sclerites grew by secretion of calcitic layers at the base (Fig. 7D; see also Bengtson 1977, 1978). Their marginal increment is indicated by (1) unquestionable growth lines in such genera as *Deltacoleus* and *Plicacoleus* (Fig. 9A), with a complete gradation to typical rugose sclerites (Fig. 6), and (2) morphology of muscle attachment scars at the inner surface of thick machaeridian sclerites (Figs. 5 and 9), with traces of migration from an initially apical position toward the anterior margin, as in brachiopods and bivalves; this would be incompatible with periodic molting of elytra.

In *Plumulites,* however, mineralization probably occurred instantaneously over the entire surface of the elytra. Even large-sized sclerites are

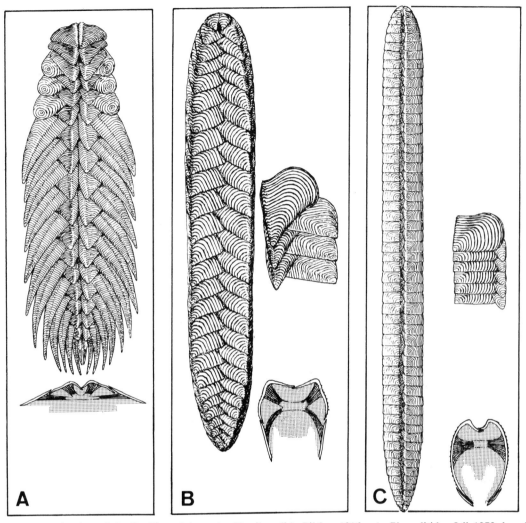

Fig. 1. Body plan of the families of the order Turrilepadida Pilsbry 1916. A. Plumulitidae Jell 1979, based mainly on *Plumulites pieckorum* Jell 1979 (see Jell 1979). B. Turrilepadidae Clarke 1896, based on *Turrilepas wrightiana* (Koninck 1857) (see Withers 1926). C. Lepidocoleidae Clarke 1896, dorsal view based on Ordovician species of *Aulakolepos* (see Withers 1926; Pope 1975), combined with lateral view of *A. ketleyanum* (Reed 1901).

very thin walled and always compressed in *Plumulites,* even those occurring with undistorted calcareous fossils, including other machaeridian sclerites (Fig. 2; see also Withers 1926). They probably were elastic, perhaps organic, and only weakly calcified during life. Their marginal growth is suggested merely by their homology to marginally growing sclerites of the other Turrilepadida.

Dorsal turrilepadid sclerites are always convex and close to isometric in outline. Bengtson (1970) proposed the term sellate for similar sclerites in the Tommotiida. Lateral turrilepadid sclerites are always flat and elongated. They correspond morphologically to mitrate sclerites of the Tommotiida.

In *Plumulites,* lateral sclerites have a medial elevation, semicircular in cross section and separated from the lateral areas by very narrow, shallow depressions. Rugae usually continue over this elevation, though with a decrease in conspicuousness. The semicircular section, morphologic separation, and abrupt appearance of the elevation in the center of thoracic sclerites of *P. bohemicus* Barrande 1872 (Figs. 2A and 3) suggest the presence of a medial canal below the elevation, similar to the central canal in elytra of the Cambrian *Thambetolepis* (Jell 1981).

Lateral margins and tips of the lateral sclerites of *Plumulites* have tubular spines (Schrenk 1978), sometimes laterally branching and always with a suture along their ventral side (Fig. 4B, C). Unlike

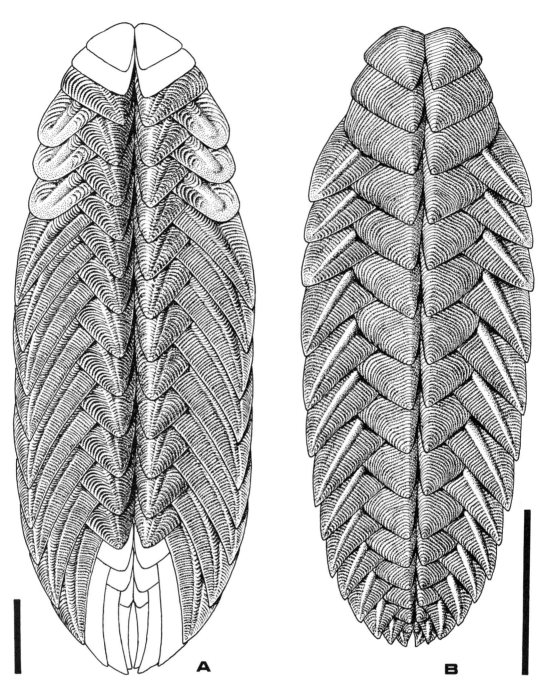

Fig. 2. Reconstruction of *Plumulites* from the Ordovician of Bohemia, based on articulated specimens. Scale bar = 10mm. A. *Plumulites bohemicus* Barrande 1872 from the Šárka Formation (Llanvirnian), Osek (cf. Fig. 3). Perhaps a few more abdominal segments should be added. B. *Plumulites folliculum* Barrande 1872 from the Letna Formation (Caradocian), Trubska. Based mainly on the lectotype NM 1425 (housed at the Narodní Museum, Prague; illustrated by Barrande 1872, pl. XX: 15-15; and Prokop 1965, fig. XIII-42; plasticine cast in Withers 1926, pl. VIII: 2).

Fig. 3. Part (A) and counterpart (B) of the lectotype of *Plumulites bohemicus* Barrande 1872 from the Šárka Formation (Llanvirnian), Osek, Bohemia (housed at the Narodní Muzeum, Prague; illustrated by Barrande 1872, pl. XX: 1-2; plasticine cast in Withers 1926, pl. VIII: 1). Abdominal segments indicated by consecutive numbers; note displaced thoracic lateral sclerite. ×2.

in other turrilepadids, the wall of the elytra was calcified also at the lateral margins in *Plumulites,* and remnants of that layer are preserved ventrally (Fig. 4), but I could not determine if there was a suture at the ventral surface. The tubular spines resemble those found in mollusks and brachiopods, but they might be parallel to the growth front, provided that marginal secretion did indeed occur in *Plumulites.* If, however, the cuticle was secreted and mineralized by the entire surface of elytra, the sutures may be analogous to cuticular border zones separating different fields in elytra of some Recent polychaetes (Pflugfelder 1933).

Notably, the marginal spines are a continuation of rugae (Fig. 4A), which suggests that the latter may reflect presence of pennately arranged internal organs similar to the lateral canals of *Thambetolepis* (see Jell 1981).

Species of *Plumulites* vary in distribution of rugae at the dorsal surface of their sclerites. The main group includes rather uniformly ornamented *P. bohemicus* Barrande 1872 from the Early Llanvirnian Šárka Formation of Bohemia (Fig. 2A), *P. peachi* (Nicholson and Etheridge 1880) from the Late Caradocian Whitehouse Group of England (Withers 1926), and *P. richo-*

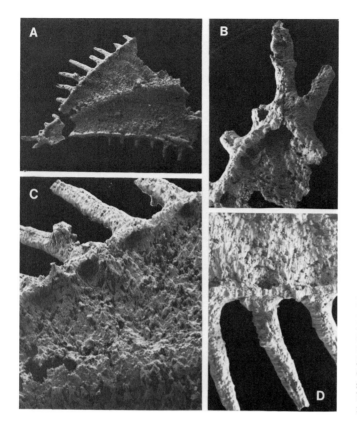

Fig. 4. Isolated lateral sclerite of *Plumulites* sp. from the erratic boulder E-231, *E. robustus* Zone, Llanvirnian (Uhakuan), Garcz, Poland (housed at the Zakład Paleobiologii, Warszawa; ZPAL V.XII/4). A. Internal side. × 60. B. Apical part; note medial ventral suture. ×300. C. Outer margin; note branching tubular spine. ×300. D. Inner margin. ×300.

rum Jell 1979 from the Early Devonian Humevale Formation of Victoria, Australia (Jell 1979). In Bohemia, *P. bohemicus* is replaced in the overlying Dobrotiva Formation by *Compacoleus compar* (Barrande 1872) (= *P. maior* in Prokop 1965, fig. XIII-45), with inner areas of the lateral sclerites having twice as many rugae as the outer areas. *Plumulites folliculum* Barrande 1872 from the Caradocian Libeň and Letna Formations of Bohemia differs from its congeners in its much shorter, densely rugose lateral sclerites (Fig. 2B). A similar ornamentation occurs in the Late Silurian *P. delicatus* Barrande 1872, with only weakly developed medial elevation in the lateral sclerites. Korejwo (1979, pl. 14: 10-11) illustrated two unidentified specimens from the Tournaisian of northern Poland that may represent dorsal and lateral thoracic sclerites of the youngest known plumulitid.

Thick dorsal sclerites of *Mojczalepas* abound in the Llanvirnian of the Baltic area and the Holy Cross Mountains, Poland. They are much thicker than in *Plumulites*, whereas the sinuous outline of rugae makes them different from *Deltacoleus*. The rugae are closely spaced, very high, and lamella shaped in *M. multilamellosa* sp. nov. (Figs. 5A, B, and 6A). In turn, *Mojczalepas* sp. *a* is ornamented with low terraces (Fig. 5C, D), and *Mojczalepas*

sp. *b* has densely spaced rugae resembling growth lines (Fig. 6B). The three species differ also in sclerite cross section. In *M. multilamellosa*, the elytra are almost rectangular in cross section; in *Mojczalepas* sp. *a*, additional angulations developed close to the inner and outer margins; in *Mojczalepas* sp. *b*, the outer area of the sclerite is convex. The paratype of *Deltacoleus crassus* Withers 1926 from the Caradocian Balclatchie Group of England (Withers 1926, pl. VIII: 6) may also belong to *Mojczalepas*, as its rugae are distinctly sinuous in outline.

Only two lateral sclerites have been found that can be attributed to *Mojczalepas*. Their morphology resembles *Plumulites* closely, but their wall is very thick (Fig. 6C, D).

The only species of *Turrilepas* that is known from articulated strobili, *T. wrightiana* (Koninck 1857), is asymmetric; its left dorsal sclerites are not mirror images of the right ones (Withers 1926). Rugae on the left sclerites are straight in the inner area and run obliquely to the inner margin, whereas the corresponding rugae on the right sclerites are gently curved and tangential to the inner margin. Such an asymmetry is unique among turrilepadids. Dorsal sclerites of *T. wrightiana* have a more complex ornamentation pattern than those of *Mojczalepas*. The lateral sclerites,

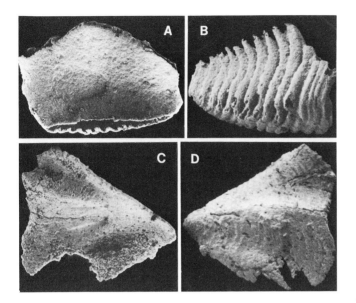

Fig. 5. Dorsal sclerites of *Mojczalepas* gen. nov. A, B. *Mojczalepas multilamellosa* sp. nov., holotype, sample MA-29. *E. reclinatus* Zone, Llanvirnian (Lasnamägian), Mójcza Limestone, Mójcza, Poland (ZPAL V.XII/2). Original calcitic wall covered with thin phosphatic(?) film. ×100. C, D. *Mojczalepas* sp. *a*, erratic boulder E-297. *E. reclinatus* Zone, Llanvirnian (Lasnamägian), Międzyzdroje, Poland (ZPAL V.XII/5). Preserved calcitic wall. ×60.

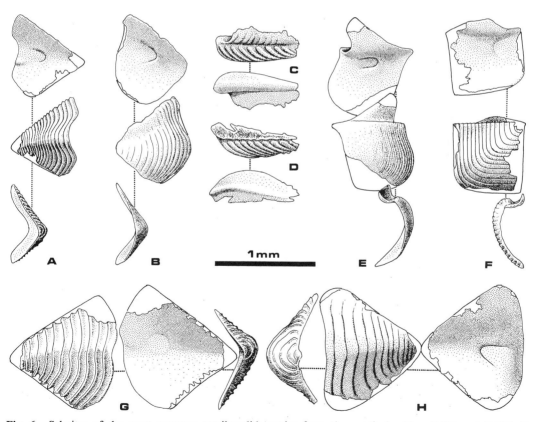

Fig. 6. Sclerites of the most common turrilepadid species from the erratic boulders E-279 and E-323. *E. reclinatus* Zone, Llanvirnian (Lasnamägian). Dorsal sclerites, except for C and D. A. *Mojczalepas multilamellosa* sp. nov. B. *Mojczalepas* sp. *b*. C, D. Lateral sclerites, perhaps *Mojczalepas* gen. nov. E. *Plicacoleus robustus* sp. nov. F. *Aulakolepos* aff. *suecicum* (Moberg 1914). G. *Deltacoleus* cf. *crassus* Withers 1926. H. *Deltacoleus* sp.

however, are flat, without any medial angulation or elevation, and with a simple pattern of rugae (Withers 1926).

In the Baltic Llanvirnian, machaeridian assemblages are dominated by *Deltacoleus.* Its dorsal sclerites have almost straight rugae, bent only at the high medial angulation. Sometimes the angulation is slightly flattened, with two secondary angulations nearby (Fig. 7A, B). This may reflect either intraspecific variability, or variation along the body, or else (the least likely) asymmetry between the right and left sclerites. There is much variation in conspicuousness and distribution of rugae (Fig. 8), as well as in the pattern of sclerite thickness. The latter character seems to have a bimodal distribution. Some specimens attain the maximum thickness in front of the muscle attachment scar, others in the center of the sclerite. Most likely these are two distinct species of *Deltacoleus,* one of which may be conspecific with *D. crassus* Withers 1926.

In *Clarkeolepis,* originally established for Middle Devonian to Early Carboniferous species, dorsal sclerites are longitudinally ornamented (Elias 1958). I have found turrilepadid dorsal sclerites ornamented with high longitudinal lamellae in an erratic boulder of Baltic origin, earliest Caradocian (Kukrusean) in age. The radially arranged lamellae cross prominent rugae and form fairly deep rectangular cells. Apart from this ornamentation, the sclerites resemble *Deltacoleus.*

Dorsal sclerites of the only known species of *Plicacoleus, P. robustus* sp. nov., are widespread but inabundant in the Llanvirnian of the Baltic area and the Holy Cross Mountains, Poland.

They are thick, elongated, convex, and almost smooth, sometimes with growth lines instead of faint rugae; a medial crest separates the convex outer area from the concave inner area (Figs. 6E and 9).

The type species of *Lepidocoleus, L. jamesi* (Hall and Whitfield 1875) from the Cincinnatian of Ohio, has relatively elongated dorsal sclerites ornamented with prominent rugae (Withers 1926; Pope 1975), which makes it similar to *Deltacoleus* and *Mojczalepas* rather than to the other species assigned by Withers (1926) to *Lepidocoleus.* This impression is reinforced by the thick wall and deeply embedded muscle attachment scar. Nevertheless, the location of the muscle attachment close to the inner side of the sclerite (Withers 1926, pl. II: 5–6) suggests that its inner concave part has been broken off. Under such circumstances, I propose to restrict the genus *Lepidocoleus* to its type species only, and to transfer the other species to the genus *Aulakolepos.*

Contrary to Withers (1926, 1933) and Bengtson (1977), *Aulakolepos ketleyanum* (Reed 1901) from the Wenlockian of England has small but fully developed lateral sclerites (see Bengtson 1977, fig. 3). They are approximately four times narrower than the corresponding dorsal sclerites, triangular in shape, and completely flat. This is indisputable evidence of the machaeridian nature of the Lepidocoleidae.

The oldest known *Aulakolepos* comes from the Late Llanvirnian (Lasnamägian) of the Baltic area (Fig. 6F). Its dorsal sclerites resemble the Ashgilian *A. suecicum* (Moberg 1914) from the same area (Withers 1926). Other Late Ordovician, Sil-

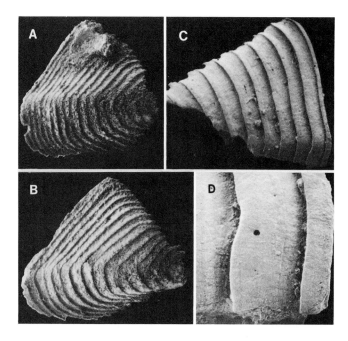

Fig. 7. Dorsal sclerites of *Deltacoleus* cf. *crassus* Withers 1926. A. Specimen ZPAL V.XII/6, erratic boulder E-279. *E. reclinatus* Zone, Llanvirnian (Lasnamägian), Międzyzdroje, Poland. × 55. B. Specimen ZPAL V.XII/7, same boulder. ×70. C. Specimen ZPAL V.XII/8, erratic boulder E-276. *P. originalis* Zone, Arenigian (Volkhovian), Międzyzdroje, Poland. Original calcitic wall replaced by a green, glauconitelike mineral. ×100. D. Same specimen. ×450.

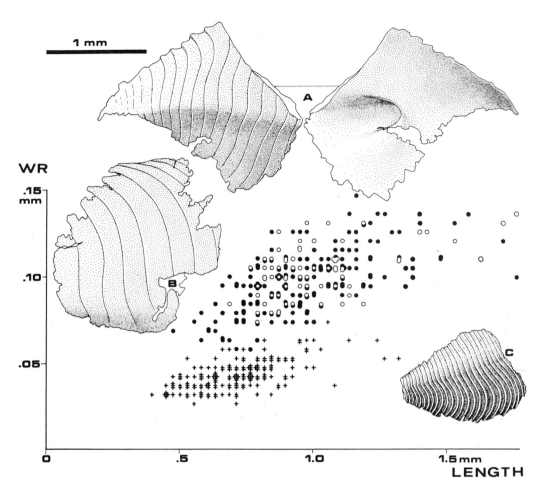

Fig. 8. Density of rugae (WR, average distance between the rugae) versus sclerite length for the most abundant turrilepadid species from the erratic boulder E-323, *E. reclinatus* Zone, Llanvirnian (Lasnamägian), Rozewie, Poland. Circles (and A), *Deltacoleus* sp.; dots (and B), *D.* cf. *crassus* Withers 1926; crosses (and C), *Mojczalepas multilamellosa* sp. nov.

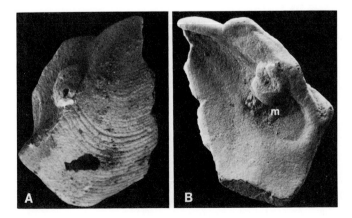

Fig. 9. External (A) and internal (B) views of a dorsal sclerite of *Plicacoleus robustus* sp. nov.. Holotype, sample MA-29, *E. reclinatus* Zone, Llanvirnian (Lasnamägian), Mójcza Limestone, Mójcza, Poland (ZPAL V.XII/3). Note growth lines and muscle attachment scar (m). × 100.

urian, and Devonian species of the genus have much shorter and wider sclerites. Their outer area and medial angulation are ornamented with rugae that continue, sometimes in form of growth lines, in the inner area. There is no evidence of a dorsal ligament, as claimed by Wolburg (1938).

The Ludlowian of the Carnic Alps, Austria, yielded some elongated sclerites with pseudoporous wall and a short posterior duplicature (Fig. 10). They are here described as *Carnicoleus gazdzickii* sp. nov. Their attribution to the Turrilepadida might be disputable, although it is compatible with the available data. Bengtson (1977, 1978) found some tubercles and depressions at the inner surface of Silurian *Aulakolepos*. No duplicature has been reported in *Aulakolepos*, but its presence might be expected because of a significant overlap of sclerites in the strobilus and by analogy to the process of mineralization in *Plumulites*. The pseudopores penetrating the sclerite wall in *Carnicoleus* do not open externally. They resemble muscle attachment scars, especially those left by the pallial muscles of some bivalves. Interpretation of *Carnicoleus* as a *Solen*-like bivalve, however, is contradicted by the absence of adductor muscle scars, presence of duplicature, and development with no indications of metamorphosis, although much change in sclerite shape occurred early in its ontogeny (Fig. 10E). In turn, interpretation of *Carnicoleus* sclerites as *Anatifopsis*-like plates of carpoid echinoderms is

contradicted by their imporous external surface with distinct growth lines and by the mirror-image symmetry.

PALEOCOLOGY

Little is known about the mode of life of the Turrilepadida. Withers (1926) refuted their interpretation as cirripedes but nevertheless envisaged the strobilus attached to the substrate, with its gaping side oriented upward, as in the barnacles. Wolburg (1938) interpreted *Aulakolepos* as a segmented clam with adductor muscles and a dorsal ligament, but this interpretation is incompatible with sclerite morphology. Jell (1979) demonstrated that *Plumulites* was free-living, with dorsoventrally flattened body and elytra confined to its dorsal side. The rugae probably mechanically strengthened the elytra. The body shape and thin sclerite walls of *Plumulites* suggest much mobility of the animal. Its strobilus did not allow for much lateral flexibility, but the body could be rolled up, as in the trilobites. Most probably *Mojczalepas* also belonged to epifauna, as the very high lamellae on its sclerites would be strongly disadvantageous for burrowing.

The majority of turrilepadids, however, have terracelike rugae, with their sharp edges directed posteriorly, which is an adaptation to burrowing in loose sediment. The almost completely smooth surface and clamlike shape of the sclerites of *Pli-*

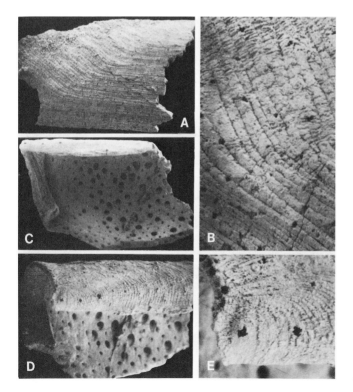

Fig. 10. *Carnicoleus gazdzickii* sp. nov. Ludlowian, Orthoceras Limestone, Valentin Törl, Carnic Alps, Austria. A. Specimen ZPAL V.XII/9. External view. ×70. B. Same specimen. ×300. C. Holotype, ZPAL V.XII/1. Internal view. ×70. D. Holotype. Dorsal view. ×100. E. Holotype. Apex, note ontogenetic changes in shape. ×300.

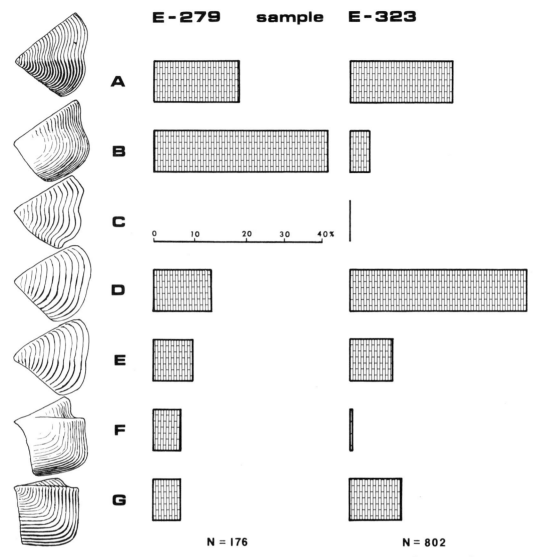

Fig. 11. Relative abundance of turrilepadid dorsal sclerites in fossil assemblages from *E. reclinatus* Zone; same samples as in Fig. 6. A. *Mojczalepas multilamellosa* sp. nov. B. *Mojczalepas* sp. *b.* C. *Mojczalepas* sp. *a.* D. *Deltacoleus* cf. *crassus* Withers 1926. E. *Deltacoleus* sp. F. *Plicacoleus robustus* sp. nov. G. *Aulakolepos* aff. *suecicum* (Moberg 1914).

cacoleus robustus may also be an adaptation to infaunal life.

There are very few data on the facies distribution of the Turrilepadida. *Plumulites* has been reported mostly from open-sea facies, such as the shales with siliceous concretions in the Šárka Formation of Bohemia. *Deltacoleus* frequently occurs in the red cephalopod limestone of the Baltic area, representative of the offshore facies of the Baltic carbonate platform. A little further onshore the gray cephalopod limestone of the Baltic area was deposited, rich in trilobites and gastropods; it contains abundant *Deltacoleus*, *Aulakolepos*, *Mojczalepas*, and *Plicacoleus*. The latter two gen-

era, however, occur most commonly in shallow-water glauconitic limestones of Oland, Sweden, and in organodetrital limestones with ooids at Mójcza in the Holy Cross Mountains, Poland.

Relative abundance of particular species could be reliably determined only for the Llanvirnian gray cephalopod limestone (Fig. 11). The assemblages comprise at least six well-defined turrilepadid and lepidocoleid species; some plumulitids could also occur in the original community. Similarly high diversity of the Turrilepadida was reported by Bengtson (1979) from the Silurian Vattenfallet section of Gotland, Sweden. With high abundance of machaeridian sclerites taken also into account, this implies that in many Early Paleozoic communities the Turrilepadida were not less important than trilobites or gastropods in the vagile benthos.

PHYLOGENETIC RELATIONSHIPS OF THE TURRILEPADIDA

The Turrilepadida range from the Tremadocian (Kobayashi and Hamada 1976) to the Pennsylvanian (Chronic in Bengtson 1978). They did not give origin to any extant animal group. Their possible relatives include the Cambrian Tommotiida and Sachitida and the Middle Paleozoic Hercolepadida.

The order Tommotiida includes animals with body covered with numerous sclerites that are prominently ornamented and lamellar in microstructure. These forms can be arranged into a morphologic series, from high-conical, almost symmetric sclerites of *Lapworthella* through *Kelanella* and *Bengtsonia* and up to low, almost flat, strongly asymmetric and dimorphic sclerites of *Tannuolina* (see Bengtson 1970; Matthews 1973; Matthews and Missarzhevsky 1975; Bischoff 1976; Missarzhevsky and Grigoreva 1981; Yuan and Zhang 1983). The sclerites are composed of calcium phosphatic lamellae, frequently separated by empty spaces; in *Tannuolina,* they are interconnected by tubuli. Such a loose distribution of phosphatic lamellae resembles the pattern observed in some acrotretid brachiopods (Poulsen 1971) and phosphatized arthropod remnants. The spaces between lamellae could have been originally filled with organic tissue. Secondary phosphatization cannot be ruled out, as the majority of tommotiid sclerites are associated with abundant phosphatized fossils or phosphorites (Bengtson 1970). This is in fact suggested by deposition of apatite prismatic layers on lamellae of the tommotiid *Sunnaginia* and subsequent filling of the spaces by loose, coarse apatite crystals (Landing et al. 1980).

Morphologically, turrilepadid sclerites can be easily derived from the Tommotiida. The tommotiid sclerites supported pyramidal protrusions of the soft body. The animal probably had rows of scleritized horns along the body. In some species of *Camenella* and *Dailyatia,* however, the sclerites were secreted on flattened protrusions of the body (Bengtson 1970; Matthews and Missarzhevsky 1975; Bischoff 1976), very much like the plumulitid elytra. The sclerites are prominently ornamented dorsally but their lower side is almost smooth. In fact, sellate sclerites of *Camenella* closely resemble dorsal sclerites of *Plumulites,* while mitrate sclerites of the former resemble lateral sclerites of the latter (Bengtson 1970). This interpretation is supported by the range of variation in shape and size which is smaller in the sellate than in the mitrate sclerites of *Camenella,* for

the corresponding dorsal sclerites are largely uniform in *Plumulites,* whereas the lateral sclerites are differentiated into thoracic and variable abdominal ones (cf. Jell 1979).

Thus, phylogenetic derivation of the Plumulitidae from the Tommotiidae appears plausible, although the time gap between the early Middle Cambrian (Bischoff 1976) and the Tremadocian (Kobayashi and Hamada 1976) calls for caution.

Another Cambrian group of elytra-bearing animals, the Siphogonuchitidae, is closely related to the Wiwaxiidae (see Bengtson and Missarzhevsky 1981) and included here in the order Sachitida. The best known sachitid species is *Wiwaxia corrugata* (Matthew 1899) from the Middle Cambrian Burgess Shale (Conway Morris 1982: W; Chapter 13 by Briggs and Conway Morris, this volume). Its oval body was covered with leaf shaped elytra arranged into 20 longitudinal rows and 8 or 9 segments (Fig. 12A). According to Conway Morris (1982: W), a radulalike organ with two rows of organic teeth occurs in *Wiwaxia.* Wiwaxiid elytra have a very complex internal structure, which is best known in the Early Cambrian *Thambetolepis delicata* Jell 1981 from Australia (Jell 1981). Their external surface is flat, with shallow longitudinal depressions separated by sharp ridges. Internally, they are subdivided into pennately arranged tubular compartments. The axial compartment is wider and opens to the proximal, basal part of the elytron. It seems that the scleritization proceeded from the elytron surface inward, which resulted in subdivision of the retreating soft body into a featherlike internal organ. This mode of secretion would be rather unusual if mollusk affinities of the wiwaxiids were accepted, as proposed by Conway Morris (1982: W). More primitive sachitids probably lacked internal compartmentalization of elytra; the arrangement of their elongated, angular sclerites, however, was similar to that in *Wiwaxia* (see Matthews and Missarzhevsky 1975; Bengtson and Missarzhevsky 1981; Bengtson and Conway Morris 1984).

As noted above, the pennate organization of *Thambetolepis* elytra (Jell 1981) resembles the pattern of medial elevation, rugae, and marginal spines in the lateral elytral of *Plumulites.* Of course, this is not conclusive evidence for phylogenetic relationship between the plumulitids and wiwaxiids but, on the other hand, Bengtson's and Missarzhevsky's concept (1981) of Coelosclerito-phora including the wiwaxiids and the "hexacti-nellid sponges" Chancelloriidae is questionable. The "spicules" of *Chancelloria* originated similarly to *Wiwaxia* sclerites (Sdzuy 1969). As judged after the few articulated specimens (Rigby 1978), however, its body plan was different from bilaterally symmetric wiwaxiids. *Chancelloria* was almost certainly sedentary, with radial symmetry of the body and triradiate symmetry of the

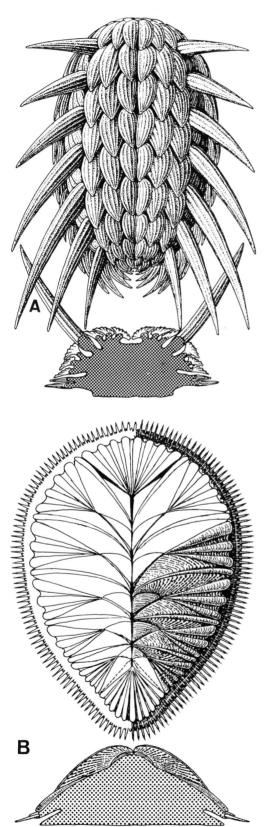

"spicules." This resembles the Early Cambrian *Anabarites* (see Abaimova 1978), which might link the Chancelloriidae to an enigmatic group including *Pirania muricata* Walcott 1920 from the Middle Cambrian Burgess Shale, *Amgaella amgaensis* Korde 1957 and *Yakutina aciculata* (Korde 1957) from the Middle Cambrian of Yakutia. These organisms used to be interpreted as either dasycladacean algae (Korde 1963) or sponges (Conway Morris 1982). The Coeloscleritophora appear thus to be polyphyletic.

Possible relatives of the Turrilepadida include also barnacle-like *Hercolepas signata* (Aurivillius 1892) from the Early Silurian Visby Beds of Gotland, Sweden (Aurivillius 1892). Provided that *Protobalanus hamiltonensis* Hall and Clarke 1888, as described by Van Name (1925), belongs indeed to the same group, which is here proposed as the Hercolepadida ord. nov., their body plan can be tentatively reconstructed (Fig. 12B). The oval body is covered with calcareous dorsal sclerites arranged in four rows, possibly with unpaired, bilaterally symmetric cephalic and caudal sclerites (Van Name 1926). Sclerites of the inner rows are scalloplike in shape, with umbones and auricles contacting along the medial commissure. Sclerites of the outer rows are triangular and rather flat. All sclerites are conspicuously ornamented with radial ribs, concentric rugae, and perhaps also punctae. The most peculiar feature of both *Hercolepas* and *Protobalanus* is a corona of small needle shaped spines surrounding the strobilus.

Hercolepadid dorsal sclerites resemble sellate sclerites of *Camenella* in shape, while the lateral sclerites are comparable to tommotiid mitrate sclerites. They resemble plumulitid sclerites in structure, but their orientation in the strobili is different. Sclerite tips are directed posteriorly in the Turrilepadida but almost medially in *Hercolepas*. When taken in conjunction with the corona of marginal spines, this may suggest sessile life habits of the hercolepadids.

I propose to assign the Sachitida, Tommotiida, Turrilepadida, and Hercolepadida to the class Machaeridia Withers 1926.

The most morphologically parsimonious model of machaeridian phylogeny (Fig. 13) suggests that their common ancestor had numerous conical protoelytra with cuticularized surfaces. They

---

Fig. 12. A. Dorsal view and cross section of reconstructed *Wiwaxia corrugata* (Matthew 1899) from the Middle Cambrian Burgess Shale, British Columbia (modified after Bengtson & Conway Morris 1984); enlarged. B. Composite dorsal view of *Hercolepas signata* (Aurivillius 1892) (shadowed) fitted into the strobilus of *Protobalanus hamiltonensis* Hall and Clarke 1888 (see Van Name 1925, 1926); also hypothetical cross section, enlarged.

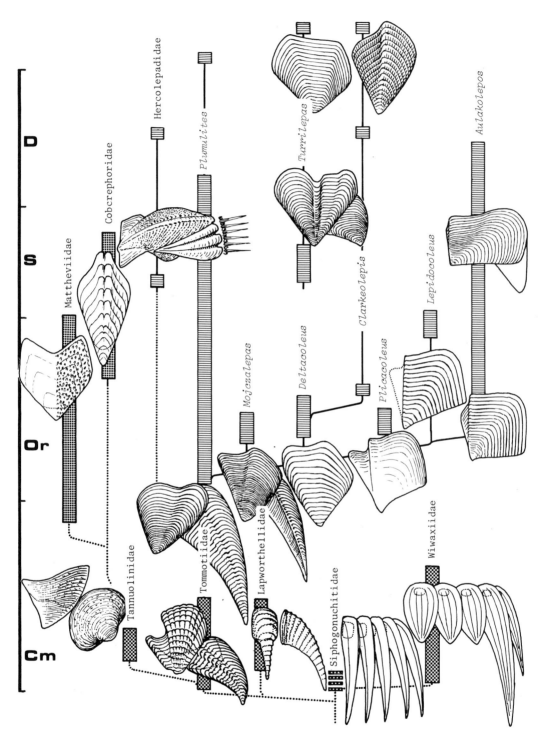

Fig. 13. Proposed relationships among the considered taxa, and their stratigraphic distribution; note that the alleged chitons Cobcrephoridae had phosphatic valves (Bischoff 1981).

could be low, as in *Tannuolina,* or elongated, as in the Siphogonuchitidae, but because of considerable sclerite mineralization in *Tannuolina* and *Camenella* and their complex ornamentation and microstructure, the latter interpretation seems more likely. Flattening of protoelytra into typical leaf shaped elytra probably took place independently in the Wiwaxiidae and Tommotiidae. The latter group may have given origin to all the post-Cambrian machaeridians, as indicated by the number of longitudinal rows of elytra. Subsequently, dorsal calcification of elytra replaced scleritization of their entire surface.

PHYLOGENETIC AFFINITIES OF THE MACHAERIDIA

The Machaeridia were originally interpreted as an extinct group of barnacles. This hypothesis was refuted by Withers (1926; but see Bischoff 1976), who suggested their echinoderm affinities. Bather (in Withers 1926) considered carpoids as the closest relatives of the Turrilepadida, while Pope (1975) interpreted lepidocoleid strobili as fragmentary mitrate spines. Microstructural counterevidence to the latter interpretation was presented by Bengtson (1977, 1978), who proposed earlier a close relationship between the Turrilepadida and Tommotiida, and suggested their annelid affinities (Bengtson 1970). This concept was supported by Jell (1979).

With monophyly of the Machaeridia taken for granted, their typical representative has a bilaterally symmetric body covered by (pseudo)metamerically arranged dorsal protuberances or elytra, each of them with internal canals (known in *Thambetolepis* and inferred in *Plumulites*) that may branch pennately. Metamerically arranged dorsoventral muscles attached to the dorsal wall of each elytron occur in more advanced machaeridians. Body segments tend to differentiate into tagmae. Spines or spicules occur outside the strobilus in the Hercolepadida. Jaw apparatus has been reported in *Wiwaxia.*

The major problem with phylogenetic affinities of the Machaeridia is that the available data allow for more than one coherent interpretation. Potential relatives of the Machaeridia can be sought in the Annelida as well as in the mollusk subphylum Amphineura. A direct phylogenetic relationship of machaeridians to any known representative of those two groups seems unlikely, however. The following speculations are merely aimed at exploring alternative interpretations of the machaeridian body plan within the annelid and mollusk anatomic frameworks.

COMPARISONS TO THE ANNELIDA

Bengtson (1970) compared the Tommotiida to the Early Cambrian onychophore *Xenusion auers-waldae* Pompeckj 1927 (Jaeger and Martinsson 1967) and suggested that the dorsal humps of *Xenusion* could bear machaeridian-like sclerites. These humps seem to be homologous to tubercles in the cuticular rings of another Cambrian onychophore, *Aysheaia pedunculata* Walcott 1911 (Whittington 1978), whereas the latter closely resemble the tubercles of Recent terrestrial onychophores. Cloud and Bever (1973) illustrated, as the trace fossil *Plagiogmus* sp., two Early Cambrian specimens that may represent a connecting link between *Xenusion* and *Aysheaia.* They provide evidence for the onychophoran nature of *Xenusion* and also indicate that more than just two longitudinal rows of dorsal humps may occur in marine Onychophora. Still, the morphologic gap between the onychophorans and machaeridians is very wide.

Wiwaxiid and plumulitid sclerites do not significantly differ in shape and structure from elytra of the Recent Polychaeta. The elytron originates as a flat extension of dorsal tubercle in the Polychaeta. The tubercle, or elytrophore, contains intestinal caecum of the segment (Pettibone 1953). There are only two longitudinal rows of elytrophores in the Recent amphinomiid polychaetes, with elytra actually present at every second segment. In elytra-bearing segments, the cirri associated with elytrophores are reduced. It is a matter of dispute whether elytra are homologous to cirri (Duncker 1906). If not, there is nothing implausible in a supposition that some extinct polychaetes had more than two rows of elytrophores. Their intestinal caecum might potentially protrude into the elytron to develop into a structure like the internal canal of *Thambetolepis.* The elytrophore has muscles (Duncker 1906) that might be transformed into dorsoventral muscles of the Turrilepadida. Moreover, the jaw structure of *Wiwaxia* (Conway Morris 1982: W) might represent the polychaete jaw apparatus, which may be hardly discernible from the mollusk radula in the fossil record (Kielan-Jaworowska 1966). *Wiwaxia* has no setae that are diagnostic of the Polychaeta. The phyllodocid polychaetes, however, also lack setae, while the marginal spines of *Hercolepas* might in fact represent modified setae. The Machaeridia could thus be interpreted as related to the Polychaeta.

Such an interpretation, however, encounters functional problems. With an *Aulakolepos*-like skeleton, parapodia could not function as a locomotory organ. The only mechanism of propulsion available to such an annelid would be snakelike lateral swinging of the body. This mechanism could hardly work in *Lepidocoleus* or *Turrilepas,* with their short bodies. The problem with locomotion could be overcome by the assumption of a footlike crawling and digging organ. The latter concept, however, is incompatible with the body plan of the phylum Annelida.

COMPARISONS TO THE AMPHINEURA

It is now generally accepted that the Mollusca are acoelomate and evolved from crawling flatworms (Salvini-Plawen 1982), while the origin of the Annelida was preceded by development of the coelom in their ancestors, who peristaltically burrowed in soft mud. Thus, the Machaeridia cannot be related to both polychaetes and polyplacophorans, but they might be related to both turbellarians and polyplacophorans (Runnegar et al. 1979).

Leaf shaped protuberances of the body filled with intestinal caeca occur in polyclad flatworms (Hyman 1951). They are not metamerically arranged, but both metamerization and cuticularization are compatible with the body plan of the phylum Platyhelminthes. Fossil record of the flatworms is practically nonexistent. The only fossil that could be related to the Recent *Turbellaria* is the Late Precambrian *Dickinsonia*. With its large, flat, and contractible body, concentric muscular pattern, and medially separated "segments" resembling in their distribution the intestinal caeca and color bands of large-sized polyclad flatworms, *Dickinsonia* fits quite well into the body plan of the Platyhelminthes, perhaps better than into any other phylum (but see Runnegar 1982; see also Chapter 6 by Fedonkin, this volume). If this in-

terpretation were correct, some prototurbellarians would have existed in the Late Precambrian and they could have given origin to the Machaeridia. The Mollusca must also have evolved from Late Precambrian flatworms. It is the Polyplacophora that appear to be the most primitive mollusks (Runnegar et al. 1979). The problem is, then, if the functional analogy between machaeridian elytra and chiton valves reflects their homology. Superficially, there is little similarity between the unpaired, flat valves of Recent chitons, penetrated by numerous aesthetae, and the paired, imporous, pyramidal sclerites of early machaeridians. This dissimilarity, however, turns out to be less pronounced when the earliest chitons are considered.

As shown by Runnegar et al. (1979), the Late Cambrian *Matthevia variabilis* Walcott 1885 was a chiton. Ventral morphology of the valves of the Early Ordovician *Septemchiton aequivoca* (Robson 1913) from the Šárka Formation of Bohemia very closely resembles *Matthevia* (Fig. 14). The armor of that species is intermediate in morphology between *Matthevia* and other early chitons. *Matthevia*, however, had very thick, conical valves (Fig. 14B), perhaps in response to the high environmental energy. *Septemchiton*, in turn, lived in deeper water (Rolfe 1981) and had roof

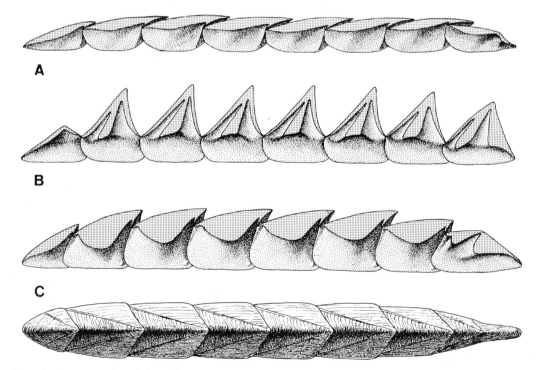

Fig. 14. Reconstruction of the earliest polyplacophorans; not to scale. A. *Chelodes* sp. Based on Runnegar et al. (1979) and specimens from the Mójcza Limestone, Mójcza, Poland. B. *Matthevia variabilis* Walcott 1885. Based on Runnegar et al. (1979). C. *Septemchiton aequivoca* (Robson 1913). Longitudinal section and dorsal view. Based on several articulated but incomplete specimens from the Šárka Formation, Osek, Bohemia (housed at the Narodní Museum, Prague).

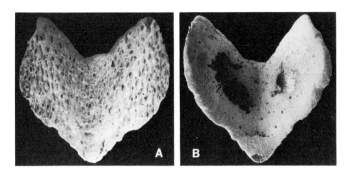

Fig. 15. External A. and internal B. views of an intermediate valve of *Septemchiton* aff. *aequivoca* (Robson 1913), sample MA-4. Late Caradocian, Mójcza Limestone, Mójcza, Poland (ZPAL V.XII/10). Note small duplicature. ×100.

shaped valves with flat lateral areas and narrow posterior duplicature (Fig. 15). The duplicature is homologous to the posterior wall of the pyramidal valves of *Matthevia*. Even the most primitive chitons, such as the Cambrian *Praeacanthochiton* (Runnegar et al. 1979) or the Ordovician *Septemchiton* (Fig. 15), show peculiar microornamentation that, by analogy to Carboniferous chitons (Hoare et al. 1983), may be taken to reflect a system of aesthetae in the external layer of the shell (see Haas and Kriesten 1974).

Generally, the plumulitids and most tommotiids had externally smooth sclerites completely lamellar in microstructure. There are, however, at least two exceptions. First, *Tannuolina* had some tubes that penetrated the lamellae and opened at the external surface of the sclerite (Bischoff 1976; Bengtson 1977). Second, *Carnicoleus,* interpreted here as a lepidocoleid machaeridian, had pseudopores (Fig. 10). Its external ornamentation, however, differed from the chitons. The main dorsoventral muscle attachment scars are arranged in two longitudinal rows in the chitons. This is the pattern observed also in the Turrilepadida. One might thus think of the roof shaped valves of *Septemchiton* having originated each by fusion of two dorsal sclerites of *Tannuolina,* although the embryonic development of Recent polyplacophorans (cf. Haas et al. 1979) does not support this hypothesis.

Nevertheless, it is not implausible that the chitons evolved from a machaeridian. Corroborating evidence might in fact be sought in the perinotal(?) spicules of *Hercolepas,* with its otherwise machaeridian strobilus, and in the radula of *Wiwaxia.* But the available data do not allow us to determine if the class Machaeridia is to be placed in the Annelida, parallel to the Polychaeta, or rather in the mollusk subphylum Amphineura, next to the Polyplacophora and Aplacophora.

SYSTEMATIC PALEONTOLOGY

CLASS MACHAERIDIA Withers 1926

Diagnosis: Bilaterally symmetric animals with body covered dorsally with metameric, scleri-

tized elytra arranged into longitudinal rows. Intestinal caeca(?) protruded in elytra in early forms. In later forms, strongly mineralized sclerites propelled by dorsoventral muscles.

ORDER SACHITIDA He 1980

Emended diagnosis: Machaeridians with elongated, usually leaf shaped elytra in about 20 rows; 8 or more segments of the body. Elytra entirely sclerable into pennately arranged compartments.
Families: Siphogonuchitidae Qian 1977; Wiwaxiidae Walcott 1911.
Distribution: Early to Middle Cambrian.

ORDER TOMMOTIIDA Missarzhevsky 1970

Emended diagnosis: Machaeridians with elytra in four(?) rows, pyramidal in shape, covered with thick mineralized sclerites; elytra of allegedly inner rows sellate, those of outer rows mitrate in shape.
Families: Lapworthellidae Missarzhevsky 1966; Tommotiidae Missarzhevsky 1970; Tannuolinidae Fonin and Smirnova 1967.
Distribution: Early to earliest Middle Cambrian.

ORDER HERCOLEPADIDA ord. nov.

Diagnosis: Machaeridians with oval body covered with four rows of elytra. Elytra calcified dorsally, with complex ornamentation and tips directed medially. Numerous marginal spines around the body.
Family: Hercolepadidae fam. nov.
Diagnosis: As for the order.
Genera: *Hercolepas* Withers 1915; *Protobalanus* Clarke, 1888.
Distribution: Early Wenlockian to Late Eifelian.

ORDER TURRILEPADIDA Pilsbry 1916 nom. corr.

Diagnosis: Machaeridians with elongated body covered with four rows of dorsally calcified ely-

tra. Two first segments of the body have only two inner elytra each; the following three segments have lateral elytra modified. Lateral elytra reduced in advanced forms.
Distribution: Tremadocian to Westfalian.

FAMILY PLUMULITIDAE Jell 1979

Emended diagnosis: Dorsoventrally flattened body with about 20 segments; elytra thin walled; lateral elytra large.
Genera: Plumulites Barrande 1872; *Compacoleus* Schallreuter 1985.
Distribution: Tremadocian to Tournaisian.

FAMILY TURRILEPADIDAE Clarke 1896

Emended diagnosis: Box-shaped body with about 30 segments, covered with angular dorsal elytra and smaller, flat lateral elytra; elytra thick walled, each with a distinct muscle scar.
Genera: Turrilepas Woodward 1865; *Deltacoleus* Withers 1926; *Clarkeolepis* Elias 1958; *Mojczalepas* gen. nov.; *Spinacoleus* Schallreuter 1985 (*-Rugacoleus* Schallreuter 1985).
Distribution: Arenigian to Namurian.

Mojczalepas gen. nov.

Type species: M. multilamellosa sp. nov.
Diagnosis: Elytra similar to *Plumulites* but with thick, calcified dorsal wall. Dorsal elytra have concave inner and convex outer areas, and relatively deep and wide medial sinus at the anterior margin. Lateral elytra have thick, narrow sclerites with prominent medial elevation.
Species included: Type species, *Mojczalepas* sp. *a, Mojczalepas* sp. *b.*
Distribution: Arenigian to Llanvirnian of Baltic area, Llanvirnian of Holy Cross Mountains, Poland; possibly also Caradocian of England.

Mojczalepas multilamellosa sp. nov. (Figs. 5A, B, 6A, and 8C)

Holotype: ZPAL V.XII/2, Fig. 5A, B.
Type horizon and locality: E. reclinatus Zone, Llanvirnian, Mójcza Limestone, Mójcza, Holy Cross Mountains, Poland.
Diagnosis: A species of *Mojczalepas* with sclerites ornamented with densely spaced, high rugae. Outer area of dorsal sclerites with semicircular lobe at the anterior margin, inner area with medial shallow sinus and marginal narrow lobe.
Remarks: M. multilamellosa differs from its congeners in its very prominent and simple pattern of ornamentation. The lateral sclerites (Fig. 6C, D) derived from the erratic boulder E-323 may belong to this species.

Distribution: Llanvirnian of Baltic area and Holy Cross Mountains, Poland.

FAMILY LEPIDOCOLEIDAE Clarke 1896

Emended diagnosis: Laterally compressed body with up to 60 segments, completely covered with large dorsal elytra. Dorsal elytra with large, convex outer area and narrow, concave inner area separated from each other by medial rib. Lateral elytra reduced in size or lacking.
Genera: Lepidocoleus Faber 1886; *Aulakolepos* Wolburg 1938; *Plicacoleus* gen. nov.; *?Carnicoleus* gen. nov.
Distribution: Llanvirnian to Givetian.

Plicacoleus gen. nov.

Type species: P. robustus sp. nov.
Diagnosis: Dorsal elytra have thick-walled sclerites with very convex outer area separated by prominent crest from wing shaped inner area. Sclerite width subequal to its length. Sclerites are almost smooth, with indistinct growth lines in the only known species.
Remarks: The peculiar medial crest on dorsal sclerites distinguishes *Plicacoleus* among the machaeridians. The sclerites have sharp posterior margin without any duplicature or marginal spines.
Species included: Type species only.
Distribution: Llanvirnian of Baltic area and Holy Cross Mountains, Poland.

Plicacoleus robustus sp. nov. (Figs. 6E, and 9A, B)

Holotype: ZPAL V.XII/3, Fig. 9A, B.
Type horizon and locality: E. reclinatus Zone, Llanvirnian, Mójcza Limestone, Mójcza, Holy Cross Mountains, Poland.
Diagnosis: As for the genus.
Distribution: As for the genus.

Carnicoleus gen. nov.

Type species: C. gazdzickii sp. nov.
Diagnosis: Sclerites elongated, semicylindrical in shape, ornamented with growth lines; sclerite wall relatively thin, penetrated with numerous pseudopores. Posterior end of the sclerite sharply truncated, with short duplicature. Inner area of the sclerite narrow, separated by weak angulation from outer area.
Remarks: This is an enigmatic form with uncertain affinities. Though generally lepidocoleid in shape, its sclerites differ from the other Machaeridia in their pseudoporous, perhaps originally aragonitic (now phosphatized) wall and peculiar duplicature.
Species included: Type species only.

Carnicoleus gazdzickii sp. nov. (Fig. 10A–D)

Holotype: ZPAL V.XII/1, Fig. 10C, D.
Type horizon and locality: Ludlowian, Orthoceras Limestone, Valentin Törl, Carnic Alps, Austria.
Diagnosis: As for the genus.
Name derivation: After Dr. Andrzej Gaździcki who collected the sample.
Distribution: Type locality only.

Acknowledgments

My most sincere thanks are due to Stefan Bengtson, who introduced me to the subject, made available machaeridian specimens, and provided many valuable comments. I am grateful also to John Repetski and the staff of the Smithsonian Institution, Washington, D. C., who made available the specimens of *Wiwaxia*. SEM micrographs were taken at Nencki's Institute of Experimental Biology, Warszawa, Poland.

REFERENCES

Abaimova, G. P. 1978. Anabaritidy—drevneishye iskopaemye s karbonatnym skeletom [Anabaritids—the oldest fossils with calcareous skeleton]. *Trudy SNIIGGIMS* **260**, 77–83.

Aurivillius, C. W. S. 1892. Über einige ober-silurische Cirripeden aus Gotland. *Bih. K. Sven. Vetenskapsakad. Handl.* **18,4**(3), 1–24.

Barrande, J. 1872. *Système Silurien du Centre de la Bohème. Ière Partie: Recherches Paléontologiques*, vol. 1, suppl. *Trilobites, Crustacés divers et Poissons, Planches.* Prague, 35 pl.

Bengtson, S. 1970. The Lower Cambrian fossil *Tommotia. Lethaia* 3, 363–392.

———. 1977. Aspects of problematic fossils in the Early Palaeozoic. *Acta Univ. Upsal. Abstr. Uppsala Dissert. Fac. Sci.* **415**, 1–71.

———. 1978. The Machaeridia—a square peg in a pentagonal hole. *Thalassia Yugosl.* **12**, 1–10.

———. 1979. Machaeridians, *Hercolepas. Sver. Geol. Unders. Arsbok* **73**(3), 211–212.

——— and Conway Morris, S. 1984. A comparative study of Lower Cambrian *Halkieria* and Middle Cambrian *Wiwaxia. Lethaia* **17**, 307–329.

——— and Missarzhevsky, V. V. 1981. Coeloscleritophora, a major group of enigmatic Cambrian metazoans. *U.S. Geol. Surv. Open-File Rep.* **81-743**, 19–21.

Bischoff, G. C. O. 1976. *Dailyatia*, a new genus of the Tommotiidae from the Cambrian strata of S.E. Australia (Crustacea, Cirripedia). *Senckenb. Leth.* **57**, 1–33.

———. 1981. *Cobcrephora* n.g., representative of a new polyplacophoran order Phosphatoloricata with calciumphosphatic shells. *Senckenb. Leth.* **61**, 173–215.

Cloud, P. and Bever, J. E. 1973. Trace fossils from the Flathead Sandstone, Fremont County, Wyoming, compared with Early Cambrian forms from California and Australia. *J. Paleont.* **47**, 883–885.

Conway Morris, S. (ed.). 1982. *Atlas of the Burgess Shale.* Palaeontological Association, Nottingham, England.

Duncker, H. 1906. Über die Homologie von Cirrus und Elytron bei den Aphroditiden (Ein Beitrag zur Morphologie der Aphroditiden). *Z. Wiss. Zool.* **81**(2/3), 13–343.

Elias, M. K. 1958. Late Mississippian fauna from the Redoak Hollow Formation of southern Oklahoma, part 4, Gastropoda, Scaphopoda, Cephalopoda, Ostracoda, Thoracica, and Problematica. *J. Paleont.* **32**, 1–57.

Haas, W. and Kriesten, K. 1974. Studien über das Mantelepithel von *Lepidochitona cinerea* (L.) (Placophora). *Biomineralization* **7**, 100–109.

———, ———, and Watabe, N. 1979. Notes on the shell formation in the larvae of the Placophora (Mollusca). *Biomineralization* **10**, 1–8.

Hoare, R. D., Mapes, R. H., and Atwater, D. E. 1983. Pennsylvanian Polyplacophora (Mollusca) from Oklahoma and Texas. *J. Paleont.* **57**, 992–1000.

Hyman, L. H. 1951. *The Invertebrates: Platyhelminthes and Rhynchocoela. The Acoelomate Bilateria*, vol. 2. McGraw-Hill, New York.

Jaeger, H. and Martinsson, A. 1967. Remarks on the problematic fossil *Xenusion auerswaldae. Geol. För. Stockh. Förhandl.* **88**, 435–452.

Jell, P. A. 1979. *Plumulites* and the machaeridian problem *Alcheringa* 3, 253–259.

———. 1981. *Thambetolepis delicata* gen. et sp. n., an enigmatic fossil from the Early Cambrian of South Australia. *Alcheringa* **5**, 85–93.

Kielan-Jaworowska, Z. 1966. Polychaete apparatuses from the Ordovician and Silurian of Poland and a comparison with modern forms. *Palaeont. Polon.* **16**, 1–152.

Kobayashi, T. and Hamada, T. 1976. Occurrence of the Machaeridia in Japan and Malaysia. *Proc. Jpn. Acad.* **52**, 371–374.

Jorde, K. B. 1963. Triby Cambroporellae, Amgaellae i Seletonellae. *In* Orlov, A. (ed.). *Osnovy Paleontologii. Vodorosli, mochoobraznye, psilofitovye, plaunovidnye, chlenistostebelnye, paporotniki.* Izdatelstvo AN SSSR, Moskva, pp. 219–220.

Korejwo, K. 1979. Biostratigraphy of the Carboniferous sediments from the Wierzchowo area (western Pomerania). *Acta Geol. Polon.* **29**, 457–473.

Landing, E., Nowlan, G. S. and Fletcher, T. P. 1980. A microfauna associated with Early Cambrian trilobites of the *Callavia* Zone, northern Antigonish Highlands, Nova Scotia. *Can. J. Earth Sci.* **17**, 400–418.

Matthews, S. C. 1973. Lapworthellids from the

Lower Cambrian *Strenuella* Limestone at Comley, Shropshire. *Palaeontology* **16**, 139–148.

——— and Missarzhevsky, V. V. 1975. Small shelly fossils of late Precambrian and early Cambrian age: a review of recent works. *J. Geol. Soc.* **131**, 289–304.

Missarzhevsky, V. V. and Grigoreva, N. V. 1981. Novye predstaviteli otryada Tommotiida. *Paleont. Zh.* **1981** (4), 91–97.

Pettibone, M. H. 1953. *Some Scale-Bearing Polychaetes of Puget Sound and Adjacent Waters.* Univ. of Washington Press, Seattle.

Pflugfelder, O. 1933. Zur Histologie der Elytren der Aphroditiden. *Z. Wiss. Zool.* **143**(4), 497–537.

Pope, J. K. 1975. Evidence for relating the Lepidocoleidae, machaeridian echinoderms, to the mitrate carpoids. *Bull. Am. Paleont.* **67**, 385–406.

Poulsen, V. 1971. Notes on an Ordovician acrotretacean brachiopod from the Oslo region. *Bull. Geol. Soc. Denmark* **20**, 265–278.

Prokop, R. 1965. Třída Machaeridia. *In* Špinar, Z. (ed.). *Systematická Paleontologie Bezobratlých.* Academia, Praha, pp. 997–999.

Rigby, J. K. 1978. Porifera of the Middle Cambrian Wheeler Shale from the Wheeler Amphitheater, House Range in western Utah. *J. Paleont.* **52**, 1325–1345.

Rolfe, W. D. I. 1981. *Septemchiton*—a misnomer. *J. Paleont.* **55**, 675–677.

Runnegar, B. 1982. Oxygen requirements, biology, and phylogenetic significance of the Late Precambrian worm *Dickinsonia,* and the evolution of the burrowing habit. *Alcheringa* **6**, 223–239.

———, Pojeta, J., Jr., Taylor, M. E., and Collins, D. 1979. New species of the Cambrian and Ordovician chitons *Matthevia* and *Chelodes* from Wisconsin and Queensland: evidence for the early history of polyplacophoran molluscs. *J. Paleont.* **53**, 1374–1394.

Salvini-Plawen, L. von, 1982. A paedomorphic origin of the oligomerous animals? *Zool. Scripta* **11**, 77–81.

Schallreuter, R. 1985. Mikrofossilien aus Geschieben IV. Machaeridier. *Geschiebesammler* **18** (4), 157–171.

Schrenk, E. 1978. Machaeridia aus silurischen Geschieben. *Geschiebesammler* **11**(4), 5–22.

Sdzuy, K. 1969. Unter- und mittelkambrische Porifera (Chancelloriida und Hexactinellida). *Paläont. Z.* **43**, 115–147.

Van Name, W. G. 1925. The supposed Paleozoic barnacle *Protobalanus* and its bearing on the origin and phylogeny of the barnacles. *Am. Mus. Novitates* **197**, 1–8.

———. 1926. A new specimen of *Protobalanus,* supposed Paleozoic barnacle. *Am. Mus. Novitates* **227**, 1–6.

Whittington, H. B. 1978. The lobopod animal *Aysheaia pedunculata* Walcott, Middle Cambrian, Burgess Shale, British Columbia. *Phil. Trans. R. Soc. Lond. (B.)* **284**, 165–197.

Withers, T. H. 1926. *Catalogue of the Machaeridia.* British Museum (Natural History), London.

———. 1933. The machaeridian *Lepidocoleus ketleyanus* (Reed ex Salter MS). *Ann. Mag. Nat. Hist.* **61** (ser. 10–11), 162–163.

Wolburg, J. 1938. Beitrag zum Problem der Machaeridia. *Paläont. Z.* **20**, 289–298.

Yuan, K. and Zhang, S. 1983. Discovery of the *Tommotia* fauna in SW China. *Acta Palaeont. Sinica* **22**(1), 31–41.

11. THE PHYLUM CONULARIIDA

LOREN E. BABCOCK AND RODNEY M. FELDMANN

Conulariids are a group of extinct metazoans, possessing a four-sided, bilaterally symmetric, pyramidal exoskeleton (Fig. 1A). Because they have no recent analogs and because they superficially resemble representatives of several major animal groups, they have been referred to, or associated with, "Vermes," Cnidaria, Mollusca, Hemichordata, and Conodonta. They have also been placed within a separate phylum, Conulariida, along with a variety of unrelated Paleozoic conical or tubular fossils. Attempts at a satisfactory assignment of conulariids to a phylum have been hindered by four factors: (1) inadequate knowledge of hard-part morphology, (2) lack of documented information on soft-part morphology, (3) lack of consensus on what constitutes a conulariid, and (4) lack of consensus on morphologic terminology. These interrelated problems have conspired to make conulariids a poorly understood group at all taxonomic levels. Herein we define the hard-part morphology of conulariids, present what information is available relative to their soft-part anatomy, define appropriate morphologic terminology, and propose placement of conulariids, as here restricted, in an independent phylum, Conulariida.

The genus *Conularia* was proposed by Miller (in Sowerby 1821) to embrace two species, *C. quadrisulcata* and *C. teres,* from the Carboniferous and Silurian of Great Britain. *Conularia quadrisulcata* was designated the type species of the genus. *Conularia teres* was subsequently re-identified as an orthocone nautiloid (Barrande 1867).

The early interpretations fixed the placement of the conulariids in the phylum Mollusca until early in the twentieth century. Most influential textbooks, including Grabau and Shimer (1910) and Eastman (1913), considered them to be pteropods. Hall (1847), Dana (1849), and others considered conulariids to be cephalopods, although Hall subsequently (1859, 1876, 1879) adopted the view that they were pteropods. Pelseneer (1889) regarded conulariids as opisthobranch gastropods but not as pteropods.

Although it was early recognized that conulariids have a unique morphology, Neumayr (1879) noted that hypotheses on the systematic position of conulariids usually were based on inadequate analyses of morphology. Citing a four-sided pyramidal shape and transverse raised markings, he

erected the group (= order?) Conulariden within the Mollusca as coordinate with the Pteropoda. Miller and Gurley (1896) erected the molluscan order Conularida to embrace flexible conical and pyramidal shells of Paleozoic pelagic animals, which were composed of calcium phosphate, and which were either smooth or with longitudinal or transverse markings. This definition, used by most later workers with only subtle modifications, is so broad that several unrelated groups have been associated with conulariids. Among these are hyoliths, which are probably mollusks, *Styliolina* (see Chapter 5 by Yochelson and Lindemann, this volume) *Tentaculites, Coleolus, Sphenothallus,* and a medusoid cnidarian *Conchopeltis.* The inclusion of these last two nonconulariid genera has resulted in considerable misinterpretation of the natural history of the conulariids.

Moore et al. (1952) identified elongate conoidal hard-part structures as usually diagnostic of worms. They raised the Conularida to phylum level. Respelled Conulariida, the group included conulariids, *Sphenothallus,* tentaculitids, and styliolinids. That conulariids were worms had been suggested previously by Ruedemann (1896a,b) and Weller (1925).

The current idea that conulariids are medusoid cnidarians was independently proposed by Kiderlen (1937) and Knight (1937). The cnidarian affinities of conulariids were supported by Moore and Harrington (1956a,b), Sysoev and Chudinov (1962), Tasch (1973), and Lehmann and Hillmer (1983). Werner (1966, 1967, 1969) suggested that the living *Stephanoscyphus* might represent a primitive scyphozoan link with conulariids. Other recent suggestions that conulariids are conodont-supporting organisms (Bischoff 1978) or that they are protochordates (Termier and Termier 1949, 1953) seem to have gathered little support. Sinclair (1948a), and more recently, Kozłowski (1968), Clarkson (1979), Fedonkin (1983), Oliver (1984), and Babcock and Feldmann (1984) have suggested that conulariids are not appropriately assignable to any described phylum.

SYSTEMATIC PLACEMENT

Conulariids pose two systematic and taxonomic problems. The first relates to the range of morphologic variation of organisms within a single taxonomic unit. The second relates to the placement of the conulariids within the Metazoa.

When organisms such as the conulariids are classified, there are three possible outcomes. They may be assignable to a previously defined taxon because they possess the "typical" morphology of that group. Another possibility is that they may not conform in all regards to any previously de-

fined taxon, but they may closely resemble a certain group in many respects. In this case, it may be appropriate to redefine and expand the existing taxon to embrace the forms being studied. Finally, they may differ from all known organisms to such an extreme extent that their inclusion in an existing taxon would not express a probable biologic relationship. In this case, a new taxon

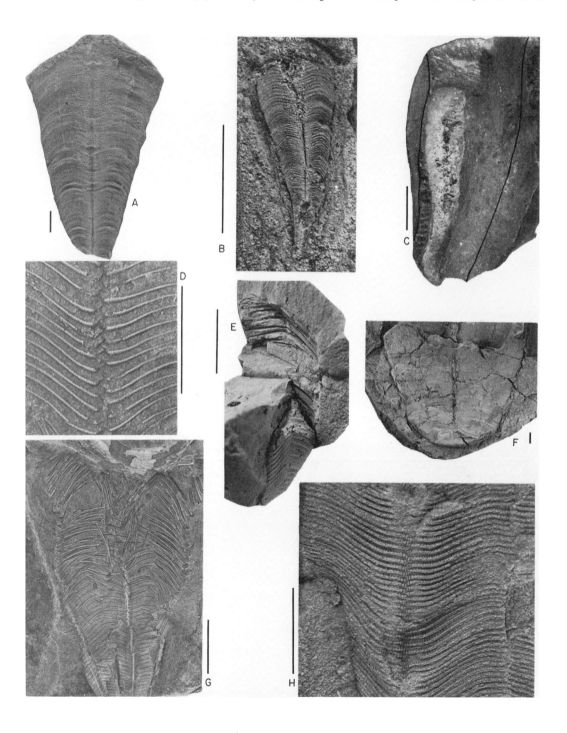

should be erected. It is this latter approach that we feel most appropriately conveys the uniqueness of the conulariids.

A heterogeneous assemblage of organisms producing tubular hard parts has, at one time or another, been brought together as "conulariids." The result of this amalgamation has been that the definition of the group has been broadened to the extent that it can no longer be considered biologically unified. Many of these organisms have recently been reassigned to more appropriate taxa; the hyoliths to the Mollusca (Yochelson 1964; Marek and Yochelson 1976), *Sphenothallus* to "Vermes" (Mason and Yochelson 1985), and *Conchopeltis* to the Cnidaria (Oliver 1984). Tentaculitids, styliolinids, *Coleolus,* and most species referable to such genera as *Conulariina* and *Metaconularia,* are so vastly different from conulariids that, although their affinities remain unknown, the groups must be separated. Removal of these organisms from association with the conulariids permits a redefinition that results in an internally consistent group of organisms that are separate and distinct from other such groups at the phylum level.

As restricted herein, the phylum Conulariida Miller and Gurley 1896 comprises organisms with a four-sided, steeply pyramidal, bilaterally symmetric exoskeleton composed of a multilayered, flexible integument of calcium phosphatic material. The skeleton is supported by arching, often spined rods that traverse the four sides and that are articulated along grooves at each corner. The apex is complete in young individuals, but in adults the convergent end of the pyramid may be truncated by a convex apical wall. The aperture is simple and open. Internal hard-part structures are unknown.

MORPHOLOGY

Morphologic terms

The literature on conulariid morphology includes the important review papers of Slater (1907), Kiderlen (1937), Richardson (1942), Sinclair (1948a), and Moore and Harrington (1956a,b). Nevertheless, misunderstanding of conulariid morphology arises from terms that are improperly defined, or undefined, terms that are ambiguous, and terms that imply systematic affinities. We propose a set of morphologic terms to denote unique morphologic features of conulariids but not to connote our views on their systematic placement.

Adapertural spine: Long spine projecting from near the adapertural side of a rod, in the direction of the aperture.
Adapical spine: Short spine projecting from near the internal adapical side of a rod, in the direction of the apex.
Angulated circular curve: Style of rod articulation in which two abutting rods on a face form a broad arcuate, adapically concave ridge, interrupted by a slight adapertural point at the midline, and by gentle adapertural turns in the vicinity of the facial margins.
Apertural constriction: Exoskeletal constriction located nearest the aperture.
Apertural termination: Rounded or bluntly subtriangular extension of exoskeleton on each face at widest end of exoskeleton.

←——————————————————————————————

Fig. 1. Scale bars = 5 mm. A. *Conularia multicostata* Meek and Worthen. Corner view, showing general shape and well-pronounced exoskeletal constrictions. Cuyahoga Formation, Early Mississippian, Sciotoville, Ohio. USNM 50128. B. *Conularia desiderata* Hall. Juvenile specimen preserving the apex. Note the exoskeletal constrictions. A small, bulbous feature appears immediately below the apex; this is produced by marks made in matrix in the process of preparing the specimen. Skaneateles Formation, Pompey Shale Member, Hamilton, New York. USNM 395827. C. *Paraconularia subulata* (Hall). View of interior showing remains of presumed soft-part structures. Black ink line denotes margin of the exoskeleton. The specimen is not coated with ammonium chloride. Borden Formation, Nancy Member, Early Mississippian, Daniel Boone National Forest, Bath County, Kentucky. USNM 395828. D. *Paraconularia subulata* (Hall). Detailed view of rod articulation along midline. Cuyahoga Formation, Orangeville Member, "Sunbury" Submember, Early Mississippian, South Chagrin Reservation, Bentleyville, Ohio. USNM 395829. E. *Paraconularia byblis* (White). Specimen showing flaps resembling "apertural flaps" at base of specimen (not visible in this view) and about midway up the specimen. Position of the aperture is indicated in the counterpart, located in the upper half of the photograph. Borden Formation, Nancy Member, Early Mississippian, Daniel Boone National Forest, Bath County, Kentucky. USNM 395830. F. *Conularia multicostata* Meek and Worthen. Apical region, showing apical wall. Cuyahoga Formation, Wooster Member?, Loudonville, Ohio. USNM 50647. G. *Paraconularia subulata* (Hall). Specimen showing rods embedded in integument in apical region and rods lacking an integumental matrix in apertural region. Note that adjacent rods on a face act as independent units. A portion of an orbiculoid brachiopod that has attached to the apertural end of the conulariid is visible at the top of the photograph. Cuyahoga Formation, Orangeville Member, "Sunbury" Submember, Early Mississippian, South Chagrin Reservation, Bentleyville, Ohio. USNM 395831. H. *Conularia desiderata* Hall. View from midline of specimen having integument draped closely over the framework of rods and spines. Note articulation of rods at the corner groove and the prominent adapertural spines. Ludlowville Formation, Wanakah Shale Member, "*Nautilus* bed," Middle Devonian, Lake Erie shore, Wanakah, New York. USNM 395832.

Aperture: Opening at widest end of exoskeleton.

Apex (biologic apex): Narrowest termination of exoskeleton, where the four faces join at a closed point. Compare with *hypothetical apex*.

Apical angle: Hypothetical angle formed by one face of the exoskeleton; measured at the intersection of two lines each identified by tracing positions on the exoskeleton tangential to the facial margins and defining the maximum angle of separation. See *major apical angle* and *minor apical angle*.

Apical wall: Broadly rounded, adapically convex, portion of integument lacking rods that completely covers the apical end of the exoskeleton when the apex itself is missing.

Central cavity: Region located internal to the four faces of the exoskeleton.

Corner angle: Longitudinal line in the marginal region of a face connecting points of greatest inflection of the rods.

Corner groove: Longitudinal invagination of exoskeleton connecting points where pairs of rods from adjacent faces cross near the marginal terminations of those rods.

Exoskeletal constriction: Depression, restricted in the longitudinal direction, traceable on all four faces of the exoskeleton in the same relative position. Compare with *apertural constriction*.

Exoskeleton (skeleton): Four-sided pyramidal structure, open at the widest end and closed at the narrowest end, comprising rods joined by integument.

Face: One of four sides of the exoskeleton crossed by ridges; it is delimited by the aperture, by the apex or the apertural wall, and by two corner grooves. See *major face* and *minor face*.

Gothic arch: Style of rod articulation in which two adjacent rods on a face form ridges that meet at an obtuse, adapically concave angle at the midline and proceed away from the midline along lines subtly curved adapically.

Hypothetical apex: Point in space where two lines, traced along the mean direction of the corner angles, meet; the hypothetical apex may or may not coincide with the position of the (biologic) apex.

Inflected circular curve: Style of rod articulation in which two adjacent rods on a face form a broadly arcuate, adapically concave ridge except in the vicinity of the facial margins, where they turn gently adaperturally.

Inflected gothic arch: Style of rod articulation in which two adjacent rods on a face form ridges that meet at an obtuse, adapically concave angle at the midline and proceed away from the midline along lines subtly curved adapically except in the vicinity of the facial margins, where they turn gently adaperturally.

Integument: Multilayered, presumably flexible structure composed of calcium phosphate and protein, within which rods and spines were embedded and held in position.

Interridge area: Roughly transverse band of integument located between two facial ridges.

Interridge crest: Raised area, usually a linear ridge, located in an interridge area and positioned at a right angle to a ridge; formed by integument covering an adapertural or adapical spine.

Interridge furrow: Low area, usually linear, located in an interridge area, and between two interridge crests.

Interrod area: Open region located between two rods; exposed only when integument is absent.

Major apical angle: Apical angle subtended by a major face.

Major face: Wider of two adjacent faces.

Margin (facial margin): Longitudinal edge of a face, or a line connecting points where two faces meet in a corner groove.

Midline: Longitudinal line connecting points where either two adjacent rods on a face meet, or central to the facial terminations of each pair of adjacent rods if the rods do not meet. The midline can be expressed as either a thin groove or a raised line if the integument is preserved. The midline is pigmented in some specimens.

Minor apical angle: Apical angle subtended by a minor face.

Minor face: Narrower of two adjacent faces.

Node: Minute, subcircular, raised surface on a rod or ridge.

Ridge: Raised line crossing a face from a corner groove to the midline area, and formed by integument covering a rod.

Rod: Narrow, elongate structure that is subcircular in cross section, composed of calcium phosphate, and embedded within the integument; it is thickened near the marginal termination, and tapers very gradually to a blunt point at the facial termination.

Rod angle: Angle subtended by a line connecting the two most distant points of a rod along a longitudinal line and a line constructed perpendicular to the facial margin at the point where that ridge intersects the corner angle.

Spine: Solid, narrow, short or elongate structure, projecting from, and whose axis is at a right angle to, a rod; tapers gradually to a sharp point distally. See *adapical spine* and *adapertural spine*.

Exoskeleton

When preserved in three dimensions, the exoskeleton of a conulariid has a four-sided elongate pyramidal shape. The profile may be modified by the development of one or more exoskeletal constrictions (Fig. 1A). The exoskeleton is generally < 10 cm in length, but in a few species, may attain a length > 20 cm (Fletcher 1938; Lamont 1946). The conulariid exoskeleton expands gradually and uniformly from a closed apex to an open apertural end (Fig. 1B). Each of the four faces of the exoskeleton is crossed transversely by thin ridges, and each of the four corners is invaginated by a longitudinal groove. An indistinct line runs longitudinally down the middle of each face (Fig. 2E–K).

The thin walls of the exoskeleton are made up of a multilayered calcium phosphate and protein integument (Fig. 3H). The precise number of layers and the extent to which this number is consistent from species to species has yet to be determined. Rods are support structures that cross each face transversely; they are composed of calcium phosphate and are subcircular in cross section. An individual rod articulates proximally with an adjacent rod in a corner groove; thus, rods on adjacent faces alternate in position at the corner groove (Fig. 1D, H). The points of articulation of the rods may be nodose swellings. Each rod crosses one-half of each face transversely until its distal end abuts with, or crosses, the distal end of an adjacent rod (Fig. 3A, C).

A rod may be equipped with numerous spines that project adaperturally or adapically (Figs. 1H and 3B, D, F, G). Adapertural spines are longer than those that project adapically. The function of a spine seems to have been to interlock between two spines of an immediately adjacent rod (Fig. 3D), thus providing a stronger framework to support the integument. Most rods have spines, but some conulariids have integument supported by rods alone (Fig. 2H). Rods may be ornamented on the external side of the exoskeleton by numerous minute nodes, arranged in a single row along a rod (Fig. 3B,C). These structures are unrelated to the spines.

The manner of rod articulation has long been used as a diagnostic character at the species level (Hall 1859; Barrande 1867; Holm 1893). Herein, we recognize four modes of rod articulation that produce patterns of ridges useful as species-level taxonomic criteria (Fig. 2A–J). Stylized examples of each pattern are presented in Fig. 2A–D. As we define them, modes of rod articulation are purely for descriptive purposes and are not intended to imply phyletic relationships. Indeed, the manner of rod articulation may change through the ontogeny of a single individual (Fig. 2E, K). Rod angles typically decrease with the addition of new exoskeletal material.

A complex of structures are produced on the external surface of the exoskeleton when integument is draped over a framework of rods and spines. The majority of species of Devonian and Mississippian conulariids of North America are based largely or solely upon the extent to which the external surface of the exoskeleton mirrors the nature of the framework.

A difference in the acute apical angle between adjacent sides of a conulariid, mostly attributed to compression, has been noted by numerous authors, including Barrande (1867), Hall (1879), Slater (1907), Bouček (1939), and Sinclair (1948a). Our studies of compressed and presumably uncompressed materials indicate that opposite sides of a conulariid exoskeleton are paired. In cross section, a conulariid is rectangular. Each face subtends an apical angle equal to that of the face opposite it, but different from either adjacent face (Fig. 4A, B). This suggests that conulariids are bilaterally symmetric, rather than tetramerally symmetric, metazoans. Rhomboid shaped conulariids may exist, but most thought to have a rhomboidal cross section probably were described from subtly compressed specimens.

The apertural region of conulariids has been the subject of much speculation. "Flaps" or "lappets," partially or wholly closing the apertural region, were first described by Miller (in Sowerby 1821), and subsequently by Richter and Richter (1930), Kowalski (1935), Kiderlen (1937), Sinclair (1948a,b), Moore and Harrington (1956b), and Branisa (1965). Kiderlen (1937) and Moore and Harrington (1956b) proposed elaborate mechanisms for closure of the apertural region involving the infolding of a flexible exoskeleton. They assumed the line of flexure to be a straight line normal to the midline. Moore and Harrington (1956b) suggested that, in order for a conulariid to have been so flexible in the apertural region, the line of flexure at the base of each "apertural flap" was chitinophosphatic, while the remainder of the exoskeleton was phosphatic.

Specimens exhibiting closed or partially restricted apertures are common. In > 70 specimens with well-preserved "apertural flaps," none were used to close the tube. There is no evidence of a line of flexure. Typically, the line along which a flap is developed is not straight (Figs. 1E, and 2E, F, K). Instead, the line bends in a pattern that emulates the style of rod articulation. No two adjacent faces on the same specimen necessarily fold inward at the same position (Fig. 2F, K). Within the same species, there is no consistency from individual to individual either in the placement of a line of flexure or in the mode of closure, as defined by Kiderlen (1937) and Moore and Harrington (1956b). Furthermore, in the specimen shown in Fig. 1E, an apertural termination is present at the top of the figure; about midway down from the top occur two additional infolded faces. At the bottom of the figure, not observable in this view, another face is infolded into a "flap." If these flaps were to occur at the aperture, they would clearly be regarded as "apertural flaps"; however, their presence at various places on the exoskeleton indicates that they are taphonomic phenomena. Richter and Richter (1930), noted the extreme flexibility of Conularia tulipa. In their opinion, a

conulariid exoskeleton was flexible enough to
have collapsed under its own weight. We suggest
that the infoldings of exoskeleton commonly
found in the apertural region, and less commonly
elsewhere on a conulariid, are taphonomic structures resulting from collapse of the exoskeleton
after death. The exoskeleton may not be as weak
as suggested by Richter and Richter, but it is cer-

tainly not as rigid, for example, as a molluscan
shell.

The apex of a conulariid has been variously interpreted in the literature. Since the work of Kiderlen (1937), however, conulariids have been
thought of as metazoans having a sharp point in
the juvenile state. The point was presumably attached by an attachment disc to a hard substra-

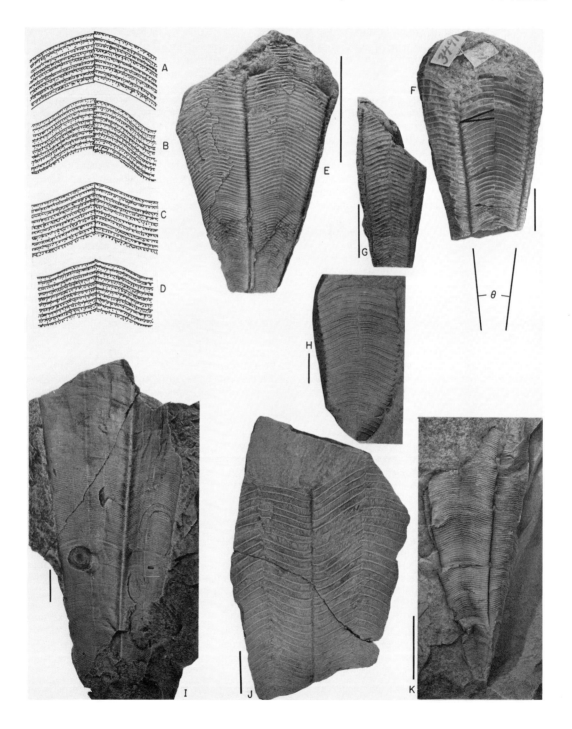

tum. Support for Kiderlen's argument was provided by supposed conulariids, described by Ruedemann (1896a,b), but which are here considered tubes of *Sphenothallus.* Pointed apices of conulariids are exceedingly rare. In > 1200 specimens examined, only 2 had complete, closed, blunt points (Fig. 1B).

Conulariids with their presumed attachment discs have never been demonstrated. Small, round, black, and presumably chitinous bodies attached to bryozoans or brachiopods may be misidentified as conulariid attachment discs. Often, the presumed base of a tube connected to such a structure is preserved. Invariably, such tubes are circular or subcircular in cross section. Sinclair (1948a) indicated that these problematic fossils resemble basal attachment discs of *Sphenothallus,* such as those described by Ruedemann (1896a,b).

Conulariids having rounded apical ends have been illustrated many times (Hall 1876, 1879; Barrande 1867; Kiderlen 1937; Moore and Harrington 1956a,b). We did not observe any specimens with rounded apices. All illustrations of specimens showing such structures are drawings, and, although we have not examined Barrande's or Kiderlen's materials, our studies of conulariid integument and rods lead us to think that rounded apices are conjectural. Alternatively, a rounded apical end may have been a stylized way of illustrating a compressed or expanded apical wall.

Smooth, imperforate apical walls have been noted by many authors (Miller, in Sowerby 1821; Hall 1876, 1879; Slater 1907; Richter and Richter 1930; Sinclair 1948a; Moore and Harrington 1956b; Babcock and Feldmann 1984). These structures have been termed septa, apical diaphragms, or Schotten. None of these terms seems appropriate, either because of the genetic impli

cations or because of an incomplete description of the morphology. Apical walls are single units of integument that cover the convergent end of a conulariid close to the apical terminus. An apical wall is not located at the apical terminus itself but seems to be attached to the interior of the faces on the exoskeleton slightly adaperturally of this region (Fig. 1F). Apical walls appear to lack support from rods or other structures, and may be slightly bowed adapically. Multiple apical walls have been reported in a single individual (Eichwald 1860; Slater 1907; Sinclair 1948a), but we have not confirmed the observation.

Although the structure is rarely well preserved, an imperforate apical wall seems to be a feature of nonjuvenile conulariids. References to conulariids with centrally perforated apical walls include Slater (1907), Richardson (1942), and Swartz and Richardson (1945). Figure 4A, far left, shows the apical wall of a compressed or collapsed individual. A subcircular structure is located centrally on the apical wall, but it does not penetrate the wall. On the basis of observations of soft-part morphology discussed below, we interpret this subcircular structure as the compression of a thin, flexible apical wall against the apical portion of some soft-part morphologic feature.

The term "septum" has been applied to three separate morphologic features in conulariids: apical walls (e.g., Miller, in Sowerby 1821; Slater 1907), ridges (Slater 1907), and T-shaped structures radiating inward from the walls at their midlines (Wiman 1895). None of these seems to be an appropriate application of the term. Most commonly, the word "septum" as applied to conulariids means large T-shaped structures of the exoskeleton that project inward from the midlines. These supposed hard-part structures were described in *Conularia loculata* by Wiman (1895)

Fig. 2. Scale bars = 10 mm.   A. Gothic arch style of rod articulation, diagrammatic.   B. Inflected circular curve style of rod articulation, diagrammatic.   C. Inflected gothic arch style of rod articulation, diagrammatic. D.   Angulated circular curve style of rod articulation, diagrammatic.   E. *Conularia pyramidalis* Hall. Syntype, showing gothic arch rod articulation in apical region and inflected gothic arch articulation elsewhere. Note also an "apertural flap" on the major face. New Scotland Limestone, Early Devonian, Clarksville, New York. AMNH 33018.   F. *Paraconularia subulata* (Hall), showing inflected circular curve style of rod articulation. Also shown are pen lines denoting an apical angle (bottom of figure), and a rod angle (on the figure). An "apertural flap" is preserved on the ?minor face, shown here to the left on the figure. Cuyahoga Formation, Meadville Member?, Early Mississippian, Richfield, Ohio. NYSM 3491.   G. *Paraconularia planicostata* (Dawson). Holotype, portion of specimen showing inflected gothic arch style of rod articulation in the lower portion of the photograph and gothic arch style in the apertural region. Early Mississippian, Irish Cove, Nova Scotia. RM(MU) 2749.   H. *Paraconularia* cf. *P. subulata* (Hall), showing inflected circular curve style of rod articulation. Note that this specimen lacks both integument and spines. Heath Formation, Bear Gulch Limestone Member, Late Mississippian, Fergus County, Montana. CM 35000.   I. *Conularia undulata* Conrad. External mold showing inflected circular curve style of rod articulation. Subcircular pits on specimen are attachment sites of orbiculoid brachiopods. Skaneateles Formation, Middle Devonian, Cazenovia, New York. AMNH 41093.   J. *Paraconularia* sp., showing inflected gothic arch style of rod articulation. Note how integument is wrinkled about the adapertural spines in the interridge areas, producing well-defined interridge crests and interridge furrows. Wellsville Formation, Late Devonian, Wellsville, New York. CM 35001.   K. *Conularia elegantula* Meek. Holotype, showing inflected gothic arch style of rod articulation in apical half of the specimen and angulated circular curve style of rod articulation in the apertural half. A poorly developed "apertural flap" is exhibited on the major face. Delaware Limestone?, Early Devonian, Delaware, Ohio. AMNH CU 282G.

and were thought by Kiderlen (1937) to be homologous to septa composed of endodermal tissue in living scyphomedusans. These structures have not been observed in any specimens other than those of Wiman, and unfortunately his material was illustrated only by drawings and the specimens are lost (W. A. Oliver, Jr., personal communication). Wiman's observations cannot be replicated, and we suspect he may have been misled by some taphonomic feature.

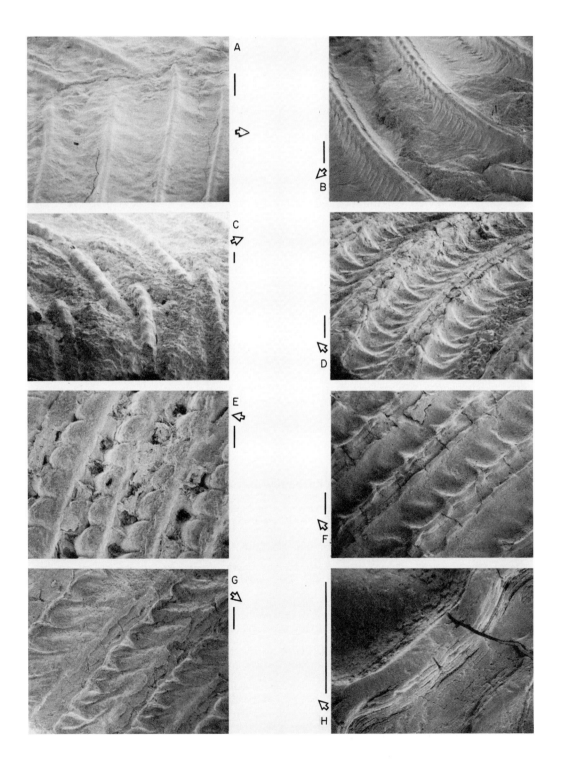

Soft-part morphology

Much speculation has surrounded the study of the soft parts of conulariids. Since the work of Kiderlen (1937), Moore and Harrington (1956a,b), and Werner (1966, 1967, 1969), conulariids have been interpreted as tentacled creatures. Support for such interpretation is weak.

Remains of presumed soft parts from a North American conulariid are here illustrated for the first time (Fig. 1C). At present, our knowledge of soft parts is limited to eight specimens that exhibit contracted masses and vague tubular structures replaced by iron oxide. The tubular structures are always located near one corner groove and extend along most of the length of the exoskeleton. Near the aperture, the tubes turn abruptly and exit the exoskeleton at an angle of nearly 90° to the long axis of the body. The tubular structures, presumably representing remains of organ systems, appear to be reduced in size when compared to expected living organs and show no details of soft-part anatomy. These structures may be "brown bodies" or contracted masses of internal tissues. If so, they may be evidence that such a specimen as shown in Fig. 1C was encysted at the time of death. A more detailed and certain interpretation of the soft parts must await more thorough examination of the material.

OCCURRENCES AND PALEOECOLOGY

Conulariids are known exclusively from marine rocks and are most common in low-diversity faunas. Upon death, the delicate conulariid exoskeleton probably could have withstood very little movement. Conulariids are found in environments of very slow deposition such as North American midcontinental black shales, or in rocks representing very rapid sedimentation such as turbidites or tempestites.

Conulariids have been reported from the Late Precambrian to the Recent (Caster 1957). This range, however, includes a variety of taxa now referable to other groups. Occurrences here considered valid include specimens from Middle Ordovician through, perhaps, Late Triassic.

Fossils referred to as conulariids have been identified from all continents. However, the sole occurrence of a conulariid from Antarctica (Cordini 1955) has been questioned and may represent plant material (Dalziel et al. 1981).

Many species of conulariids seem to have been geographically widespread. For example, the Late Mississippian species, *Paraconularia chesterensis* (Worthen) ranges from Alabama to Nevada and occurs in rocks of various lithologies. The occurrence of this and other conulariid species in vastly dissimilar lithologies and environments of deposition, and their great geographic distribution, suggest that they may have been planktonic. The bilaterally symmetric body plan, however, may suggest that some species at least were weakly nektonic.

Conulariids are usually referred to as solitary animals. Clusters are rare, but a few have been illustrated by Slater (1907, pl. 2, fig. 1), Hall (1876, pl. 28, fig. 1; 1879, pl. 24, fig. 1), Sinclair (1940, pl. 2, fig. 5), and Babcock and Feldmann (1984, p. 17; herein, Fig. 4A,C). In all these, individual specimens are shown radiating from a centroid. Apices are almost invariably pointing inward in pre-

---

Fig. 3. Scanning electron micrographs. All views are of the external surfaces of specimens. Arrows point in the direction of the aperture. Scale bars = 0.1 mm. A. *Paraconularia subulata* (Hall). View along midline, showing rods alternating at their facial terminations. Integument is draped loosely over the rods and spines, producing ridges and subtle interridge crests and interridge furrows. Cuyahoga Formation, Orangeville Member, "Sunbury" Submember, Early Mississippian, South Chagrin Reservation, Bentleyville, Ohio. USNM 395833. B. *Paraconularia byblis* (White). Integument draped closely over rods, showing well-defined ridges, nodes, interridge crests, and interridge furrows. Same specimen as illustrated in Fig. 1E. Borden Formation, Nancy Member, Early Mississippian, Daniel Boone National Forest, Bath County, Kentucky. USNM 395830. C. *Paraconularia subulata* (Hall). View of rods at the midline, integument lacking. Note that the nodes are arranged in a single row on each of the rods. Borden Formation, Nancy Member, Early Mississippian, Daniel Boone National Forest, Bath County, Kentucky. USNM 395834. D. *Conularia desiderata* Hall. View of ridges, interridge crests, and interridge furrows. The rods have been broken away in this specimen. Note that the adapertural and adapical spines, which underlie the interridge crests, interdigitate in the interridge areas. Same specimen as illustrated in Fig. 1H. Ludlowville Formation, Wanakah Shale Member, "*Nautilus* bed," Middle Devonian, Lake Erie shore, Wanakah, New York. USNM 395832. E. *Conularia desiderata* Hall. View of counterpart at approximately the same position as that of the part shown in Fig. 3F. Counterpart specimens frequently preserve, in negative relief, morphologic features that appear to be lacking in the part specimens. The nodes appear in this specimen but seem to be lacking in the part, Fig. 3F. Same specimen as illustrated in Fig. 3D. F. *Conularia desiderata* Hall. View of part at approximately the same position as that of the counterpart shown in Fig. 3E. The interridge crests produced by the adapertural spines appear to be better pronounced than those produced by the adapical spines, implying that the adapertural crests are located closer to the external surface of the exoskeleton. Same specimen as illustrated in Fig. 3D. G. *Conularia desiderata* Hall. View showing ridges with rods removed. The interridge crests have their axes at acute angles to the axes of the ridges, probably produced by a shearing of the spines. Same specimen as illustrated in Fig. 3D. H. *Conularia desiderata* Hall. view of ridge with rod removed, showing a multilayered integument. Same specimen as illustrated in Fig. 3D.

Fig. 4.   Scale bars = 10 mm.   A. *Conularia congregata* Hall. Portion of disaggregated cluster comprising 13 compressed or collapsed individuals. The two nearly complete specimens at the center and at the left preserve apical walls. Three of the specimens in this figure exhibit two longitudinal lines each, located centrally. These are corner grooves; one is preserved in positive relief, the other has been impressed from the opposite side of each individual and appears in negative relief. Orbiculoid brachiopods are attached to the exoskeletons of three conulariids. Ithaca Formation, Late Devonian, Ithaca, New York. NYSM 3483.   B. *Conularia desiderata* Hall. View along corner groove of a specimen preserved in three dimensions. The major face is to the right in this figure, the minor face to the left. NYSM 3487.   C. *Paraconularia chesterensis* (Worthen). Cluster of 22 individuals disposed radially about a centroid. Arrows point to probable plant material near the center of the cluster. Borden Formation, Mississippian, Crawfordsville, Indiana. USNM 50150.

sumed undisturbed aggregations. In portions of a somewhat disaggregated cluster illustrated by Hall (1876, 1879), and later by Babcock and Feldmann (1984; herein, Fig. 4A), apical walls are visible in several specimens. Black carbonaceous fossil remains, possibly attachment stalks or plant material, are preserved among specimens illustrated here in Fig. 4C. It is possible that certain conulariids were pseudoplanktonic and were attached to, or entwined with, planktonic algae.

Whether conulariids were planktonic or nektonic, attached or free swimming, the soft-part organs in the region of the aperture of the exoskeleton probably functioned for filter feeding. There is no evidence for aggressive food-gathering behavior.

Associations in clusters of numerous specimens comprising single species serve to indicate that some, if not all, conulariids were gregarious, at least during some part of the adult phase of the life cycle. No evidence of budding or other asexual reproductive style exists.

The style of growth in conulariids is not known. Growth lines such as those seen on mollusks and brachiopods are not known. Exoskeletal constrictions (Fig. 1A, B), in exactly the same relative positions on all four faces, are possible evidences of incremental growth. It is not clear whether exoskeletal constrictions are lines indicating temporary cessations of growth. It is probable, however, that addition of rods, spines, and integument occurred at the apertural termination. When the internal cavity of an individual reached a certain volume, an apical wall was probably secreted adaperturally of the apical terminus. Sometime thereafter, exoskeletal material may have been shed from the apical region. However, as suggested above, conulariids may have had the capacity to encyst at one or more times during their life cycle, in order to survive temporarily inhospitable environmental conditions.

Epibionts on conulariids include orbiculoid brachiopods (Hall 1876, 1879; Moore and Harrington 1956a; Babcock and Feldmann 1984), encrusting bryozoans (Finks 1955), and edrioasteroid echinoderms (Barrande 1867; Moore and Harrington 1956a).

In the portion of a conulariid cluster illustrated in Fig. 4A, six orbiculoids are attached to the exoskeletons of three individuals of *Conularia congregata* Hall. Each orbiculoid is preserved with the "dorsal" valve oriented in the same direction. Moreover, the orbiculoids are uniformly attached only to two conulariid faces and only in the vicinity of the conulariid aperture. The apertural regions of these conulariids are partially dismembered. While this may be coincidental, it could indicate that attachment of the orbiculoids to this substratum was preferential. Upon falling to the seafloor, the flexible conulariid exoskeletons probably underwent some collapse, and perhaps, partial burial. Orbiculoids subsequently attached to the portions of the exoskeletons that remained above the sediment–water interface. Finally, those portions of conulariid integument, such as the apertures, which were exposed above the substratum, underwent some disintegration and dismemberment.

CONCLUSIONS

Conulariids are a group of Paleozoic to early Mesozoic marine metazoans with a unique exoskeleton consisting of a framework of articulated rods supporting a semiflexible integument of calcium phosphate and forming a four-sided, bilaterally symmetric, open pyramid. Conulariids are so strikingly different from all other taxa that they can be considered as an independent extinct phylum. Thus defined, a variety of other tubular remains previously considered conulariids can better be assigned to the Mollusca, the Cnidaria, the "Vermes," or other groups.

Although knowledge of the soft-part anatomy of conulariids is limited, there is evidence, in the preservation of clusters of individuals, of a gregarious, solitary life-style. This occurrence, coupled with a lack of evidence for asexual reproduction or colonial development, implies a sexual mode of reproduction. No mechanisms for aggressive food gathering are known. Preservation of conulariids in a variety of sedimentary rock types, typically associated with few benthic invertebrate remains, suggests that they may have been planktonic or pseudoplanktonic during most of their life cycle.

Acknowledgments

Ellis L. Yochelson suggested this study, offered many helpful discussions, and reviewed this paper. Roger J. Cuffey, Alan Stanley Horowitz, Charles E. Mason, and Jon Mortin discussed with us conulariid biology and provided reference materials. Mitchell J. Ciccarone and Gerald J. Kloc collected some specimens shown herein. For the loan of museum specimens, we thank Ingrid Birker, John L. Carter, Frederick J. Collier, Niles Eldredge, and Ed Landing. William A. Oliver, Jr., reviewed this manuscript. This work was supported in part by American Association of Petroleum Geologists Grant-in-Aid No. 582-12-01 and by a Grant-in-Aid of Research from Sigma Xi, The Scientific Research Society. Contribution 296, Department of Geology, Kent State University, Kent, Ohio.

ABBREVIATIONS

Specimens are listed by catalog numbers with the repository names abbreviated as follows: AMNH, American Museum of Natural History, New York, New York; AMNH CU, American Museum of Natural History, Columbia University Collection, New York, New York; CM, Carnegie Museum, Pittsburgh, Pennsylvania; NYSM, New York State Museum and Science Service, Albany, New York; RM (MU), Redpath Museum (McGill University), Montreal, Quebec; USNM, United States National Museum of Natural History, Washington, D.C.

REFERENCES

Babcock, L. E. and Feldmann, R. M. 1984. Mysterious fossils. *Earth Sci.* **37**(3), 16–17.

Barrande, J. 1867. *Système Silurien du Centre de la Bohème. Ière Partie: Recherches Paléontologiques,* vol. 3. *Classe des Mollusques, Ordre des Pteropodes.* Prague and Paris, 179 pp., 16 pl.

Bischoff, G. C. O. 1978. Internal structures of conulariid tests and their functional significance, with special reference to Circonulariina n. suborder (Cnidaria, Scyphozoa). *Senckenb. Leth.* **59**, 275–327.

Bouček, B. 1939. Conularida. *In* Schindewolf, O. H. (ed.). *Handbuch der Paläozoologie,* vol. 2A, pp. A113–A131.

Branisa, L. 1965. Los Fosiles Guias de Bolivia. I. Paleozoico. *Serv. Geol. Boliv. Bol.* **6,** 282 pp., 78 pl.

Caster, K. E. 1957. Problematica. *In* Ladd, H. S. (ed.). *Treatise on Marine Ecology and Paleoecology,* vol. 2. *Geol. Soc. Am. Mem.* **67,** 1025–1032.

Clarkson, E. N. K. 1979. *Invertebrate Palaeontology and Evolution.* George Allen and Unwin, London. 323 pp.

Cordini, I. R. 1955. Contribucion al conocimiento del Sector Antarctico Argentino. *Publ. Inst. Antart. Argent.* **1,** 273–277.

Dalziel, I. W. D., Elliot, D. H., Jones, D. I., Thompson, J. W., Thompson, M. R. A., Wells, N. A., and Zinsmeister, W. J. 1981. The geological significance of some Triassic microfossils from the South Orkney Islands, Scotia Ridge. *Geol. Mag.* **118**, 15–25.

Dana, J. D. 1849. *Geology. United States Exploring Expedition During the Years 1838, 1839, 1840, 1841, 1842. Under the Command of Charles Wilkes, U.S.N.,* vol. 10. C. Sherman, Philadelphia, 756 pp.

Eastman, C. R. 1913. *Text-book of Paleontology* (adapted from the German of Karl A. von Zittel), vol. 1. 2nd ed. MacMillan and Co., London, 839 pp.

Eichwald, C. E. 1860. *Lethaea Rossica, ou Paléontologie de la Russie,* vol. 1, *L'Ancienne Période,* part 2. Stuttgart, pp. 681–1657.

Fedonkin, M. A. 1983. Organic world of the Vendian. Stratigraphy and paleontology. *Itogi Nauki Tekhniki. VINITI AN SSSR* **12**, 127 pp. [in Russian].

Finks, R. M. 1955. *Conularia* in a sponge from the west Texas Permian. *J. Paleont.* **29**, 831–836.

Fletcher, H. O. 1938. A revision of the Australian Conulariae. *Aust. Mus. Rec.* **20**, 235–255.

Grabau, A. W. and Shimer, H. W. 1910. *North American Index Fossils,* vol. 2. A. G. Seiler and Co., New York. 909 pp.

Hall, J. 1847. *Palaeontology of New York,* vol. 1, *Containing Descriptions of the Organic Remains of the Lower Division of the New York System (Equivalent of the Lower Silurian Rocks of Europe).* C. Van Benthuysen, Albany, N.Y., 338 pp., 99 pl.

———. 1859. *Geological Survey of New York. Palaeontology,* vol. 3, *Containing Descriptions and Figures of the Organic Remains of the Lower Helderberg Group and the Oriskany Sandstone.* C. Van Benthuysen, Albany, N.Y., 532 pp.

———. 1876. *Geological Survey of the State of New York. Palaeontology. Illustrations of Devonian Fossils: Gastropoda, Pteropoda, Cephalopoda, Crustacea and Corals of the Upper Helderberg, Hamilton and Chemung Groups.* Weed, Parsons and Co., Albany N.Y., 7 pp., 74 pl.

———. 1879. *Geological Survey of the State of New York. Palaeontology,* vol. 5, part 2, *Containing Descriptions of the Gastropoda, Pteropoda, and Cephalopoda of the Upper Helderberg, Hamilton, Portage, and Chemung Groups.* Charles Van Benthuysen and Sons, Albany, N.Y., 492 pp., 120 pl.

Holm, G. 1893. *Sveriges Kambrisk-Siluriska Hyolithidae och Conulariidae. Sveriges Geologiska Undersokning Afhandlingar och uppsatser,* series C, no. 112. 172 pp., 6 pl.

Kiderlen, H. 1937. Die Conularien. Über Bau und Leben der ersten Scyphozoa. *N. Jb. Mineral.* **77**, 113–169.

Knight, J. B. 1937. *Conchopeltis* Walcott, an Ordovician genus of the Conulariida. *J. Paleont.* **11**, 186–188.

Kowalski, J. 1935. Les Conulaires. Quelques observations sur leur structure anatomique. *Bull. Soc. Sci. Natur. de l'Ouest Fr.* **5** (ser. 5), 281–293.

Kozłowski, R. 1968. Nouvelles observations sur les Conulaires. *Acta Palaeont. Polon.* **13**, 497–535.

Lamont, A. 1946. Largest British *Conularia. Quarry Manager's J.* **29**, 569–570.

Lehmann, U. and Hillmer, G. 1983. *Fossil Invertebrates.* Cambridge Univ. Press, Cambridge, 350 pp.

Marek, L. and Yochelson, E. L. 1976. Aspects of the biology of Hyolitha (Mollusca). *Lethaia* **9**, 65–82.

Mason, C. and Yochelson, E. L. 1985. Some tubular fossils (*Sphenothallus:* "Vermes") from the Middle and Late Paleozoic of the United States. *J. Paleont.* **59**, 85–95.

Miller, S. A. and Gurley, W. F. E. 1896. New species of Palaeozoic invertebrates from Illinois and other states. *Ill. State Mus. Nat. Hist. Bull.* **11**, 50 pp.

Moore, R. C. and Harrington, H. J. 1956a. Scyphozoa. *In* Moore, R. C. (ed.). *Treatise on Invertebrate Paleontology,* part F, *Coelenterata.* Geological Society of America and Univ. of Kansas Press, Lawrence, pp. F27–F38.

———— and ————. 1956b. Conulata. *In* Moore, R. C. (ed.). *Treatise on Invertebrate Paleontology,* part F, *Coelenterata.* Geological Society of America and Univ. of Kansas Press, Lawrence, pp. F54–F66.

————, Lalicker, C. G., and Fischer, A. G. 1952. *Invertebrate Fossils.* McGraw-Hill, New York, 766 pp.

Neumayr, M. 1879. Zur Kenntnis der Fauna des untersten Lias in den Nordalpen. *Kaiserlich-königlichen Geol. Reichsanstalt* **7**(5), 46 pp.

Oliver, W. A., Jr. 1984. *Conchopeltis:* its affinities and significance. *Palaeontogr. Am.* **54**, 141–147.

Pelseneer, P. 1889. Sur un nouveau *Conularia* du Carbonifère et sur les prétendus "Pteropodes" primaires. *Soc. Belg. Geol. Paleont. Hydrolog. Mem.* **3**, 124–136.

Richardson, E. S., Jr. 1942. A Middle Ordovician and some Lower Devonian conularids, with two orthoceratids, from central Pennsylvania. Unpubl. M.S. thesis, Pennsylvania State Univ., State College.

Richter, R. and Richter, E. 1930. Bemerkenswert erhaltene Conularie und ihre Gattungsgenossen im Hunsrückschiefer (Unterdevon) des Rheinlandes. *Senckenb. Leth.* **12**, 152–171.

Ruedemann, R. 1896a. Note on the discovery of a sessile *Conularia,* art. 1. *Am. Geol.* **17**, 158–165.

————. 1896b. Note on the discovery of a sessile *Conularia,* art. 2. *Am. Geol.* **18**, 65–71.

Sinclair, G. W. 1940. A discussion of the genus *Metaconularia* with descriptions of new species. *R. Soc. Canada, Sect. IV, Trans., Ser. 3* **34**, 101–121, pl. 1–3.

————. 1948a. The biology of the Conularida.

Unpubl. Ph.D. Dissert., McGill Univ., Montreal.

————. 1948b. Aperture of *Conularia. Bull. Geol. Soc. Am.* **59**, 1352.

Slater, I. L. 1907. *A Monograph of British Conulariae.* Palaeontographical Society, London, 41 pp.

Sowerby, J. 1821. *The Mineral Conchology of Great Britain; or Coloured Figures and Descriptions of Those Remains of Testaceous Animals or Shells, Which Have Been Preserved at Various Times, and Depths in the Earth,* vol. 3, part 46. W. Arding Co., London.

Swartz, F. M. and Richardson, E. S., Jr. 1945. New structures in Early Devonian Conularidae. *Bull. Geol. Soc. Am.* **56**, 1206.

Sysoev, B. A. and Chudinov, I. I. 1962. Subclass Conulata. *In* Orlov, Yu. A. (ed.). *Osnovy Palaeontologii: Sponges, Archaeocyathids, Coelenterates, Vermes.* Izdatelstvo AN SSSR, Moscow, pp. 187–191 [in Russian].

Tasch, P. 1973. *Paleobiology of the Invertebrates.* Wiley, New York, 946 pp.

Termier, H. and Termier, G. 1949. Position systématique et biologie des Conulaires. *Rev. Sci.* **3300**, 711–722.

———— and ————. 1953. Les Conularides. *In* Piveteau, J. (ed.). *Traité de Paléontologie,* vol. 3, *Onychophores, Arthropodes, Echinodermes, Stomocordes.* Masson et Cie, Paris, pp. 1006–1013.

Weller, S. 1925. A new type of Silurian worm. *J. Geol.* **33**, 540–544.

Werner, B. 1966. *Stephanoscyphus* (Scyphozoa Coronatae) und seine direkte Abstammung von den fossilen Conulata. *Helgoländer Wiss. Meeresunters.* **13**, 317–347.

————. 1967. *Stephanoscyphus* Allman (Scyphozoa Coronatae), ein rezenter Vertreter der Conulata? *Paläont. Z.* **41**, 137–153.

————. 1969. *Neue Beiträge zur Evolution der Scyphozoa und Cnidaria.* I Sympos. Internac. Zoofilogen., Universidad de Salamanca, pp. 223–244.

Wiman, C. 1895. Palaeontologische Notizen 1–2. *Univ. Uppsala Geol. Inst. Bull.* **2**, part 11(3), 109–117.

Yochelson, E. L. 1964. Paleozoic mollusk: *Hyolithes. Science* **146**, 1674–1675.

12. EARLY STROMATOPOROIDS

BARRY D. WEBBY

Stromatoporoids are a more-or-less unified group of problematic organisms with meshlike skeletons that form abundant and distinctive fossils in many Ordovician–Devonian carbonate deposits. They are perhaps best regarded as multicellular animals with a grade of organization lying somewhere in the evolutionary plexus between Porifera and Coelenterata (Stearn 1982). However, many modern workers prefer to ally them more specifically, to a particular sponge or coelenterate group. Since the rediscovery of living sclerosponges in Jamaican waters (Hartman and Goreau 1970), the stromatoporoids have been more closely allied to sponges. Indeed, Hartman (1980, 1983) has classified all the nonspiculate Palaeozoic stromatoporoids, the so-called Mesozoic stromatoporoids, and the living genera *Astrosclera* (with siliceous spicules) and *Calcifibrospongia* in the order Stromatoporoidea of the class Sclerospongiae. Stearn (1980, 1982, 1983, 1984), however, has taken a less extreme view, regarding the Paleozoic stromatoporoids as poriferans but not representatives of the class Sclerospongiae. He preferred to raise the status of the Stromatoporoidea to the level of a separate class within the Porifera.

Coelenterate affinities are favored by a number of other workers. For example, Mori (1978; 1982, 1984) has viewed stromatoporoids as closest to scleractinian corals, Nestor (1981) has linked them with heliolitine corals, and Bogoyavlenskaya and Boiko (1979) have grouped them with hydrozoans. Another, more radical viewpoint is taken by Kaźmierczak (1976, 1980, 1981), suggesting a connection with Cyanobacteria. However, many stromatoporoid workers remain skeptical of this latter suggestion (Riding and Kershaw 1977).

The calcareous skeletons (coenostea) of the Paleozoic Stromatoporoidea are composed of domal to laminar, and columnar to digitate growth forms, and internally of a meshwork of platelike elements (cysts or laminae) and vertical rodlike pillars, or an amalgamate network; they may exhibit latilaminae, astrorhizae, and mamelons, and lack evidence of coloniality or spicules. The stromatoporoids were dominant in the narrowly adapted and environmentally sensitive reef areas in the surf zone (Newell 1971; Walker and Alberstadt 1975).

The history of the Paleozoic stromatoporoids has been outlined by Stearn (1982). There were important extinctions of stromatoporoids at the end of the Ordovician and at the end of the Frasnian. The columnar, aulacerid Labechiida which have widespread occurrences in Late Ordovician sequences of North America, Siberia, Baltoscandia, and Australia (Webby 1980), attain their largest sizes, some 3 m in height and 300 mm in diameter, in the uppermost part of the Ordovician succession on Anticosti Island (Petryk 1982). The abrupt disappearance (and probably extinction) of these massive *Aulacera* columns is associated with a major regressional phase caused by the glacial maximum at the end of the Ordovician. The cooling (and probably contributing sudden drop in sea level) presumably caused the demise of these huge, specialized columnar forms. *Stratodictyon, Cystostroma, Pseudostylodictyon,* and *Dermatostroma* are other genera with occurrences in Late Ordovician (Ashgill and equivalent) horizons (Fig. 5) but that do not appear to have survived into the Silurian.

The second major extinction event, at the end of the Frasnian, saw the collapse of the Devonian coral–stromatoporoid reef community (McLaren 1970, 1983) and the disappearance of many stromatoporoid genera, but there seems to have been a brief resurgence of the Labechiida in the Famennian and Early Carboniferous (Dong 1964; Stearn 1979; Bogoyavlenskaya 1982) and then Late Paleozoic extinction of the group.

Until a more continuous stratigraphic record is available, the so-called Mesozoic stromatoporoids and the Recent sclerosponges should be regarded as separate groups only exhibiting convergently similar features to the Paleozoic stromatoporoids. Late Paleozoic sphinctozoan sponges with diaphragms in their chambers and astrorhiza-like structures on the trabecularium (Finks 1983) may provide an alternative line of descent to that of the sclerosponges for the Mesozoic stromatoporoid-like forms. As Stearn (1982) has remarked, the reappearance of the stromatoporoids in mid-Mesozoic time suggests that they are "not the direct descendants of the Paleozoic stromatoporoids but arose independently."

PROBLEMATIC EARLY RECORDS

Cambrian stromatoporoid-like structures

A group of Soviet specialists (Yavorsky 1932, 1940, 1947; Khalfina 1960a, 1971; Vlasov 1961;

Khalfina and Yavorsky 1967, argued that Ordovician stromatoporoids were derived from certain Cambrian stocks. The supposed "ancestral" faunas come from the Early Cambrian of the Altai-Sayan region, where they are associated with archaeocyaths. Some of these forms are exclusively Cambrian genera (*Altaicyathus* Vologdin, *Korovinella* Khalfina, *Praeactinostroma* Khalfina, and *Cambrostroma* Vlasov); others have been linked to Ordovician or later stromatoporoids, for example *Clathrodictyon* Nicholson and Murie, *Rosenellina* Radugin, and *Stromatocerium* Hall.

Taxonomic relationships of the exclusively Cambrian genera remain confused. While some authors (e.g., Vlasov 1967; Hill 1972) have viewed *Korovinella* as a junior synonym of *Altaicyathus* (see Webby 1979a), others, such as Fonin (1982), have placed *Altaicyathus* (with its junior synonym *Abakanicyathus* Konyushkov) and *Korovinella* in separate families of the archaeocyath suborder Archaeosyconina—*Altaicyathus* in family Archaeosyconidae Zhuravleva and *Korovinella* in family Korovinellidae Khalfina. As the best known form, *Korovinella* typically exhibits a complete meshwork of horizontal and vertical elements—porous horizontal laminae and short rounded vertical pillars and canals (Fig. 1B–D). The porous nature of the laminae (see Fig. 1D) clearly allies *Korovinella* to the irregular archaeocyaths. *Praeactinostroma* has discontinuous rodlike horizontal elements, vertical pillars, and canals producing an incomplete reticulate mesh (Fig. 1A). It appears to have closer similarities to later stromatoporoids, but may be merely a less well-calcified offshoot of *Korovinella*. *Cambrostroma* resembles *Korovinella* and has been regarded as a junior synonym by Nestor (1966a). *Korovinella* not only developed a stromatoporoid-type habit but also exhibits close similarities to sphinctozoan sponges such as the Mesozoic genus *Verticillites* (A. Yu. Zhuravlev, personal communication, 1984). The presence of perforate laminae and the gap of about 70 Ma in the con-

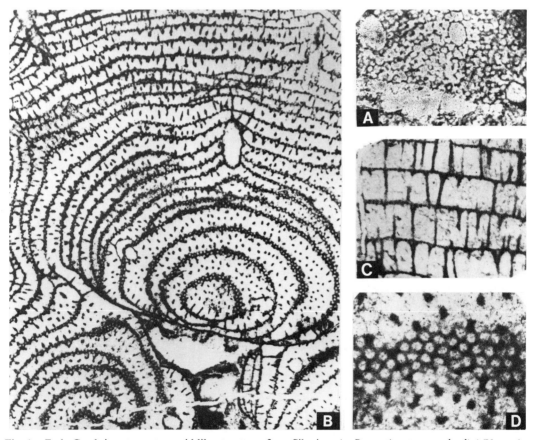

Fig. 1. Early Cambrian stromatoporoid-like structures from Siberia. A. *Praeactinostroma vologdini* (Yavorsky 1932). Transverse section. ×10. Adapted from Khalfina (1960b, p. 143, pl. Cm-XIII, fig. 3). B–D. *Korovinella sajanica* (Yavorsky 1932) B. Oblique section of specimen A-28-32, Palaeontological Institute, Moscow. About × 5. C. Vertical section. ×10. Adapted from Khalfina (1960b, p. 142, pl. Cm-XIII, fig. 1b). D. Detailed transverse section. ×30. Adapted from Khalfina (1960b, pl. Cm-XIV, fig. 1b).

tinuity of the record are factors that weigh against the suggestion that *Korovinella* and its allies are related to the Paleozoic Stromatoporoidea.

Of the forms that may be linked to Ordovician and later stromatoporoids, Radugin's supposed *Rosenellina* (1936) has not been described or illustrated, and Vlasov's *Clathrodictyon formozovae* (1961) is based on specimens that are too fragmentary to allow its taxonomic position to be verified (Nestor 1966a). Vlasov's species may merely represent the growth of dissepiment-like exothecal tissue of an associated archaeocyath. This leaves the two species of *Stromatocerium* described by Khalfina and Yavorsky (1974), which do indeed show the typical meandriform pillar structures of some Ordovician members of the genus (Galloway and St. Jean 1961). These represent the most convincingly stromatoporoid-like morphologies but seem more likely to be convergently similar to Ordovician stromatoporoids than direct antecedents of them.

Another group of problematic fossils from the Siberian and Mongolian Early Cambrian (Sayutina 1980, 1982) may have relationships with stromatoporoids. These are the sheetlike and branching forms assigned to the family Khasaktidae Sayutina. Of the laminate forms, *Vittia* Sayutina has superposed denticles forming short pillars, cysts, and lamina (Fig. 2B) as in the stromatoporoid genus *Pachystylostroma* (Fig. 2C). But the structures are much finer and they occur in much smaller, sheetlike masses than in the typical representatives of *Pachystylostroma* and *Labechia* of the *L. prima* group (Fig. 2A). *Khasaktia* Sayutina is another small laminate (sometimes saucer or channel shaped) form with cysts and laminae but has no pillars or denticles. It resembles *Pseudostylodictyon* except for the saucer shaped growth form and the laminae sometimes being flexed into ridge- and groovelike folds.

None of the branching khasaktids closely resemble cylindrical Ordovician stromatoporoid genera. *Edelsteinia* Vologdin has much more slender branches (1–4 mm in diameter) and finer internal morphologic features (mamelonlike columns from 0.05 to 0.2 mm in diameter and spaced up to 0.2 mm apart). *Drosdovia* Sayutina is a small branching form with perforate laminae, possibly again suggesting links with archaeocyaths. *Rackovskia* Vologdin, with its chainlike, branching structure, seems more likely to have coelenterate relationships.

Interpreted as members of a relatively homogeneous group, the Cambrian khasaktids overlap rather than fall entirely within the normal range of growth form, size, and morphologic features of the Ordovician stromatoporoids. They are smaller, and some genera exhibit features, for example, the chainlike branching structures of *Rackovskia* and the perforate laminae of *Drosdovia,* which are not known among the Ordovician La-

bechiida. Only *Vittia* and *Khasaktia* appear to have connections with the Labechiida (Sayutina 1980). But even *Khasaktia* is in doubt because the ridge- and groovelike plications are more suggestive of algal mat-type growth patterns. While the khasaktids as a whole do not seem to be altogether satisfactorily accommodated in the Stromatoporoidea, *Vittia* and perhaps *Khasaktia* could be viewed as ancestral stocks from which Labechiida and later stromatoporoids evolved (Fig. 10). The less likely alternative is to view the khasaktids as an early attempt to evolve framework skeletons of stromatoporoid type by animals not necessarily in direct descent to the Ordovician stromatoporoids.

No other stromatoporoid-like structures are known from the Cambrian except a problematic calcareous sediment-stabilizing organism informally named *Yoholaminites* from Middle Cam-

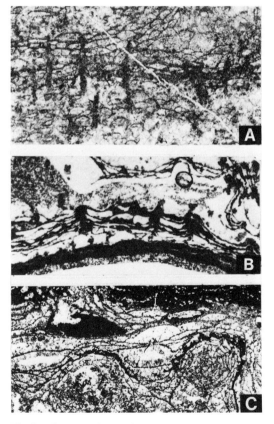

Fig. 2.   Comparative vertical sections of representative Early Cambrian khasaktids (B) and Middle Ordovician labechiids (A, C). A. *Labechia* aff. *prima* Kapp and Stearn 1975. Middle Ordovician, Cashions Creek Limestone, Tasmania. ×10. After Webby (1979b, figs. 2B, C). B. *Vittia vallaris* Sayutina 1980. Early Cambrian, Siberian Platform. ×20. Adapted from Sayutina (1980, pl. 3, fig. 1). C. *Pachystylostroma surculum* Webby 1979c. Middle Ordovician, Mjøsa Limestone, Norway. ×20. After Webby (1979c, fig. 2A).

brian carbonates of British Columbia (McIlreath and Aitken 1976). This form has not been adequately described or illustrated but seems more likely to represent a stromatolite than a stromatoporoid.

Pulchrilamina, the Early Ordovician stromatoporoid-like organism

In the Early Ordovician (Arenig) carbonate buildups of west Texas and southern Oklahoma, there are conspicuous occurrences of a problematic structure superficially resembling some algal stromatolites. Named *Pulchrilamina spinosa* by Toomey and Ham (1967), it forms one of the three major biotic components in the upper part of individual "reef" mounds (Toomey 1970, 1981; Toomey and Nitecki 1979; Toomey and Babcock 1983). The laminae of these domelike calcified masses exhibit horizontal spar and sediment-layered alterations with conspicuous mainly vertical "spines" arising in spar-filled layers and extending upward into the directly overlying sediment (Figs. 3A, B, D, and 4A, B, D). The vertical "spines" are reminiscent of the pillars of some labechiids. The randomly spaced vertical to near-vertical spines have been interpreted as originally hollow, now infilled by "mosaic" calcite (Toomey and Nitecki 1979).

Toomey and Ham (1967) thought that *Pulchrilamina* might be a "primitive coelenterate" with affinities to the Stromatoporoidea. While it shared with algal stromatolites the ability to trap layers of mud, this "colonial" organism differed in being able to secrete the spiny layers. Threads of the alga *Girvanella* were noted by Toomey and Nitecki (1979) as occurring on some laminae of *Pulchrilamina*.

Toomey (1970) and Toomey and Nitecki (1979) noted that in the west Texas mounds, *Pulchrilamina* occupies a similar ecologic niche as do the laminate stromatoporoids of the Middle Ordovician (Chazyan) and later Ordovician–Devonian reefs. They are the dominant elements of reef community in the shallowest waters of the surf zone (see also Alberstadt et al. 1974; Walker and Alberstadt 1975). Individually the *Pulchrilamina* masses may be up to 500 mm high and 300 mm across in the tops of these buildups.

Pratt and James (1982) have interpreted *Pulchrilamina* in Early Ordovician buildups of the St. George Group in Newfoundland as an encrusting sponge rather than a coelenterate. The problematic form has also been reported from the younger (pre-Chazyan Whiterockian) Table Head Group of Newfoundland (Klappa and James 1980).

Sayutina (1980), following Nestor (1978), has regarded *Pulchrilamina* as the earliest member of the stromatoporoid family Lophiostromatidae (order Labechiida). Stromatoporoid affinities

have also been stressed by Stock (1983), who reported that Toomey and Nitecki's fig. 13 (1979) showed specimens of *Pulchrilamina* with cysts and remnants of pillars as in the Labechiida. Similarly, Webby (1984a,b) has referred to *Pulchrilamina* as stromatoporoid-like with the generalized growth form, latilaminae, cysts, and pillars of typical Ordovician labechiids.

The apparently hollow spar-filled vertical spines are very similar to the "hollow pillars" of many labechiids (Kapp and Stearn 1975). In thin sections of *Pulchrilamina,* spines range from 0.02 to 0.07 mm in diameter and up to 0.6 mm in height; they taper slightly upward (Figs. 3D, and 4B, D), have a rounded cross section (Fig. 3C) and a spacing of from 0.05 to 0.2 mm apart. They only differ from the pillars of typical labechiids (Fig. 4F) in being more slender and in protruding relatively further upward above the tops of "latilaminae" into overlying sediments. Occasionally a few spines may be tilted away from the vertical (Fig. 4B), suggesting that they grew in a more loosely aggregated skeleton than in many labechiids, or the horizontal tissue support was less well calcified. Alternatively there was some early diagenetic dissolution of finer horizontal skeletal tissue causing the spines to become relatively unsupported. As they protrude upward into the sediment, as in the pillars of some labechiids (Fig. 4F), the spines are unlikely to have originated as hollow structures; they would seem to represent spar-filled replacements or originally solid carbonate or siliceous structures.

The originally siliceous lithistid sponge *Archaeoscyphia* occurs with *Pulchrilamina* in the same beds of the Early Ordovician buildups of west Texas, and has been affected by dissolution and replacement by calcite (Rigby 1966; Palmer and Fürsich 1981). Also in the sediment associated with skeletons of *Pulchrilamina* are calcified monactinal sponge spicules with axial filaments indicating they were originally siliceous. The spines in the *Pulchrilamina* skeleton show no traces of such axial filaments to support the view that the organism was originally a siliceous sponge.

The growth form is characteristically of the labechiid type, showing well-developed alternations of spar-filled "latilaminae" and sediment layers. The latilaminae are usually from 0.1 to 3.0 mm thick (exceptionally up to 8 mm thick) with a ragged edge zone resulting from the interplay between phases of growth and influxes of mud (Figs. 3A–D, and 4A–D), as in stromatoporoids. In some areas of the spar-filled skeleton, usually in narrow, lenticular patches toward the tops of latilaminae, it is possible to recognize undulating rows of long, low cysts spaced from 0.025 to 0.030 mm (rarely up to 0.04 mm) vertically (Figs. 3A, B, D, E, and 4C–E). The cysts illustrated by Toomey and Nitecki (1979, fig. 13) have similar

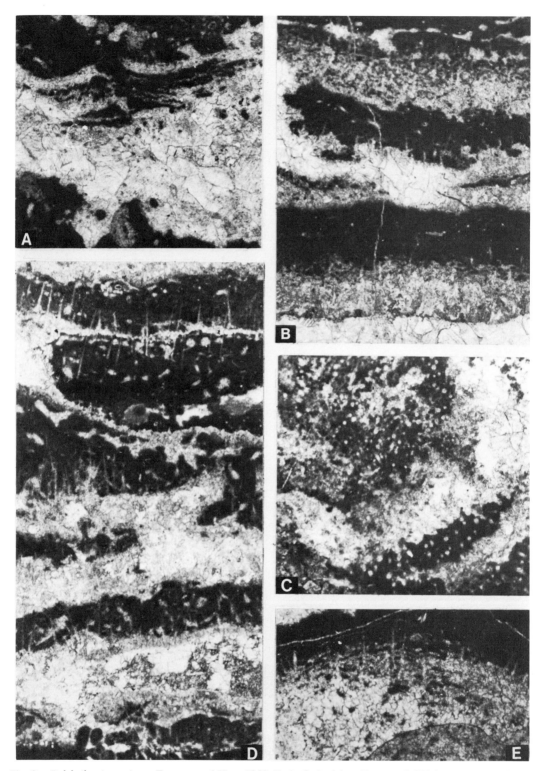

Fig. 3. *Pulchrilamina spinosa* Toomey and Ham 1967. Early Ordovician, Texas and Oklahoma. A–D. ×20. E. ×40. A. Vertical section. PP22795, Unap Mountain section, "450 ft" above base of the Kindblade Formation, Wichita Mountains, Oklahoma. B. Vertical section of specimen from "mound rock," McKelligon Canyon Formation, South Franklin Mountains at El Paso, west Texas (D. V. LeMone collection). C. Transverse section, PP22845. D. Vertical section, PP22843. C, D. From mound horizons (subunit B1) in McKelligon Canyon Formation, South Franklin Mountains at El Paso, west Texas. E. Vertical section, PP22967. From Mill Creek section "450 ft" above the base of the Kindblade Formation, Arbuckle Mountains, Oklahoma.

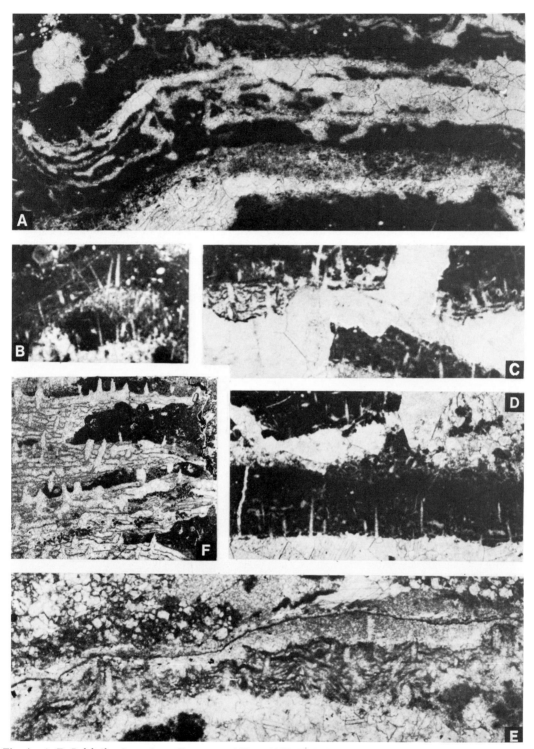

Fig. 4. A–E. *Pulchrilamina spinosa* Toomey and Ham 1967. Early Ordovician, Texas and Oklahoma. A, B, D. ×20. C, E. ×40. A. Vertical section, PP22963. Mill Creek section "450 feet" above the base of the Kindblade Formation, Arbuckle Mountains, Oklahoma. B. Vertical section of specimen (MC-38-MB, D. V. LeMone collection) from McKelligon Canyon Formation, South Franklin Mountains at El Paso, west Texas. C, D. Vertical sections of specimen from "mound rock," McKelligon Canyon Formation, South Franklin Mountains at El Paso, west Texas (D. V. LeMone collection). E. Vertical section, PP22849. From mound horizons (subunit B1) in McKelligon Canyon Formation, South Franklin Mountains at El Paso, west Texas. F. *Stromatocerium bigsbyi* Webby 1979b. Vertical section, Middle Ordovician, Cashions Creek Limestone, Tasmania. ×5. After Webby (1979b, fig. 5A).

153

vertical spacing and variable lateral continuity. Under moderately high magnification the cysts may be distinguished from patches of *Girvanella* strands with diameters of about 0.02 mm, and which lie between the latilaminae of the *Pulchrilamina* skeleton.

The bulk of the *Pulchrilamina* skeleton appears to have been modified by dissolution and replacement by a variety of diagenetic alteration processes. The vertical pillarlike structures at the tops of the latilaminae are the most conspicuous feature of the skeleton (Figs. 3D, and 4B, D), with horizontal elements only rarely well preserved (Figs. 3E, and 4C, E). This contrasts with the coenosteal fabric of most Chazyan stromatoporoids (Kapp and Stearn 1975; Webby 1979b), which is dominantly of horizontal tissue with vertical elements such as pillars usually relatively poorly developed. Only the most common Chazyan stromatoporoid from Tasmania, *Stromatocerium bigsbyi,* has pillars with considerable vertical continuity and prominence (Fig. 4F). Nevertheless, it seems difficult to exclude Early Ordovician *Pulchrilamina* from an assignment with the stromatoporoids of the family Labechiidae Nicholson, especially now that pre-Chazyan Whiterockian representatives of the Labechiida have been recognized in China (Dong 1982) and Malaysia (Webby et al. 1985). With its laminar, distinctly latilaminate coenosteum, long slender pillars, and long, low cysts, the genus *Pulchrilamina* exhibits no significant dissimilarities from other members of the Labechiidae. It seems to be most closely allied to the *Labechia prima* group, although it exhibits a greater continuity of vertical elements than typical representatives of this group. It should not be retained in the family Lophiostromatidae Nestor, as recommended by Nestor (1978), because it differs markedly from *Lophiostroma* Nicholson and *Dermatostroma* Parks. *Lophiostroma* has a coenosteum almost completely filled with poorly differentiated, thickened skeletal tissue, and *Dermatostroma* is a thin, encrusting form.

MIDDLE–LATE ORDOVICIAN RADIATIONS

Labechiida Kühn 1927

The classification of the order Labechiida adopted herein is slightly amended from that proposed by Stearn (1980). Four families, the Labechiidae, the Rosenellidae Yavorsky, the Aulaceridae Kühn, and the Lophiostromatidae Nestor, are included in the order.

Until recently it was widely accepted that the North American Chazyan stromatoporoid fauna (Galloway and St. Jean 1961; Nestor 1964, 1966a; Bogoyavlenskaya 1969; Kapp 1974; Kapp and Stearn 1975) and its counterpart in Tasmania (Webby 1979a,b, 1980) included the earliest members of the group. Kapp and Stearn (1975) recognized the morphologically simple genus *Pseudostylodictyon* as appearing first and giving rise to two main branches, one producing the various species of *Labechia* and the other, species of *Pachystylostroma*. However, the Chazyan fauna from the Cashions Creek Limestone of Tasmania (Burrett, in Webby et al. 1981) includes occurrences of *Stromatocerium, Labechia,* and *Stratodictyon,* whereas *Pseudostylodictyon* is not recorded in Australian successions until about Blackriveran (= Early Caradoc) time. *Stromatocerium* is a morphologically complex form, yet it is the earliest and most abundant to appear.

The relatively "advanced" features of the Tasmanian Chazyan stromatoporoid fauna, and the conspicuous roles *Labechia* and *Pachystylostroma* played in the North American Chazyan reef community (Kapp, 1975) lead to the conclusion that the Labechiida may have had a pre-Chazyan history (Fig. 5).

Dong (1982) seems to have confirmed this presumption by recording a Whiterockian fauna of labechiids, rosenellids, and aulacerids from North China. The fauna from the Majiagou Formation of northern Anhui includes representatives of *Labechia,* a possible *Pseudostylodictyon* or *Rosenella, Cryptophragmus, Ludictyon,* and *Aulacera.* Conodont faunas studied by An (1981) and An et al. (1983) imply that the Majiagou Formation of North China spans an interval from Whiterockian to earliest Chazyan (i.e., from Early to Late Whiterockian of Ross et al. 1982). This is consistent with Flower's correlations (1976) based on nautiloids. Other members of the Labechiida previously reported by Yabe and Sugiyama (1930), Ozaki (1938), and Kobayashi (1969) as coming from the "Toufangian fauna" of Liaoning and Shandong provinces of North China may also come from the Majiagou Formation or equivalents, and may also include some pre-Chazyan Whiterockian faunas.

The earliest stromatoporoid-bearing horizons in Southeast Asia are more precisely determined from associated nautiloids and conodonts. In the Middle Ordovician Lower Setul Limestone of Malaysia there are labechiids and rosenellids of definite pre-Chazyan Whiterockian age (Webby et al. 1985). The fauna of *Labechia, Cystostroma?,* and *Rosenella* occurs in association with the nautiloids *Wutinoceras robustum, Chaohuceras?* sp., *Armenoceras chediforme,* and *Tofangoceras nanpiaoense,* and the conodonts *Histiodella* and *Scolopodus.*

These Asian occurrences of sheetlike to domal *Labechia, Cystostroma?,* and *Rosenella,* and cylindrical *Ludictyon* and *Cryptophragmus,* seem to record the first major diversification of the group in pre-Chazyan Whiterockian (Early–Middle Llanvirnian) times, possibly from a *Pulchrilam-*

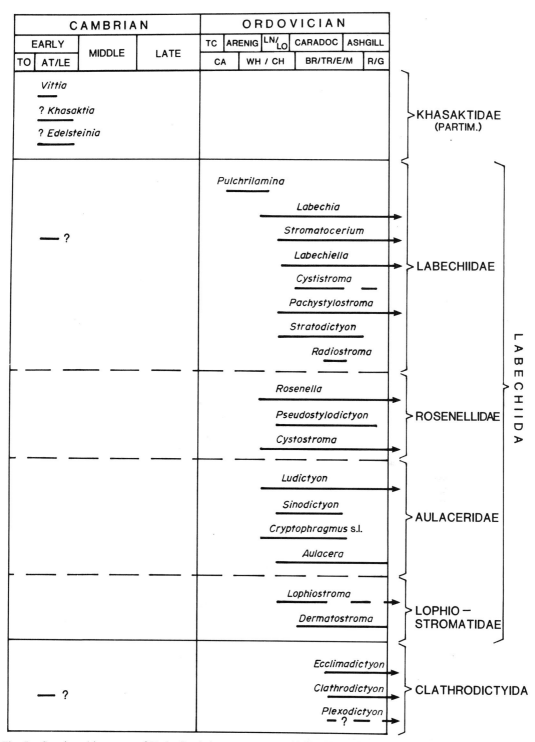

Fig. 5. Stratigraphic ranges of Early Cambrian stromatoporoid-like khasaktid and known Ordovician stromatoporoid genera (Labechiida and Clathrodictyida). Post-Ordovician ranges of genera indicated by arrows. Abbreviations: TO, Tommotian; AT/LE, Atdabanian–Lenian; TC, Tremadoc; LN/LO, Llanvirn–Llandeilo; CA, Canadian; WH/CH, Whiterockian–Chazyan; BR/TR/E/M, Blackriveran–Trentonian–Edenian–Maysvillian; R/G, Richmondian–Gamachian.

ina-like ancestor (Fig. 5). There may have been two main lines of descent from *Pulchrilamina,* to more fully developed "mature" meshworks of *Labechia* with pillars and cysts, and to simpler, "immature" cystose structures of *Rosenella* and *Cystostroma.* Cylindrical labechiids such as *Ludictyon* and *Cryptophragmus* appear to have been derived by extension of mamelonlike upgrowths off a *Rosenella* or *Cystostroma* base. Previously the North American Chazyan occurrences of *Cystostroma* (Galloway 1957) or *Pseudostylodictyon* (Kapp and Stearn 1975) were taken as ancestral stocks. However, Kapp and Stearn (1975) have already noted that *Cystostroma* occurs in the Chazyan material as bands at the bases of the coenostea of species of *Labechia* and therefore only represents an "immature" stage of growth of a *Labechia.* Other examples of *Labechia* and *Cystostroma*-like alternations have been recorded (Webby 1979b, p. 241).

Although the precise stratigraphic positions of other members of the Labechiida reported from the Majiagou Formation of North China (part of the "Toufangian fauna" of Kobayashi 1969) are not known, the general indications of age based on correlations with the North American Midcontinent conodont assemblages (An 1981; Harris et al. 1979) suggest that uppermost horizons are no younger than Early Chazyan. *Pseudostylodictyon, Labechiella, Lophiostroma, Sinodictyon,* and *Aulacera* are additional elements of the fauna. *Pseudostylodictyon* and *Labechiella* may be viewed as "immature" and "mature" types with predominantly laminar rather than cystose tissue, and *Lophiostroma* as an offshoot of *Labechia,* with its "overmature" thickened tissue. By analogy with growth patterns in early heliolitine corals (Webby and Kruse 1984), the cystose and laminar forms *(Cystostroma, Pseudostylodictyon)* may be interpreted as reflecting relatively faster growth than the meshwork forms *(Labechia, Labechiella).* The much thickened forms *(Lophiostroma)* may in contrast represent a slower rate of growth. *Sinodictyon* and *Aulacera* may have been derived from other cylindrical forms like *Ludictyon* and/or *Cryptophragmus.* Indeed it now seems likely that by the end of the Chazyan, more than two-thirds of labechiid genera had made their appearances (Fig. 5).

The Labechiida achieved their greatest diversity and widest geographic spread through the Middle–Late Ordovician interval (Webby 1980). Their significance as contributions to Middle–Late Ordovician reefs has been outlined by Webby (1984a,b). Most of the cylindrical forms, including the giant, columnar *Aulacera,* had become extinct by the end of the Ordovician (Fig. 5). Only *Ludictyon* lingered on into the Early Silurian (Dong and Yang 1978). The hemispherical–laminar forms also show a decline in importance through the Silurian and Devonian, but enjoyed a brief resurgence (Dong 1964; Stearn 1979) after the collapse of the reef community and extinction of some other stromatoporoid groups in the Late Devonian. The revival of the Labechiida was, however, short-lived. By the end of the Tournaisian this conservative, long-ranging group had disappeared.

Clathrodictyida Bogoyavlenskaya 1969

Only three genera of the Clathrodictyida are known from the Ordovician, namely, *Clathrodictyon* Nicholson and Murie, *Ecclimadictyon* Nestor, and *?Plexodictyon* Nestor (Fig. 5). Representatives of the group first appeared in the shallow waters of an offshore volcanic "island arc" in eastern Australia during Trentonian (Middle Caradoc) times, taking almost to the end of the Ordovician to achieve a circumequatorial distribution (Webby 1980, 1984b). *Ecclimadictyon nestori* and *Plexodictyon?* sp. are the first to be recorded in the eastern Australian succession, from the middle part of the Belubula Limestone (Webby and Morris 1976; Webby and Packham 1982), and then two species of *Clathrodictyon* follow in the upper part of the Belubula Limestone.

Stearn (1980) assigned these forms to two different families: *Clathrodictyon* to the family Clathrodictyidae and *Ecclimadictyon* and *Plexodictyon* to the Ecclimadictyidae Stearn. Essentially the Clathrodictyidae have regular laminae and pillars confined to a single interlaminar space or derived by downward flexure of the laminae, while the Ecclimadictyidae exhibit chevron-folded laminae and in some forms continuous pillars and paralaminae. Representatives of both groups show astrorhizae. However, the distinction between the two groups seems rather artificial given the associations of a *Plexodictyon*-type structure in a mamelonlike upgrowth of *Clathrodictyon plicatum* (Webby and Banks 1976, pl. 2, fig. 5), and the *Ecclimadictyon*-like tissue at the bases of the coenostea or other varieties of *C. molense* (Webby and Banks 1976, pl. 1, fig. 3). Nestor (1964) stressed the close similarities between *Clathrodictyon* and *Ecclimadictyon,* including one form with undulating *Ecclimadictyon*-like laminae as *Clathrodictyon microundulatum.* Similarly there is a complete gradation between the forms assigned to *Clathrodictyon* cf. *microundulatum* and *Ecclimadictyon amzassensis* in the Vandon Limestone (Cliefden Caves Limestone Group) of central New South Wales.

Many authors (Galloway 1957; Nestor 1966b; Bogoyavlenskaya 1969; Kaźmierczak 1971; Stearn 1982) have suggested that it is relatively easy to derive a clathrodictyid from the simple, imbricated cystose morphology of a labechiid-like *Cystostroma.* The funnel shaped pillars of some species of *Clathrodictyon* (Stearn 1966; Stearn and Hubert 1966; Mori 1978) are thought to have developed from downwardly inflexed parts of the overlying lamina, and bear no relationships to the denticles or pillars of labechiids (Fig. 6A,B).

In establishing the relationships between the Labechiida and the Clathrodictyida, the problem is not so much how to derive the laminae of clathrodictyids from the cysts of labechiids, but how to produce the orderly arrangement of rodlike pillars by downflexing of laminae. The meshwork of variably sized convex cysts of a *Cystostroma* or *Pseudostylodictyon* may have evolved into a more orderly arrangement of laterally continuous plates (zigzag or regular laminae as in *Ecclimadictyon* or *Clathrodictyon*), but the derivation of the pillars is less easy to explain. In *Ecclimadictyon* the pillars commence in downflexed laminae and descend through bars into rodlike pillars that have a very regular spacing as seen in tangential section. They therefore developed in a fundamentally different way from the pillars of labechiids. This is perhaps best emphasized by viewing upper surfaces of the coenostea: In the Labechiida (including *Pulchrilamina* and the lophiostromatids) the pillars protrude above the surface as solid tubercles or papillae (Fig. 6A), whereas in the Clathrodictyida the upper surfaces may be undulating or smooth, only rarely showing as shallow depressions the tops of the funnel shaped pillars (Fig. 6B).

Another fundamental distinction may be drawn from the internal morphology of mamelonlike columns of the two groups (Fig. 7A,B). In the growth of such columns, tissue is added more rapidly in axial areas than at the margins. In the Labechiida (including the cylindrical Aulaceridae, which have evolved from mamelonlike upgrowths) the axial areas are composed of large cysts, while the margins exhibit relatively small cysts (Fig. 7A). On the other hand, in some species of *Clathrodictyon* (e.g., *C. plicatum*) the axial area of mamelons shows less orderly *Plexodic-*

tyon-like zigzagged laminae while the margins have regular laminae (Fig. 7B).

The Clathrodictyida attained the status of a major cosmopolitan group in the Silurian, and continued to be well represented in the Devonian until its extinction during the Late Devonian faunal crisis along with the collapse of the reef community.

Cliefdenella Webby 1969

It has previously proved difficult to classify this complex "advanced" genus (family Cliefdenellidae Webby) within the Paleozoic stromatoporoids. Khalfina and Yavorsky (1973) included the family in the superfamily Tienodictyacea Bogoyavlenskaya, and Stearn (1980) grouped it in the Labechiida, adding the proviso that it might be the basis for a new order. But there is little justification for assigning forms with perforate tube pillars and complex astrorhizal canals and columns in the Labechiida, albeit tentatively.

Species of *Cliefdenella* are known from the Middle Caradoc to Early Ashgill (Trentonian–Maysvillian) of New South Wales (Webby 1969; Webby and Morris 1976), from the Ashgill of Salair, southwestern Siberia (Khalfina and Yavorsky 1974), and from the Late Ordovician of Alaska (Stock 1981). They exhibit a complex association of laminae with denticles on upper surfaces, dissepiments and a ramifying meshwork of horizontal astrorhizal canals filling interlaminar spaces, vertical astrorhizal columns and tube pillars (Figs. 6C, 8A–D, and 9C–D). Stock (1981) first identified pores in the tube pillars, and these structures have since been confirmed in the type species (Fig. 8C). The pores are crucial to the functional interpretation of *Cliefdenella* as having a water circulation system like the living sclerosponge *Calcifibrospongia*. Stock (1984) proposed that the perforated tube pillars carried the inhalant water

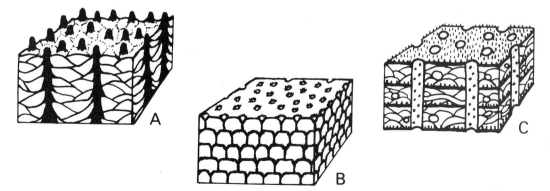

Fig. 6. Block diagrams to contrast the differing horizontal and vertical morphologic features of Ordovician Labechiida (A), Clathrodictyida (B), and the problematic genus *Cliefdenella* (C). Note the cysts and updomed pillars of the Labechiida, and laminae and discontinuous downflexed pillars of the Clathrodictyida, and the denticled laminae, vesicles, and perforate tube pillars of *Cliefdenella*.

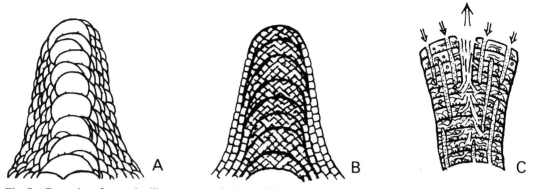

Fig. 7. Examples of mamelonlike upgrowths in Labechiida (A), Clathrodictyida (B), and the problematic genus *Cliefdenella* (C). Labechiid and clathrodictyid mamelons are presumed to have been mantled by soft tissues, with successive mineralized layers being added more rapidly in axial areas than in margins. This contrasts with *Cliefdenella* upgrowths in which new chambers are added only at the tops of the columns, and the soft tissue is inferred to be restricted to upper chambers. Arrows depict the suggested incurrent and excurrent water circulation of *Cliefdenella*.

into the galleries. Astrorhizal canals converge on astrorhizal columns and act to drain the exhalant water from the body of the animal (Fig. 7C). The poriferan affinities of *Cliefdenella* seem indisputable, but whether the genus should be retained in the Stromatoporoidea is now in doubt.

The recent discovery of Ordovician sphinctozoan sponges from central New South Wales (Webby and Rigby 1985) raises the alternative view that the cliefdenellids are sphinctozoans. Indeed, the branching, cylindrical form of *Cliefdenella*, described by Webby and Morris (1976) as *Cliefdenella* sp. (Figs. 7C, and 9C, D), has the typical growth form of sphinctozoans. The New South Wales Ordovician sphinctozoans first appear in the lower part of the Belubula Limestone (Webby and Packham 1982), stratigraphically below horizons containing the first cliefdenellids. More complex forms such as *Angullongia* Webby and Rigby (Fig. 9A, B), with chambers infilled by canals and vesicles (dissepiments), occur at higher stratigraphic levels (Angullong Tuff). Similar chambers (or galleries) with vesicles and canals occur in *Cliefdenella*. However, the angullongiids differ in having spoutlike ostia, small prosopores on external surfaces of chambers, and a central spongocoel tube with apopores surmounted by a prominent osculum. Although an angullongiid sphinctozoan seems unlikely to have given rise to the cliefdenellids, even through a line of descent that included the branching, cylindrical variety of *Cliefdenella*, the two groups may have diverged from a common ancestor earlier in the Ordovician, or in the Cambrian. Possibly angullongiids and cliefdenellids represent separate porate and aporate offshoots from the main *Amblysiphonella* lineage (Pickett and Jell 1983; Webby and Rigby 1985).

A number of Triassic sphinctozoans bear close

resemblances to *Cliefdenella*. For instance, species of *Zardinia* and *Cryptocoelia* described by Senowbari-Daryan and Schafer (1983) from the Triassic of Greece show similar developments of horizontal and vertical tissue—laminae, dissepiments (or vesicles), and perforate tube pillars (or pseudocanals)—but they lack the complex astrorhizal systems of *Cliefdenella*. A tubelike reticular mesh in the central column of another Triassic species, *Vesicocaulis depressus* (Ott 1967; Dieci et al. 1968), very much resembles the astrorhizal columns of *Cliefdenella*. These apparently close relationships may be best interpreted in terms of homeomorphy.

DISCUSSION

Of the Early Cambrian stromatoporoid-like groups, some of the khasaktids and possibly the species identified as belonging to *Stromatocerium* may have direct relationships to later stromatoporoids (Figs. 5 and 10). The second Early Cambrian group, the korovinellids, are irregular archaeocyaths with only homeomorphous resemblances to stromatoporoids.

The Early Ordovician *Pulchrilamina* is interpreted as being ancestral to the Labechiida (Fig. 5). Whether it had links with the Early Cambrian khasaktid genus *Vittia* or arose independently has yet to be ascertained. The main diversification of the Labechiida in the early–middle part of the Middle Ordovician (Whiterockian–Blackriveran) was followed in the later part of the Middle Ordovician (Trentonian) by the appearance of the Clathrodictyida representing the development of a markedly different group (Figs. 6A, B, 7A, B, and 10). There are some doubts about how in detail this group could have been derived. Indeed the distinctive morphologic differences and the

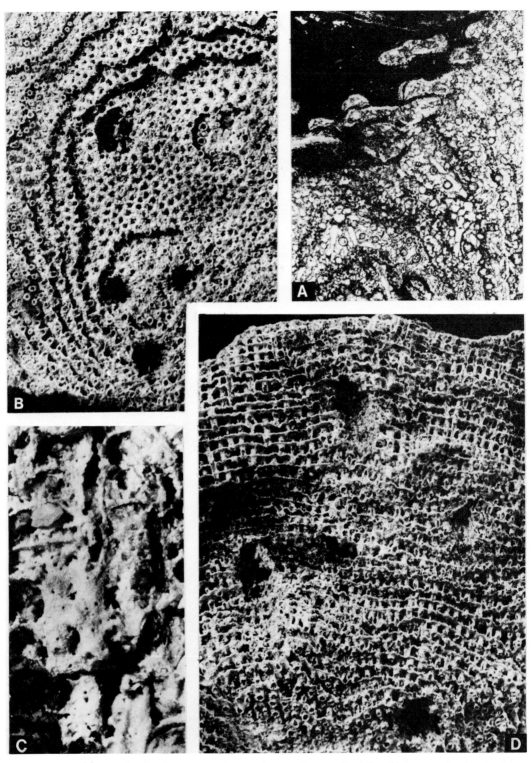

Fig. 8. A. *Cliefdenella* aff. *etheridgei* Webby 1969. Oblique section showing blisterlike outer ends of laminae sealing off successive chambers to form irregular, impervious outer margin (or dermal layer) to skeleton. SUP79150, lower part of Goonumbla Volcanics (Late Ordovician), central New South Wales. ×5. B–D. *Cliefdenella etheridgei* Webby 1969. Silicified specimen of SUP 94124, Vandon Limestone (Cliefden Caves Limestone Group) at the Island, central New South Wales. B, D. Top and oblique side views. ×3. C. Detailed side view showing perforate "tube pillars." ×15

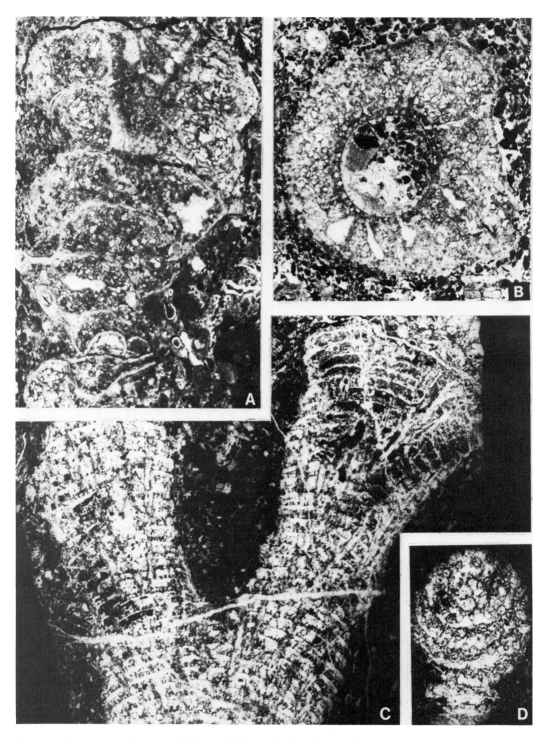

Fig. 9. A, B. *Angullongia vesica* Webby and Rigby 1985. Angullong Tuff (Late Ordovician), central New South Wales. ×5. A. Longitudinal section of paratype, SUP51004. B. Transverse section of paratype, SUP51001. C, D. *Cliefdenella* sp. Malongulli Formation (Late Ordovician), central New South Wales. Longitudinal and transverse sections. ×5. After Webby and Morris (1976, fig. 3E, F).

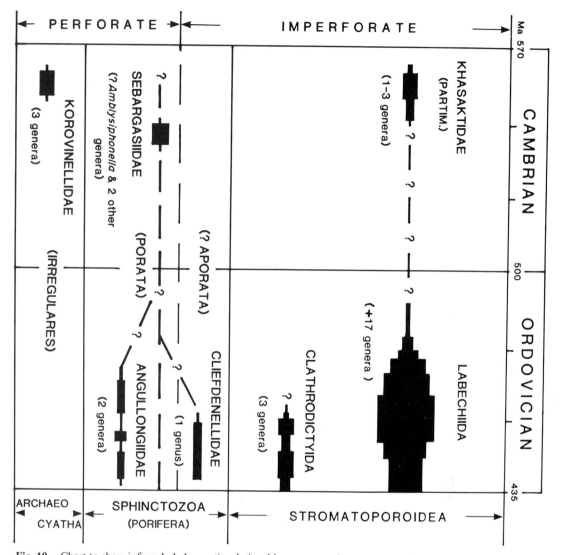

Fig. 10. Chart to show inferred phylogenetic relationships among early stromatoporoids, and stromatoporoid-like organisms of sphinctozoan sponge and archaeocyath affinities.

lack of preserved intermediate stages of development, may be inferred to suggest that Ordovician stromatoporoids were polyphyletically derived.

The Labechiida and the Clathrodictyida lie within the general morphologic field of the stromatoporoids and, following Stearn (1984), are more likely to be closely linked to sponges than to cnidarians. They show only marginal overlap with other Early Paleozoic fossil organisms, such as archaeocyaths and sphinctozoan sponges. The overlap is mainly among the few irregular Early Cambrian archaeocyaths like *Korovinella*, which have a compound, massive growth form, and some Permian–Triassic stratiform guadalupiid sphinctozoans.

Debrenne (1983) has suggested that a close relationship exists between archaeocyaths and sphinctozoan sponges, giving the recent discoveries of Middle Cambrian sphinctozoans in Australia (Pickett and Jell 1983) and the Middle–Upper Cambrian archaeocyaths from Antarctica (Debrenne et al. 1984) as evidence that no stratigraphic gap exists between the two groups. Indeed, it now seems that "there are no valid arguments against the inclusion of the Archaeocyatha within the Phylum Porifera" (Debrenne and Vacelet 1983).

The archaeocyaths can also be separated from the Stromatoporoidea on the basis of their porous tissue. The archaeocyaths and the earliest sphinctozoans (Pickett and Jell 1983) are perforate and appear to have no direct connections to the imperforate Ordovician stromatoporoids.

Zhuravleva and Miagkova (1972, 1974) introduced the kingdom Archaeata to accommodate the simple, multicellular, porate, skeletonized, cuplike organisms with double walls and a central cavity at the boundary between animals and plants. This, however, is an artificial group because it includes both plants (e.g., the Ordovician–Carboniferous receptaculitids) and animals, like the Cambrian archaeocyaths and the Silurian aphrosalpingids (see also Chapter 3 by Nitecki, this volume). Zhuravleva and Miagkova (1974) seem merely to record a homeomorphous relationship.

The only other nonspicular, Early Paleozoic, sponge-like organism of interest is the problematic genus *Lapidipanis* described by Paquette et al. (1983) from the Middle Ordovician of Newfoundland, with its large skeleton of radiating and concentrically arranged spheroidal textures. Regarded as the oldest known sclerosponge and a contemporary of the earliest stromatoporoids, its occurrence suggested to Paquette et al. (1983) that the sclerosponges and stromatoporoids diverged from a common ancestor in Early Paleozoic times.

The "compound," massive to sheetlike type of growth form exhibited by the stromatoporoids, as well as the sclerosponges, is a less common sponge morphotype than the cup, flask, vase, or simple branching forms with water circulation system differentiated into incurrent flow through the side walls and excurrent flow from the top. In the "compound," massive and encrusting stromatoporoid-type morphologies, the inflow and outflow must be differentiated essentially by the soft tissues mantling the upper surface of the skeleton, perhaps in the manner proposed by Stearn (1975).

Astrorhizae appear to have played an important role in permitting the removal of excurrent waters from galleries containing living tissue beneath the upper surface of the coenosteum, as in some demosponges and the sclerosponges. In the massive to encrusting stromatoporoids and sclerosponges, they are frequently associated with mamelons, the moundlike structures that appear to have had a function related to the need to separate incurrent and excurrent water flow on the upper, living surface of the organism. Mamelons and associated astrorhizae are well developed in some Late Ordovician clathrodictyids; see, for example, *Clathrodictyon mammillatum* of Nestor (1962, 1964).

Other groups like the Labechiida commonly exhibit mamelons but no traces of associated astrorhizal structures. Astrorhizae have only been preserved in the coenostea of a few labechiids. This is probably a reflection of an exhalant canal system that mainly remained uncalcified and of a living tissue that did not extend into the galleries beneath the growing surface of the organism (Stearn 1975, fig. 4). The possibility of a derivation of cylindrical Labechiida from the strongly mammillated members of *Cystostroma, Pseudostylodictyon,* or *Rosenella* has been discussed previously (Webby 1979a). This developmental trend may in functional morphologic terms represent a return, perhaps a reversion, to a more differentiated water circulation system, with inflow at the sides and outflow from the tops of the columns.

The problematic genus *Cliefdenella* is now regarded as a sphinctozoan sponge, possibly the earliest member of the Aporata (Fig. 10).

It is beyond the scope of this paper to assess all the various alternative views of stromatoporoid relationships. However, two recently published interpretations of stromatoporoids as Cyanobacteria and as cnidarians having an origin in common with heliolitine corals require comment. Kaźmierczak (1976, 1980, 1981) reported the presence of coccoid and rare filamentous cyanobacteria-like microfossils in the tissue of well-preserved Devonian stromatoporoid skeletons as implying that the skeletons were stromatolitic constructions. However, if such microorganisms were capable of secreting the relatively complex stromatoporoid meshwork through Early to Middle Paleozoic times, why were these cyanobacterial organisms incapable of producing similar skeletons during their early record in the Precambrian, and their later record to the present? If the cyanobacterial microorganisms can be shown to be common associates of stromatoporoid skeletons, they may be more realistically interpreted as symbionts or harmless contaminants than the producers of these meshworks.

Nestor (1981), while stressing the contemporaneity of early appearances of stromatoporoids and heliolitine corals, has downplayed the importance of the fundamental dissimilarity between the two groups, namely that the heliolitines exhibit corallites and the stromatoporoids lack them. Nestor has inferred the stromatoporoids to be cnidarians and, in a reconstruction of *Lophiostroma,* has interpreted the "zooids" as occupying positions above the upraised papillae. Yet *Lophiostroma* is an encrusting form, like its environmental counterparts among the heliolitines, *Protaraea* and *Coccoseris.* The much thickened, bacular *Coccoseris* has corallites that give some measure of protection to the zooids. It represents a low-level encruster that lived in an active environment with a comparatively slow rate of growth compared to the hemispherical *Heliolites* or *Propora* (Webby and Kruse 1984). To have "zooids" occupying a "high," unprotected position on the raised papillae of *Lophiostroma* is inconsistent with its interpretation as an encrusting form occupying an active environment.

From the remarkably complete record of heliolitine occurrences in the Ordovician succession of New South Wales (Webby and Kruse 1984), it is possible to give a full account of the early evolution of the heliolitines, including the suggestion that the *Propora*-type morphology developed in response to relatively faster growth rates by a heliolitican ancestor like *Heliolites,* losing the vertical components of its coenosclerenchyme. *Propora,* with its typical small, hemispherical, "high-level" colonies (Webby and Kruse 1984), does not exhibit any clear phylogenetic relationships with the sheetlike, latilaminate coenostea of *Cystostroma* in the New South Wales succession (Webby 1969).

Furthermore, wherever labechiids and heliolitines occur together in the same localities and horizons of the New South Wales Ordovician successions, the preservation of the heliolitines is far better than the associated labechiids. This perhaps suggests some important original compositional differences between the two groups. The labechiids, like the poorly preserved tetradiid corals, may have had a skeleton of original aragonitic composition (Semeniuk 1971), whereas the heliolitines may have been originally calcitic. In conclusion, it seems that the resemblances between labechiids and heliolitines are likely to be a reflection of convergence in response to adaptation to similar environmental niches.

Acknowledgments

I thank D. F. Toomey, M. H. Nitecki, and D. V. LeMone for kindly permitting me to study specimens of *Pulchrilamina* housed in the Field Museum of Natural History, Chicago, and in the University of Texas at El Paso. C. W. Stearn kindly reviewed the manuscript. Thin sections illustrated in Figs. 3 and 4 with the prefix PP are from Dr. Toomey's collection in the Field Museum of Natural History. Specimens shown in Figs. 8 and 9 with the prefix SUP are in the paleontology collection of Sydney University. My research on stromatoporoids has been assisted by funds from the Australian Research Grants Scheme (ARGS grants E73/151002 and E79/15763).

REFERENCES

Alberstadt, L. P., Walker K. R., and Zurawski, R. P. 1974. Patch reefs in the Carter Limestone (Middle Ordovician) in Tennessee, and vertical zonation in Ordovician reefs. *Bull. Geol. Soc. Am.* **85,** 1171–1182.

An Taixiang. 1981. Recent progress in Cambrian and Ordovician conodont biostratigraphy of China. *Geol. Soc. Am. Spec. Pap.* **187,** 209–226.

An Taixiang, Zhang Fang, Xiang Weida, Zhang Youqiu, Xu Wenhao, Zhang Huijuan, Jiang Debiao, Yang Changsheng, Lin Liandi, Cui Zhantang, and Yang Xinchang. 1983. *The conodonts of North China and the Adjacent Regions.* Science Press, Beijing, pp. 1–223 [in Chinese].

Bogoyavlenskaya, O. V. 1969. K postroyeniyu klassifikatsii stromatoporoidey. *Paleont. Zh.* **1969** (4), 12–27.

———. 1982. Stromatoporaty pozdnego devona-rannego karbona. *Paleont. Zh.* **1982** (1), 33–38.

——— and Boiko, E. V. 1979. Sistematicheskoye polozheniye stromatoporat. *Paleont. Zh.* **1979** (1), 22–35.

Debrenne, F. 1983. Archaeocyathids: morphology and affinity. *In* Broadhead, T. W. (ed.). Sponges and Spongiomorphs: Notes for a Short Course. Univ. of Tennessee, Knoxville, Dept. Geol. Sci., *Studies in Geology,* no. 7, pp. 178–190.

———, Rozanov, A. Yu., and Webers, G. F. 1984. Upper Cambrian Archaeocyatha from Antarctica. *Geol. Mag.* **121,** 291–299.

——— and Vacelet, J. 1983. Archaeocyatha: is the sponge model consistent with their structural organization? *Abstr. 4th Int. Symp. Fossil Cnidaria, Washington, D.C.,* p. 4.

Dieci, G., Antonacci, A., and Zardini, R. 1968. Le spugne cassiane (Trias mediosuperiore) della regione dolomitica attorno a Cortina d'Ampezzo. *Bull. Soc. Paleont. Ital.* **7,** 94–155.

Dong Deyuan. 1964. Stromatoporoids from the Early Carboniferous of Kwangsi and Kueichow. *Acta Palaeont. Sinica* **12,** 280–299. [in Chinese, Engl. summary].

———. 1982. Lower Ordovician stromatoporoids of northern Anhui. *Acta Palaeont. Sinica* **21,** 577–583 [in Chinese, Engl. summary].

——— and Yang Jhinzhi. 1978. Lower Silurian stromatoporoids from northeastern Guizhon. *Acta Palaeont. Sinica* **17,** 421–438 [in Chinese, Engl. summary].

Finks, R. M. 1983. Pharetronida; Inozoa and Sphinctozoa. *In* Broadhead, T. W. (ed.). *Sponges and Spongiomorphs: Notes for a Short Course.* Univ. of Tennessee, Knoxville, Dept. Geol. Sci., *Studies in Geology,* no. 7, pp. 55–69.

Flower, R. H. 1976. New American Wutinoceratidae with review of actinoceroid occurrences in Eastern Hemisphere. *Mem. New Mexico Bur. Mines Mineral. Res.* **28,** 1–12.

Fonin, V. D. 1982. Klass Irregularia. *In* Rozanov, A. Yu. (ed.). Granitsa dokembriya i kembriya v geosinklinalnikh oblastyakh (oporniy razrez Salany-Gol, MNR). *Trans. Soviet-Mongolian Paleont. Exped.* **18,** 83–109. Nauka, Moskva.

Galloway, J. J. 1957. Structure and classification of the Stromatoporoidea. *Bull. Am. Paleont.* **37,** 345–480.

——— and St. Jean, J., Jr. 1961. Ordovician Stromatoporoidea of North America. *Bull. Am. Paleont.* **43,** 1–111.

Harris, A. G., Bergström, S. M., Ethington, R. L.,

and Ross, R. J., Jr. 1979. Aspects of Middle and Upper Ordovician conodont biostratigraphy of carbonate facies in Nevada and Southeast California and comparison with some Appalachian successions. *Brigham Young Univ. Geol. Stud.* **26** (3), 7–44.

Hartman, W. D. 1980. Systematics of the Porifera. *In* Hartman, W. D., Wendt, J. W., and Wiedenmayer, F. (eds). *Living and Fossil Sponges: Notes for a Short Course. Sedimenta VIII,* Univ. of Miami, Florida, pp. 24–35.

———. 1983. Modern and ancient Sclerospongiae. *In* Broadhead, T. W. (ed.). *Sponges and Spongiomorphs: Notes for a Short Course.* Univ. of Tennessee, Knoxville, Dept. Geol. Sci., *Studies in Geology,* no. 7, pp. 116–129.

——— and Goreau, T. E. 1970. Jamaican coralline sponges: their morphology, ecology and fossil representatives. *Zool. Soc. Lond. Symp.* **25**, 205–243.

Hill, D. 1972. Archaeocyatha. *In* Teichert, D. (ed.). *Treatise on Invertebrate Paleontology.* part E, vol. 1, 2nd ed. Geological Society of America and Univ. of Kansas Press, Lawrence, pp. E1–E158.

Kapp, U. S. 1974. Mode of growth of middle Chazyan (Ordovician) stromatoporoids, Vermont. *J. Paleont.* **48**, 1235–1240.

———. 1975. Paleoecology of Middle Ordovician stromatoporoid mounds in Vermont. *Lethaia* **8**, 195–207.

——— and Stearn, C. W. 1975. Stromatoporoids of the Chazy Group (Middle Ordovician), Lake Champlain, Vermont and New York. *J. Paleont.* **49**,163–186.

Kaźmierczak, J. 1971. Morphogenesis and systematics of the Devonian Stromatoporoidea from the Holy Cross Mountains, Poland. *Palaeont. Polon.* **26**, 1–150.

———. 1976. Cyanophycean nature of stromatoporoids. *Nature* **264**, 49–51.

———.1980. Stromatoporoid stromatolites: new insight into evolution of Cyanobacteria. *Acta Palaeont. Polon.* **25**, 243–251.

———. 1981. Evidences for cyanophyte origin of stromatoporoids. *In* Monty, C. (ed.). *Phanerozoic Stromatolites.* Springer, Berlin, pp. 230–241.

Khalfina, V. K. 1960a. Stromatoporoidei iz kembriiskikh otlozhenii Sibirii. *Trudy SNIIGGIMS* **8**, 79–83.

———. 1960b. Stromatoporoidei. *In* Biostratigrafiya paleozoya Sayano-Altayskoy Gornoy oblasti Tom. I. Nizhniy Paleozoy. *Trudy SNIIGGIMS* **19**, 82–84; 141–143.

———. 1971. Stromatoporoidei. *In* Sokolov, B. S., Ivanovskiy, A. B., and Krasnov, E. V. (eds.). Morphologiya i terminologiya kishechnopolostnykh. *Trudy Inst. Geol. Geofiz. Sib. Otd.* **133**, 14–22.

——— and Yavorsky, V. I. 1967. O drevne-

ishikh stromatoporoideyakh. *Paleont. Zh.* **1967** (3), 133–136.

——— and ———. 1973. Klassifikatsiya stromatoporoidey. *Paleont. Zh.* **1973** (2), 19–34.

——— and ———. 1974. K evolutsii stromatoporoidey. *In* Sokolov, B. S. (ed.). *Drevniye Cnidaria,* I. Nauka, Novosibirsk, pp. 38–45.

Klappa, C. F. and James, N. P. 1980. Small lithistid sponge bioherms, early Middle Ordovician Table Head Group, western Newfoundland. *Bull. Can. Petrol. Geol.* **28**, 435–451.

Kobayashi, T. 1969. Stratigraphy of the Chosen Group in Korea and South Manchuria, sect. D. The Ordovician of Eastern Asia and other parts of the continent. *J. Fac. Sci. Tokyo Univ.* **17** (2), sect. 2 163–316.

McIlreath, I. and Aitken, J. D. 1976. *Yoholaminites* (Middle Cambrian), problematical calcareous sediment-stabilizing organism. *Geol. Assoc. Can. Prog. Abstr.* **1**, 84.

McLaren, D. J. 1970. Time, life and boundaries. *J. Paleont.* **44**, 801–815.

———. 1983. Bolides and biostratigraphy. *Bull. Geol. Soc. Am.* **94**, 313–324.

Mori, K. 1970. Stromatoporoids from the Silurian of Gotland, part 2. *Stockh. Contr. Geol.* **22**, 1–152.

———. 1978. Stromatoporoids from the Silurian of the Oslo Region, Norway. *Norsk Geol. Tidsskrift* **58**, 121–144.

———. 1982. Coelenterate affinities of stromatoporoids. *Stockh. Contr. Geol.* **37**, 167–175.

———. 1984. Comparison of skeletal structures among stromatoporoids, sclerosponges and corals. *Paleontogr. Am.* **54**, 354–357.

Nestor, H. E. 1962. Reviziya stromatoporoidei opisannykh F. Rozonom v 1867 godu. *Trudy Geol. Inst. AN Est. SSR* **9**, 3–23.

———. 1964. Stromatoporoidei ordovika i llandoveri Estonii. *Inst. Geol. AN Est. SSR, Tallinn,* pp. 1–112.

———. 1966a. O drevneishikh stromatoporoideyakh. *Paleont. Zh.* **1966** (2), 3–12.

———. 1966b. Stromatoporoidei wenloka i ludlowa Estonii. *Inst. Geol. AN Est. SSR, Tallinn,* pp. 1–87.

———. 1978. Evolutsiya i usoviya obitaniya paleozoiskikh stromatoporat. Dissertatsiya na soiskanie uchenoy stepeni doktora geologo-mineralogicheskikh nauk Avtoreferat, *Inst. Geol. AN Est. SSR.,* Tallinn, pp. 1–38.

———. 1981. The relationship between stromatoporoids and heliolitids. *Lethaia* **14**, 21–25.

Newell, N. D. 1987. An outline history of tropical organic reefs. *Am. Mus. Novitates* **2465**, 1–37.

Ott, E. 1967. Segmentierte Kalkschwämme (Sphinctozoa) aus der alpinen Mitteltrias und ihre Bedeutung als Riffbildner im Wettersteinkalk. *Abh. Bayer. Akad. Wiss. Mathem.-Naturwiss. Kl.* **131**, 1–96.

Ozaki, K. E. 1938. On some stromatoporoids

from the Ordovician limestone of Shantung and South Manchuria. *J. Shanghai Sci. Inst., Sect. 2* **2**, 205–223.

Palmer, T. J. and Fürsich, F. T. 1981. Ecology of sponge reefs from the Upper Bathonian of Normandy. *Palaeontology* **24**, 1–23.

Paquette, J., Stearn, C. W., and Klappa, C. F. 1983. An enigmatic fossil of sponge affinities from Middle Ordovician rocks of western Newfoundland. *Can. J. Earth Sci.* **20**, 1501–1512.

Petryk, A. A. 1982. Aulacerid ecostratigraphy of Anticosti Island, and its bearing on the Ordovician–Silurian boundary and the Upper Ordovician glacial episode. *Proc. 3rd North Am. Paleont. Conv.* **2**, 393–399.

Pickett, J. W. and Jell, P. A. 1983. Middle Cambrian Sphinctozoa (Porifera) from New South Wales. *Mem. Assoc. Australas. Palaeontol.* **1**, 85–92.

Pratt, B. R. and James, N. P. 1982. Cryptalgal-metazoan biotherms of early Ordovician age in the St. George Group, western Newfoundland *Sedimentology* **29**, 543–569.

Radugin, K. V., 1936. Nekotorye tselenteraty iz nizhnego silura gornoi Shorii. *Materialy po geologii zapadno-sibirskogo Kraya* **35**, 89–106.

Riding, R. and Kershaw, S. 1977. Nature of stromatoporoids. *Nature* **268**, 178.

Rigby, J. K., 1966. Microstructure and classification of an Ordovician sponge *Dystactospongia madisonensis* Foerste from Indiana. *J. Paleont.* **40**, 1127–1130.

Ross, R. J., Jr., Adler, F. J., Amsden, T. W., Bergström, D., Bergström, S. M., Carter, C., Churkin, M., Cressman, E. A., Derby, J. R., Dutro, J. T., Jr., Ethington, R. L., Finney, S. C., Fisher, D. W., Fisher, J. H., Harris, A. G., Hintze, L. F., Ketner, K. B., Kolata, D. L., Landing, E., Neuman, R. B., Sweet, W. C., Pojeta, J., Jr., Potter, A. W., Rader, E. K., Repetski, J. E., Shaver, R. H., Thompson, T. L., and Webers, G. F. 1982. The Ordovician System in the United States. *Int. Union Geol. Sci.* **12**, 1–73.

Sayutina, T. A. 1980. Rannekembriyskoye semeystvo Khasaktiida n. fam.—vozmozhnyye stromatoporaty. *Paleont. Zh.* **1980** (4), 13–28.

———. 1982. Klass Incertae Sedis. *In* Rozanov, A. Yu. (ed.). Granitsa dokembriya i kembriya v geosinklinalnikh oblastyakh (oporniy razrez Salany-Gol, MNR). *Trans. Soviet-Mongolian Paleont. Exped.* **18**, Nauka, Moskva, pp. 66–68.

Semeniuk, V. 1971. Subaerial leaching in the limestones of the Bowan Park Group (Ordovician) of central western New South Wales. *J. Sed. Petrol.* **41**, 939–950.

Senowbari-Daryan, B. and Schafer, P. 1983. Zur Sphinctozoen-Fauna der obertriadischen Riffkalke ("Pantokratorkalke") von Hydra, Griechenland. *Geol. et Palaeont.* **17**, 179–205.

Stearn, C. W. 1966. The microstructure of stromatoporoids. *Palaeontology* **9**, 74–124.

———. 1975. The stromatoporoid animal. *Lethaia* **8**, 89–100.

———. 1979. Biogeography of Devonian stromatoporoids. *Palaeont. Assoc. Spec. Pap.* **23**, 229–232.

———. 1980. Classification of the Paleozoic stromatoporoids. *J. Paleont.* **54**, 881–902.

———. 1982. The unity of the Stromatoporoidea. *Proc. 3rd North Am. Paleont. Conv.* **2**, 511–516.

———. 1983. Stromatoporoids: affinity with modern organisms. *In* Broadhead, T. W. (ed.) *Sponges and Spongiomorphs: Notes for a Short Course.* Univ. of Tennessee, Knoxville, Dept. Geol. Sci., *Studies in Geology,* no. 7, pp. 164–166.

———. 1984. Growth forms and macrostructural elements of the coral-like sponges. *Paleontogr. Am.* **10**, 315–325.

——— and Hubert, C. 1966. Silurian stromatoporoids of the Matapedia Temiscouata area, Quebec. *Can. J. Earth Sci.* **3**, 31–48.

Stock, C. W. 1981. *Cliefdenella alaskaensis* n. sp. (Stromatoporoidea) from the Middle/Upper Ordovician of central Alaska. *J. Paleont.* **55**, 998–1005.

———. 1983. Stromatoporoids: geologic history. *In* Broadhead, T. W. (ed.). *Sponges and Spongimorphs: Notes for a Short Course.* Univ. of Tennessee, Knoxville, Dept. Geol. Sci., *Studies in Geology,* no. 7, pp. 167–172.

———. 1984. The function of tube-pillars in *Cliefdenella* (Stromatoporoidea) inferred by analogy with *Calcifibrospongia* (Sclerospongea). *Paleontogr. Am.* **10**, 349–353.

Toomey, D. F. 1970. An unhurried look at a Lower Ordovician mound horizon, southern Franklin Mountains, west Texas. *J. Sed. Petrol.* **40**, 1318–1334.

———. 1981. Organic-buildup constructional capability in Lower Ordovician and Late Paleozoic mounts. *In* Gray, J. Boucot, A. J. and Berry, W. B. N. (eds.). *Communities of the Past.* Hutchinson Ross, Stroudsburg, Pa., pp. 35–68.

——— and Babcock, J. A. 1983. Precambrian and Paleozoic algal carbonates, west Texas–southern New Mexico. *Prof. Contrib. Colo. School Mines* **11**, 1–345.

——— and Ham, W. E. 1967. *Pulchrilamina,* a new mound-building organism from Lower Ordovician rocks of west Texas and southern Oklahoma. *J. Paleont.* **41**, 981–987.

——— and Nitecki, M. H. 1979. Organic buildups in the Lower Ordovician (Canadian) of Texas and Oklahoma. *Fieldiana: Geol.* **2**, 1–181.

Vlasov, A. N. 1961. Kembriiskie stromatoporoidei. *Paleont. Zh.* **1961** (3), 22–32.

————. 1967. O rode *Altaicyathus* Volgodin. *Paleont. Zh.* **1967** (1), 120.

Walker, K. R. and Alberstadt, L. P. 1975. Ecological succession as an aspect of structure in fossil communities. *Paleobiology* **1**, 238–257.

Webby, B. D. 1969. Ordovician stromatoporoids from New South Wales. *Palaeontology* **12**, 637–662.

————. 1979a. The Ordovician stromatoporoids. *Proc. Linn. Soc. N.S.W.* 103, 83–121.

————. 1979b. The oldest Ordovician stromatoporoids from Australia. *Alcheringa* **3**, 237–251.

————. 1979c. Ordovician Stromatoporoids from the Mjøsa district, Norway. *Norsk Geol. Tidsskrift* **59**, 199–211.

————. 1980. Biogeography of Ordovician stromatoporoids. *Palaeogeogr., Palaeoclimatol., Palaeoecol.* **32**, 1–19.

————. 1984a. Ordovician reefs and climate: a review. *In* Bruton, D. L. (ed.). *Aspects of the Ordovician System.* Universitetsforlaget, Oslo, pp. 89–100.

————. 1984b. Early Phanerozoic distribution patterns of some major groups of sessile organisms. *Proc. 27th Int. Geol. Congr.* **2**, 193–208.

———— and Banks, M. R. 1976. *Clathrodictyon* and *Ecclimadictyon* (Stromatoporoidea) from the Ordovician of Tasmania. *Pap. Proc. R. Soc. Tasm.* **110**, 129–137.

———— and Kruse, P. D. 1984. The earliest heliolitines: a diverse fauna from the Ordovician of New South Wales. *Palaeontogr. Am.* **10**, 164–168.

———— and Morris, D. G. 1976. New stromatoporoids from the Ordovician of New South Wales. *J. Proc. R. Soc. N.S.W.* **109**, 125–135.

———— and Packham, G. H. 1982. Stratigraphy and regional setting of the Cliefden Caves Limestone Group (Late Ordovician), central-western New South Wales. *J. Geol. Soc. Aust.* **29**, 297–317.

———— and Rigby, J. K. 1985. Ordovician sphinctozoan sponges from central New South Wales. *Alcheringa* **9**, 209–220.

————, VandenBerg, A. H. M., Cooper, R. A., Banks, M. R., Burrett, C. F., Henderson, R. A., Clarkson, P. D., Hughes, C. P., Laurie, J., Stait, B., Thomson, M. R. A., and Webers, G. F. 1981. The Ordovician System in Australia, New Zealand and Antarctica: correlation chart and explanatory notes. *Publ. Int. Union Geol. Sci.* **6**, 1–64.

————, Wyatt, D., and Burrett, C. 1985. Ordovician stromatoporoids from the Langkawi Islands, Malaysia. *Alcheringa* **9**, 159–166.

Yabe, H. and Sugiyama, T. 1930. On some Ordovician stromatoporoids from South Manchuria, North China and Chosèn (Corea), with notes on two new European forms. *Sci. Rep. Tohoku Imp. Univ. Ser. 2 (Geol.)* **14** (1), 47–62.

Yavorsky, V. I. 1932. Ein Stromatoporenfund im Kambrium. *Zentralbl. Min. Geol. Paläont.* **B**, 613–616.

————. 1940. Klass Hydrozoa, poryadok Stromatopory. *In Atlas rukovodyashchikh form iskopaemykh faun SSSR,* vol. 1, *Kembrii.* Vses. Nauchno-Issledovat. Inst. (VSEGEI), pp. 100–103.

————. 1947. Nekotorye paleozoiskie i mesozoiskie Hydrozoa, Tabulata i Algae. *Monogr. paleontologii SSSR* **20**(1), 1–30. VSEGEI, Leningrad-Moskva.

Zhuravleva, I. T. and Miagkova, E. J. 1972. Archaeata, novaya gruppa organizomov paleozoya. XXIV. *Mezhd. geol. Kongr., Trudy sov. geol., paleont.,* Nauka, Moskva, pp. 7–14.

———— and ————. 1974. Sravnitelenaya kharakteristika Archaeata i Stromatoporoidea. *In* Sokolov, B. S. (ed). *Drevniye Cnidaria.* vol. 1. Nauka, Novosibirsk, pp. 63–70.

13. PROBLEMATICA FROM THE MIDDLE CAMBRIAN BURGESS SHALE OF BRITISH COLUMBIA

DEREK E. G. BRIGGS AND SIMON CONWAY MORRIS

Problematic fossils are no stranger to the paleontologist, but the term Problematica has been applied in a number of different ways. Hofmann (1971), for example, defined them as "structures of unknown origin," neither clearly organic (fossils) nor inorganic (pseudofossils). Häntzschel (1975), however, applied the term to organic remains that included trace fossils and some body fossils, the latter of "doubtful or completely uncertain classificatory status." Many of the body fossils included in Häntzschel's compilation (1975) are poorly preserved, and this reason alone could be enough to explain why their affinities remain in doubt. Such a restriction, however, scarcely applies to the majority of fossils discussed in this chapter. Here the term problematica is used to denote taxa with distinctive body plans that do not obviously conform to any living phylum, and represent extinct phyla. Their perceived problematic status results at least partly from our relative ignorance of evolutionary events during the early Phanerozoic metazoan radiations.

The Burgess Shale was discovered in 1909 by Charles D. Walcott on the ridge connecting Mount Wapta and Mount Field in southern British Columbia (see Whittington 1971, 1980 for review of the history of research on the fauna). The exceptional preservation of the fossils accounts for the significance of the Burgess Shale biota as a unique glimpse of Cambrian life. About 120 genera are represented, and only 33 of these genera normally fossilize in open-shelf conditions (Conway Morris and Robison 1982). The preservation, environmental setting, and general aspect of the Burgess Shale fauna have been reviewed elsewhere (Conway Morris 1979a,b, 1982; Conway Morris and Whittington 1979; Whittington 1980, 1981, 1982) and are not repreated here.

In his preliminary descriptions of the Burgess Shale fauna, Walcott assigned all the animals to classes based on living organisms. As Yochelson (1977, p. 440) observed, "the conventional assumption of past decades was that there are no extinct phyla." An exception to this rule is the Archaeocyatha, which were already given phylum status by Okulitch (1955). By the mid-1970s some paleontologists working on Cambrian shelly fossils admitted the possibility of extinct phyla, and the taxonomic status of some groups was duly elevated: Hyolitha (Runnegar et al. 1975), Agmata (Yochelson 1977). Bengtson (1977), in his review of early Cambrian problematica, noted that only 3 of the 11 small shelly fossil groups from the base of the Tommotian (basal Cambrian) represent known phyla. It is research on the Burgess Shale, however, that has extended this view to encompass soft-bodied forms, emphasizing that a significant number of Cambrian forms cannot be assigned to living phyla. This demonstration is one of the more important results of the Burgess Shale project (Conway Morris 1979a). It was only in the early 1970s that it was realized that there are any Burgess Shale problematica, in the sense used in this paper.

The Burgess Shale yields representatives of five "shelly" phyla that occur in standard Cambrian shelf conditions: Trilobita (Arthropoda), Brachiopoda, Porifera, Mollusca, and Echinodermata. The Hyolitha are accorded phylum status by some authorities (Runnegar et al. 1975; Runnegar 1980) but still included in the mollusks by others (e.g., Marek and Yochelson 1976; Dzik 1978). Soft-bodied representatives of five phyla are also present: Coelenterata, Priapulida, Annelida, Hemichordata, and Chordata. The phylum Ctenophora may be represented by *Fasciculus vesanus* but, as discussed below, this animal may belong to an extinct major group. A small number of specimens of a new animal from an adjacent locality on Mount Stephen (locality 9 of Collins et al. 1983), which may belong to the phylum Chaetognatha, are presently under study by Collins and Briggs. Thus, about 12 of the 32 living animal phyla (*sensu* Margulis and Schwartz 1982) may be represented in the Burgess Shale (this does not necessarily imply, of course, that the remainder were absent).

Twenty Burgess Shale animals that appear to defy assignment to any living phylum are reviewed in this paper. In recent reviews some of them have been described as "miscellaneous worms" (Conway Morris 1979a) or "miscellaneous" (Whittington 1980). As no clear interrelationships exist between these various taxa, they are treated alphabetically. This review is not exhaustive in that only brief summaries are given of published species, and there are a number of undescribed forms that may represent yet more new groups. Most of these animals are only known from the Walcott Quarry, but the occurrence of some with other Burgess Shale taxa at adjacent localities in British Columbia (Collins et al. 1983) and even further afield (e.g., *Anomalocaris;* see Whittington and Briggs 1985) indicates that they probably formed part of the normal Middle Cambrian open-shelf biota of North America (Conway Morris and Robison 1982; Collins et al. 1983).

• *Amiskwia sagittiformis* Walcott 1911

This is a rare species; only five specimens are known. Although this worm (Fig. 1) has been interpreted as both a chaetognath (Walcott 1911) and a pelagic nemertean (Owre and Bayer 1962; Korotkevich 1967), neither interpretation appears tenable (Conway Morris 1977a). Some striking similarities in overall body shape, especially between *Amiskwia* and certain species of the pelagic nemerteans (i.e., dorsoventrally compressed and gelatinous body, prominent fins), are regarded as functionally convergent, representing adaptations to a pelagic mode of life. The principal reason for rejecting a nemertean relationship lies in an inability, despite extraordinary soft-part preservation, to demonstrate in *Amiskwia* the presence of either the distinctive rhynchocoel and the enclosed proboscis, or characteristically located reproductive organs. Comparisons with chaetognaths fail on account of the absence of the prominent feeding apparatus, a complex and robust organ that is unlikely to have escaped preservation given the exceptional conditions of taphonomy (Conway Morris 1977a).

Conway Morris (1977a) suggested that, like the nemerteans, *Amiskwia* may have evolved from a turbellarian ancestor, and could have possessed a grade of organization broadly comparable with nemerteans. It seems possible that *Amiskwia* belongs to an extinct phylum.

• *Anomalocaris* Whiteaves 1892

Anomalocaris is the largest Burgess Shale animal; individuals reached lengths of up to 0.5 m. Two species are known from the Burgess Shale, *A. canadensis* and *A. nathorsti.* Isolated appendages of *A. canadensis* have been known since Whiteaves (1892) described them from Mount Stephen (Briggs 1979). They were originally misinterpreted as the body of an arthropod, and the ventral spines as either limbs or pleural lobes. Briggs (1979) recognized that the specimens represent appendages, and considered that the animal was an arthropod. He pointed out that limb morphology gives little indication of affinity and concluded that the arthropod could belong to "almost any class." The only known specimen of the body of *A. canadensis* (incomplete) has recently been described (Whittington and Briggs 1985) and shows that only one pair of the segmented limbs was borne by each individual, in the head.

There are about 10 specimens showing the body of *A. nathorsti* (Fig. 3). The mouth is surrounded by a circlet of toothed plates that commonly occurs isolated, presumably because, like the appendages, it consisted of more heavily sclerotized cuticle than the rest of the animal. The circlet was originally described by Walcott (1911) as a medusoid coelenterate, *Peytoia nathorsti.* [The toothed circlet is assumed to have been present in *A. canadensis* also (Whittington and Briggs 1985), although its presence in the only near-complete specimen known cannot be unequivocally demonstrated.] *Anomalocaris nathorsti,* like *A. canadensis,* bore a pair of segmented limbs in the head (Fig. 2), but they differ in structure, with fewer segments and bearing a series of ventral spinose blades. These limbs also occur in isolation, though more rarely, and were described as "Appendage F" by Briggs (1979).

The rest of the body appears to be similar in both species of *Anomalocaris* from the Burgess Shale. Behind the mouth the head region bore 3 pairs of semicircular flaps. The trunk bore 11 pairs of triangular lateral lobes that overlapped significantly and were apparently strengthened by longitudinal rays. A lamellar structure, possibly a gill, was attached dorsal to each flap and lobe. The trunk terminated bluntly and abruptly.

Anomalocaris nathorsti, at least, had large lateral eyes. A functional analysis (Whittington and Briggs 1985) suggests that the lateral lobes were employed in swimming, probably using a metachronal propulsive wave. This, combined with morphology of the head appendages, particularly in *A. nathorsti,* suggests that *Anomalocaris* was a formidable predator. Although the segmented head limbs are arthropod-like, and the body metameric, the jaw apparatus and overlapping lobes do not suggest affinity with any arthropod or other known group. *Anomalocaris* is considered to represent an extinct phylum. Isolated limbs occur at Lower and Middle Cambrian localities in British Columbia other than the Burgess Shale (Briggs 1979; Collins et al. 1983), and in the Lower Cambrian of California (Briggs and Mount 1982) and Pennsylvania (Briggs 1979). Neither the genus, nor any obvious relative, is known outside the Cambrian of North America.

• *Aysheaia pedunculata* Walcott 1911

Walcott (1911) originally described *Aysheaia* (Fig. 4) as a polychaete, but the similarity between it and Onychophora was drawn to his attention shortly after the paper appeared (see Walcott 1931, p. 8). *Aysheaia* had an elongate, subcylindrical body bearing 10 pairs of unbranched walking limbs, and an additional pair of limbs that projected laterally from the head region. The cuticle was unmineralized and flexible, and both body and limbs were annulated. The annulations on the dorsal side of the trunk were sharp crested and bore tubercles; elsewhere they were less

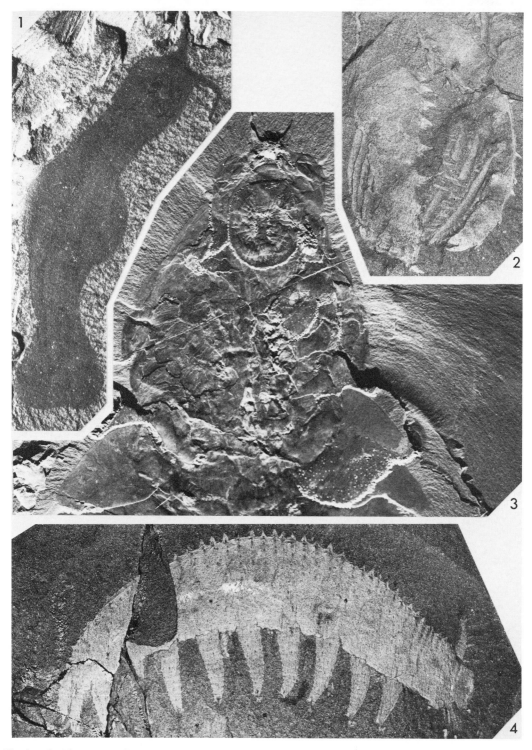

Fig. 1. *Amiskwia sagittiformis.* Dorsoventral compaction. USNM 57644, part, low-angle UV light from south-west. ×5.

Figs. 2,3. *Anomalocaris nathorsti.*

Fig. 2. Isolated pair of anterior limbs. USNM 57490, part, reflected UV light. ×2.

Fig. 3. Anterior part of body. Dorsoventral compaction. ×1. The limbs seen in Fig. 2 project on either side of the anterior end of the head and are compacted pointing down into the sediment. They flank the jaw, a subcircular structure made up of radiating plates, which has been exposed by removing the cuticle lying dorsal to it. Some of the large lateral trunk lobes that were used in locomotion are evident posteriorly. USNM 274143, part, low-angle UV light from northeast. ×1.

Fig. 4. *Aysheaia pedunculata.* Lateral compaction, right side. USNM 139206b, counterpart, reflected UV light. ×3.3.

prominent. The body consisted of at least 12 somites. There was no pronounced division between the head region or the posteriormost pair of limbs and the rest of the trunk. The mouth was situated terminally (rather than ventrally) and surrounded by about six anteriorly directed papillae. The pair of laterally projecting limbs in the head region bore movable spines, but these did not extend far enough to insert food into the mouth. About 20 specimens of *A. pedunculata* are known from the Burgess Shale, and of these about one-third are associated with sponges. Whittington (1978) suggested that the distal claws on the limbs would have provided a means of gripping a sponge; the papillae surrounding the mouth may have functioned in some preliminary external digestion of the soft tissue. *Aysheaia* probably did not kill the sponge, and this example of a predator–prey relationship may therefore be considered a kind of parasitism (Briggs and Whittington 1985). Sponge colonies would also have afforded *Aysheaia* some concealment and protection.

Hutchinson (1930) assigned *Aysheaia* to a separate extinct order of the Onychophora, and this view has prevailed in most subsequent publications. Delle Cave and Simonetta (1975) noted the similarity between *Aysheaia* and tardigrades, and concluded that it is "morphologically intermediate between the Onychophora and Tardigrada" (1975, p. 79) and warrants the status of a separate class (p. 76). Whittington (1978, p. 195) did not assign *A. pedunculata* to any higher taxon but regarded it "as the sole known representative of an early group of soft-bodied, metamerically segmented, lobopodial animals." He considered that this group might have given rise to both the Onychophora, included in the arthropod phylum Uniramia, and the Tardigrada. He nonetheless concluded that *Aysheaia* "does not fit readily into any extant higher taxon," thus placing it in the category of problematica. Whittington (in Conway Morris 1982) later assigned *Aysheaia* to "Arthropoda: phylum uncertain." Briggs (1983, p. 15) described *Aysheaia* as uniramian-like but considered that the lack of jaws and slime papillae excluded it from the group. Robison (1985), however, considers that the lack of jaws is a primitive condition and not indicative of relationships, and that the differences between the anterior appendages of *Aysheaia* and those of extant onychophorans may reflect adaptation to marine as opposed to terrestrial habitats rather than a lack of close affinity. He suggests the possibility that the annulations on *Aysheaia* represent a ringed system of vascular channels, "an important synapomorphic character indicating close phylogenetic unity of *Aysheaia* and the terrestrial Onychophora." McKenzie (1983) has drawn attention to the similarities between *Aysheaia* and a bizarre group of mainly deep-sea holothurians, the elasipodids. The similarities between *Aysheaia* and

the elasipodids are only slightly less striking than those between *Aysheaia* and the onychophorans (most significantly perhaps, elasipodids lack terminal claws on the lobopodial limbs). This serves to emphasize that the similarities between *Aysheaia* and other groups may be convergent; the role of *Aysheaia* in the ancestry of the Uniramia remains uncertain.

Aysheaia is also represented, as a new species *A. prolata* (Robison, 1985), in the Middle Cambrian Wheeler Formation of Utah. The remainder of the fossil record of onychophoran-like animals is confined to the Carboniferous. Thompson and Jones (1980) described two specimens of a lobopod animal, *Helenodora inopinata,* from the Middle Pennsylvanian Mazon Creek beds of Illinois. They concluded that "in all preserved characteristics" *Helenodora* could be classified in the Onychophora, but the lack of clear details of the head prevented them from making the assignment unequivocally. Rolfe et al. (1982) reported a "virtually identical onychophoran" to *Helenodora* from the Stephanian at Montceau-les-Mines, France.

• *Banffia constricta* Walcott 1911

Walcott's original description (1911) of this relatively abundant worm (Fig. 5) was brief. It is one of the relatively few taxa also to occur in the Raymond Quarry located some 20 m above the Phyllopod bed (Walcott Quarry), and it may also occur higher in the section. Walcott placed *Banffia* with reservations in the Gephyrea, a now-discarded polyphyletic taxon that was erected to embrace echiurans, sipunculans, and priapulids. Conway Morris (in Valentine 1977) provided a few observations, but the paleobiology and paleoecology of *Banffia* await a thorough appraisal. A prominent constriction separated an anterior section, with prominent annuli possibly arranged helically, from a slightly larger posterior section [this orientation being opposite to that proposed by Walcott (1911)]. On the assumption that *Banffia* possessed a fluid-filled body cavity, the constriction was tentatively interpreted as a hydrostatic damper that served to isolate either section, allowing them to act semiindependently (Conway Morris, in Valentine 1977). Soft-bodied worms such as polychaetes and oligochaetes use septa to compartmentalize the body cavity, as well as fluctuations in hydrostatic pressure, and the constriction in *Banffia* may represent an analogous solution to the problem of controlling pressure variations. Little is known yet of the internal anatomy, but examination of as-yet unstudied material is expected to reveal some unusual features. The mode of life of this worm, which could grow in excess of 80 mm in length, is also problematic. Despite Walcott's assignment (1911) to the now-obsolete Gephyrea, there is no convincing evidence to link *Banffia* with any of the com-

Fig. 5. *Banffia constricta.* Incomplete specimen showing part of anterior annulate section, separated by a constriction from the posterior section. ×1.5. USNM 57638, ordinary light, specimen under water.

Fig. 6. *Dinomischus isolatus.* MCZ 1083, reflected UV light. ×6.

Fig. 7. *Hallucigenia sparsa.* Lateral compaction, right side. USNM 83935, low-angle UV light from north. ×7.

Fig. 8. *Nectocaris pteryx.* Lateral compaction, right side. USNM 198667, reflected UV light. ×7.

Fig. 9. *Odontogriphus omalus.* Feeding apparatus indicated by arrow. ×2. USNM 196169a, counterpart, ordinary light, specimen under alcohol.

Fig. 10. *Oesia disjuncta.* USNM 57630, reflected UV light. ×1.5.

ponent phyla. The relationships of *Banffia* are obscure, and it may represent an extinct phylum.

• *Dinomischus isolatus* Conway Morris 1977

The body of the animal (Fig. 6) consisted of a cuplike calyx that was supported on a comparatively long stalk, terminating in a slightly bulbous holdfast and apparently lacking any rootlike extensions that would serve to embed the organism in the soft substrate. The calyx evidently housed the principal organs including a U-shaped gut with openings located on the upper surface. Elongate structures ran from the calyx wall to the gut, and if they acted as suspensory fibers this provides a strong indication that the body cavity was originally fluid filled. The top of the calyx was encircled by lanceolate platelike structures (bracts) that presumably generated feeding currents that ultimately channeled food to the mouth. This species is rare; three specimens were originally described (Conway Morris 1977c). A probable fourth specimen (ROM 43194) was discovered at Mount Stephen (locality 10 of Collins et al. 1983). This specimen shows indifferent preservation and is incomplete; in particular much of the stalk is missing owing to rock breakage. The expanded midgut or stomach, however, is evident and is surrounded by structures that may represent the suspensory fibers.

Dinomischus was compared (Conway Morris 1977c) with a variety of stalked metazoans, including pelmatozoan echinoderms, sponges, ectoprocts, and entoprocts. It appears to resemble entoprocts most closely, but the similarities are not compelling (see also Emschermann 1982), and a relationship is either remote or entirely superficial. *Dinomischus* has no obvious affinity with other metazoans and presumably belongs to an extinct phylum.

• *Fasciculus vesanus* Simonetta and Delle Cave 1978

Simonetta and Delle Cave (1978, figs. 1 and 2) briefly described this unusual metazoan on the basis of a single specimen, and tentatively assigned the genus to the coelenterates. In addition, two specimens are known from the Burgess Shale, but the discovery of a further suite of specimens from a new soft-bodied locality adjacent to the Burgess Shale (locality 9, Collins et al. 1983) provides sufficient material for an extended redescription of the organisms (Conway Morris and Collins, in preparation). Collins et al. (1983, fig. 4) drew attention to the similarity of *Fasciculus* with the Ctenophora (see also Conway Morris 1979a, table 1), a group with an otherwise negligible fossil record (Stanley and Stürmer 1983; Conway Morris 1985b). Prominent elongate bands with multiple transverse structures have a striking resemblance to the comb rows of ctenophores. However, the number of bands exceeds

the total of eight found in all modern ctenophores. Other features of the internal anatomy of *Fasciculus* await detailed study, and the allocation of this genus to the ctenophores may be the subject of revision.

• *Hallucigenia sparsa* Walcott 1911

A redescription of this bizarre creature (Fig. 7) depicted it with seven pairs of stiltlike legs, supporting an elongate trunk with a cylindrical cross section (Conway Morris 1977b). From the dorsal surface of the trunk there arose seven flexible tentacles in a row, succeeded posteriorly by a compact group of shorter tentacles. Each of the longer tentacles terminated in a bifid snapperlike structure. In common with the spines they are composed of a bright reflective mineral that appears to have replaced a well-cuticularized material—in the case of the snappers, significantly tougher than the rest of the tentacle. The tentacles appear to have extended into the trunk to join an internal longitudinal structure that might represent the alimentary canal. Posteriorly the trunk terminated by recurving upward and forward, while anteriorly it expanded into a poorly defined area that may represent a head. Although *Hallucigenia* is relatively abundant, few specimens are well preserved and the great majority are only recognizable on account of the row of spine pairs that evidently possessed a greater resistance to decay (see above).

A unique specimen (MCZ 1084) appears to show a number ($\geq$ 18) of individuals associated with a large mass of indeterminate organic matter. This example may represent gregarious feeding by *Hallucigenia* on a corpse. *Hallucigenia* is reconstructed therefore as an epifaunal scavenger, crossing the seafloor on its spines, which may have operated with a metachronal movement. The spines terminate in sharp points and apparently lack "snowshoe" adaptations (see Thayer 1975) that would be comparable, for example, to the distal splay of short spines on the appendages of various arthropods that inhabited the soft substrates of the Devonian Hunsrückschiefer (Stürmer and Bergström 1978; Seilacher 1962). However, in the absence of any evidence for substantial hard parts that could have acted as a counterweight, the presumed near-neutral buoyancy of *Hallucigenia* suggests that even progress across muddy substrates could be feasible unless there was a deep layer of flocculent debris. Indeed, the calculated lightness of *Hallucigenia* led La Barbera (in preparation) to suggest that either the animal was restricted to areas of very low current velocity (around <2.5 cm/second) or used its tentacles to help maintain stability.

While no animal like *Hallucigenia* is known either living or in the fossil record, there are various organisms that may provide more-or-less appropriate analogs. These include deep-sea organisms such as tripod fish, which, as their name suggests, rest and possibly "walk" with two elongate pectoral spines and a tail spine, as well as some elasipodid holothurians that move in herds over the seabed supported by elongate tube feet (e.g., Herring and Clarke 1971; Heezen and Hollister 1971; Lemche et al. 1976). Interestingly, in view of the preceding comments, they have only slight negative buoyancy, so that small disturbances send them swirling into the water (Barham et al. 1967).

The original interpretation of *Hallucigenia* as a polychaetous annelid (Walcott 1911, 1931) is untenable, and its ludicrous appearance makes it difficult to offer any coherent suggestions regarding its relationships and origins. Even in the best preserved specimens the region interpreted as the head is very poorly defined, perhaps because of leakage of body fluids during decay. The possibility that *Hallucigenia* represents part of a larger animal is difficult to reconcile with the MCZ specimen mentioned above, in that the individuals of *Hallucigenia* show no coherent arrangement. A specimen recently located by D. H. Collins (ROM 43045) consists of a jumbled mass of spines over a small area of shale, and cannot be obviously restored into a more complex organism. If *Hallucigenia* represents only part of an organism with the "head" representing the zone of detachment, this might enable it to occupy more hydrodynamically vigorous environments than allowed by La Barbera's (in preparation) calculations. Although there is no evidence to support such a reconstruction, a composite animal conceivably could have consisted of several *Hallucigenia* radiating from a central region, an arrangement vaguely reminiscent of an ophiuroid echinoderm. Conway Morris (1977b) earlier pointed out a possible comparison between *Hallucigenia* and echinoderms, but present information strongly suggests that this Cambrian animal represents an extinct phylum.

• *Nectocaris pteryx* Conway Morris 1976

The only known specimen (Fig. 8) was evidently noted by Walcott, but it was almost 50 years after his death that a description was published (Conway Morris 1976b). The specimen is preserved in lateral aspect, the body divided into a head region and elongate trunk. The head bore a prominent eye, presumably paired, a shieldlike unit covering the posterior portion, and anteriorly directed structures that may represent simple appendages. The trunk appears to have been segmented and bore narrow fins that were supported by fin rays. The anatomy of *Nectocaris* is consistent with a nektonic existence, while the large eyes, anterior (?) appendages, and streamlined shape are indicative of predatory activity, with food perhaps being captured by sudden lunges.

The cephalic shield recalls a crustacean carapace, but there is little other evidence to support a relationship with either the crustaceans or other arthropods. In conversation with one of us (S. C. M.), several workers have suggested that *Nectocaris* could represent a chordate. The fin rays recall a similar arrangement in the cephalochordates (amphioxus). However, other features that would support unequivocally a chordate relationship such as gill slits and notochord are not apparent. Present evidence suggests *Nectocaris* to be the only known representative of a major extinct group.

• *Odontogriphus omalus* Conway Morris 1976

Only one specimen is known (Fig. 9), representing an animal with an apparently dorsoventrally compressed body rich in gelatinous tissue and without obvious appendages or fins (Conway Morris 1976a). Transverse structures on the body may represent either annuli or segments. The alimentary canal appears to have been straight. The mouth was subterminal, located on the ventral surface, and was surrounded by a looped structure that is interpreted as a feeding apparatus. Within this apparatus there were toothlike structures, now preserved as flattened impressions or molds.

The affinities of *Odontogriphus* are obscure. The arrangement of the feeding apparatus bears a striking resemblance to the lophophore of brachiopods, phoronids, and ectoprocts, suggesting that it too may have been tentacular, although remains of any tentacles are rather equivocal. The three phyla listed here may share a common ancestry from a phoronid-like worm (e.g., Farmer et al. 1973; Valentine 1975; Rowell 1982), and it is conceivable that *Odontogriphus* was also so derived and could be included within the superphylum Lophophorata (but see Chapter 16 by Dzik, this volume). Although the original composition is not known, in size and shape the toothlike structures within the feeding apparatus resemble certain Cambrian conodonts, and *Odontogriphus* was identified tentatively as a possible conodontophorid (Conway Morris 1976a). More recent research has rendered this interpretation less likely (see Chapter 15 by Aldridge and Briggs, this volume). The Middle Cambrian age of *Odontogriphus* means that comparisons are best drawn with either protoconodonts or paraconodonts, as euconodonts did not appear until the Upper Cambrian (Bengtson 1976, 1983). However, the evidence that euconodonts evolved from paraconodonts (Bengtson 1976, 1983), and the documentation of a convincing euconodont animal from the Lower Carboniferous of Scotland (Briggs et al. 1983), which bears no significant resemblance to *Odontogriphus,* suggest that this Burgess Shale animal is not related to the euconodont–paraconodont stock. Striking similarities between at least some protoconodonts and the grasping spines of chaetognaths (Szaniawski 1982, 1983) suggest that the affinities of *Odontogriphus* are not close to the protoconodonts either; evidence of whether protoconodonts are related to the paraconodont–euconodont lineage is still equivocal (Bengtson 1983). It appears unlikely, therefore, that *Odontogriphus* represents a conodont animal in any accepted sense (Bengtson 1983), and the nature of the toothlike structures and their relationship to the rest of the feeding apparatus remain obscure. Even if *Odontogriphus* is a lophophorate, it appears to belong to a major new group.

• *Oesia disjuncta* Walcott 1911

Walcott (1911) described this worm (Fig. 10) as an annelid, suggesting that it inhabited a fragile tube. This interpretation was disputed by Lohmann (1922, 1933–1934), who on the basis of Walcott's illustrations presented detailed comparisons in favor of *Oesia* belonging to the Larvacea (or Appendicularia), a remarkable group of neotenous tunicates otherwise unknown from the fossil record. Notwithstanding Tarlo's support (1960; see also Howell 1962), the notion that *Oesia* is an annelid is almost certainly erroneous. The minute hooks at the anterior end that Walcott (1911) identified are apparently wanting, and there are few other features consistent with an annelid relationship (Conway Morris 1979b). Preliminary restudy of *Oesia* has not yet confirmed unequivocally Lohmann's suggestions (1922, 1933–1934), and it is more prudent to regard this worm as *incertae sedis* pending a complete reappraisal.

• *Opabinia regalis* Walcott 1912

There are about 30 known specimens of this species (Fig. 11). *Opabinia* was the first animal to be described as part of the present restudy (Whittington 1975) that clearly could not be assigned to a living phylum. Its morphology was sufficiently remarkable to provoke loud laughter from a Palaeontological Association audience in Oxford in 1972.

The head of *Opabinia* bore five pairs of eyes, and a long flexible proboscis divided at its distal extremity into opposing bundles of spines, which could presumably be brought together to pick up food. The trunk was divided into 15 divisions, each bearing a lateral lobe. The lobes overlapped posteriorly, and each bore a lamellar structure dorsally that probably functioned as a gill. A short posterior section of the trunk was not divided and bears three dorsolaterally directed thin blades forming a tail fan. La Barbera (in preparation) has developed Whittington's suggestion (1975) that *Opabinia* might have been able to swim with a metachronal movement of the lobes. La Barbera suggests a hydraulic mode of propulsion similar to that employed by some branchiopod crustaceans. Water would have been drawn into the space between successive lobes as they moved apart, and forced out posteriorly as they were brought closer together during the cycle. The tail fan would have functioned in stabilizing the animal during swimming.

Walcott (1912) considered *Opabinia* to be a branchiopod crustacean, but the resemblance is superficial. As Whittington (1975) pointed out. Størmer's (1944) placement of the animal in the now-discredited class Trilobitoidea was based on Raymond's misinterpretation (1935) of the lobes as pleura and his suggestion that trilobite-like limbs were present. The characters shared by *Opabinia* with arthropods and annelids, among other groups, are primitive, and it may represent an extinct phylum.

• *Pollingeria grandis* Walcott 1911

This enigmatic genus (Fig. 12) was described by Walcott (1911) in conjunction with the lepidote metazoan *Wiwaxia* (see below), with the scalelike remains being interpreted as dispersed sclerites of a *Wiwaxia*-like creature. The type material was reillustrated in a very brief redescription (Conway Morris 1985a), the main purpose of which was to refute any alliance with *Wiwaxia*. Walcott's description (1911) was based on material from the Burgess Shale, but additional specimens have been found at other localities on Fossil Ridge (between Mounts Wapta and Field), Mount Field and Mount Stephen (Collins et al. 1983). Although the scales of *Pollingeria* may occur in abundance over limited areas of shale, they never form a coherent assemblage consistent with their representing the now dispersed scaly armor of a metazoan. As Walcott (1911) noted, their shape is foliate, generally tapering at one end. The scales show a considerable degree of morphologic variation that either may be original or represent different degrees of contraction. The nature of the irregularly meandering structures on the scales, which in life probably had positive relief, is problematic. Walcott (1911) interpreted them as created by the meanderings of a commensal worm, but this interpretation seems unlikely. Superficially they resemble lengths of alimentary canal filled with fine-grained detritus, but their irregular distribution and apparent failure to form a connected whole militate against this interpretation. In some specimens what appear to be the meandering strands are partially detached from the body, presumably as a result of decay. Each scale, which in life may have been more inflated, may represent a separate organism. The affinities of *Pollingeria* remain enigmatic.

• *Portalia mira* Walcott 1918

The only known specimen (part and counterpart) occurs on a somewhat weathered surface and is relatively poorly preserved (Fig. 13). It awaits detailed restudy following Walcott's very brief descriptions (1918, 1931). It is partially superim-

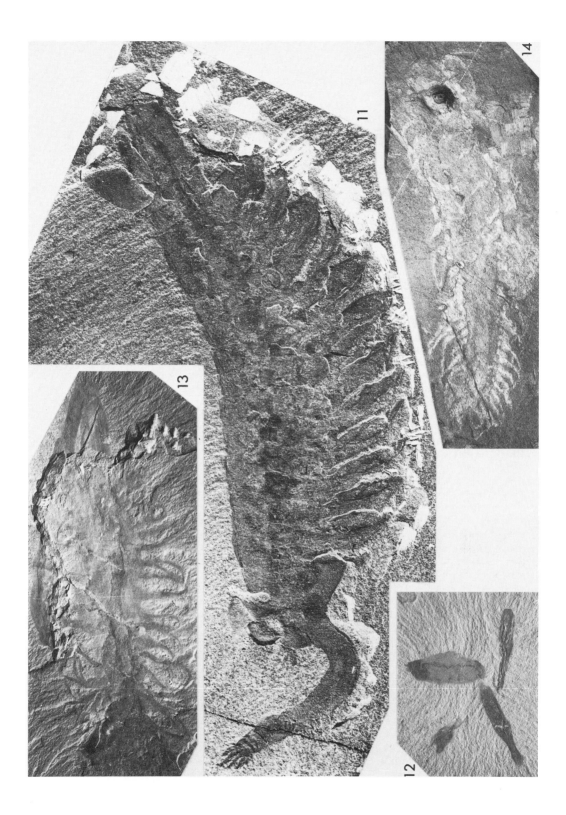

Fig. 11. *Opabinia regalis.* Lateral compaction, left side. USNM 57683, low-angle UV light from northwest. ×3.

Fig. 12. *Pollingeria grandis.* Scales. USNM 57641, low-angle UV light from northeast. ×2.

Fig. 13. *Portalia mira.* Partially superimposed by a specimen of *Mackenzia costalis.* USNM 83927, low-angle UV light from northeast. ×1.

Fig. 14. *Redoubtia polypodia.* USNM 83924, reflected UV light. ×1.5.

posed by a specimen of the possible actiniarian *Mackenzia costalis* [misidentified as *Miskoia* (= *Louisella,* see Conway Morris 1977d) by C. E. Resser, in Walcott 1931], presumably fortuitously (Walcott 1918). The body of *Portalia* was relatively long and contains a central strand that could represent the alimentary canal. Arising from the body along most of its length are elongate tentacular structures that appear to show bifurcation and further division, although whether this is original or preservational (e.g., decay) is not certain. At one end (? anterior) there is an indistinct area that may include the head. The affinities of *Portalia* remain unresolved (Conway Morris 1979b). Walcott (1918, 1931) proposed that it was a holothurian, an approach cautiously supported by Croneis and McCormack (1932) but decisively rejected by Frizzell and Exline (1955) and Madsen (1957), the latter arguing that *Portalia* was a sponge. Størmer's (1944) comment that he had "examined a specimen of the genus *Portalia* (similar to *Protocaris*) which has indications of gill-blades on the appendages" (see also Rolfe 1969) refers to a specimen of the arthropod *Odaraia alata,* a number of which were labeled *Portalia alata* in the Walcott collections (Briggs 1976, p. 12; 1981, p. 572). Durham (1971, 1974; N.B. in earlier paper *Portalia* misspelled *Pourtalia*) noted that *Portalia* could not be rejected as a holothurian without further study, but its status is probably best regarded as problematic pending its redescription.

• *Priscansermarinus barnetti* Collins and Rudkin 1981

This species was described on the basis of a unique slab preserving over 60 specimens from talus near the Walcott Quarry. Since then a few specimens have been found at two localities on Mount Field (Collins et al. 1983). An undescribed specimen from the Middle Cambrian of Utah may represent a new species. The animal consisted of two parts, an irregularly ovoid body covered by a carapace, and a thick stalk. The stalk was flexible, similar in length to the carapace, and usually tapers slightly distally terminating in a wider "disc."

Collins and Rudkin (1981) described *Priscansermarinus* as a "probable lepadomorph barnacle." They deduced the presence of the carapace based mainly on the darker area "covering most of the body." This gray area per se provides no substantive evidence for the outline or even presence of the suggested plates (Briggs 1983, p. 7); Collins and Rudkin (1981, p. 1014) suggested that they did not cover all of the capitulum at maturity. Reexamination of the material suggests that the ovoid outline of the body is defined by a carapace, apparently of two valves, which obscures any detail of structures within it. The preserved configuration results mainly from differences in orientation to the bedding. The specimen ROM 36064c (Collins and Rudkin 1981, fig. 1c, pl. 2, fig. 4) preserves a slight ridge around part of the perimeter that might represent the true margin of a valve (the valves are folded at the margins of the other specimens). The occurrence of a large number of specimens on a single slab suggests that *Priscansermarinus* may have been gregarious. There is no evidence that the specimens were attached to anything, although this remains a possible function for the terminal "disc."

Collins and Rudkin (1981) discussed and eliminated an affinity with a number of groups with an oval body and protruding stalk or foot: holothurians, tunicates, mollusks, and brachiopods! Their conclusion, that the fossil represents a probable barnacle, is however equivocal (Briggs 1983, pp. 7, 8). No evidence of appendages has been observed. Confirmation of Collins and Rudkin's interpretation "must await discovery of specimens with identifiable cirriped structures, such as cirri, within the capitulum" (Collins and Rudkin 1981, p. 1011).

The only fossil similar to *Priscansermarinus* is *Cyprilepas holmi* Wills 1962, from the Upper Silurian of Estonia, which Wills (1963) interpreted as a lepadomorph cirripede. The specimens were discovered attached to the cuticle of the eurypterid *Baltoeurypterus.* They consist of a chitinous bivalved shell and cylindrical stalk. As the specimens were extracted by dissolving the matrix in acid, the possibility that the valves were calcareous cannot be verified. No appendages or other soft parts are known. The affinities of *Cyprilepas,* like *Priscansermarinus,* are therefore equivocal. The preservation of a hexagonal pattern on the chitin of the carapace in some specimens, and traces of papillae on the stalk of others, are consistent with its interpretation as a cirripede, however, as is the direct evidence of attachment. The earliest unequivocal cirripedes occur in the Carboniferous (Schram 1975, 1982).

• *Redoubtia polypodia* Walcott 1918

The history of research on this species (Fig. 14) parallels closely that on *Portalia mira;* Walcott (1918, 1931) provided only brief descriptions and assigned the animal to the holothurians. The originally illustrated specimen (USNM 83924) (Walcott 1918, fig. 5) was redepicted by Walcott (1931) together with another specimen (USNM 83925) whose identity even at the time was regarded by C. E. Resser as questionable. There seems little doubt that USNM 83925 is not related to *Redoubtia,* and other unpublished material may provide further insights into its anatomy and affinities. An apparently superimposed fossil on USNM 83925 was identified as *?Hymenocaris* (= *Canadaspis*) by Resser (in Walcott 1931), and while this determination is suspect, this superposition may be due to chance.

Walcott's (1918, 1931) assignment of *Redoubtia* to the holothurians won the guarded support of Croneis and McCormack (1932), but has left most other workers (e.g., Frizzell and Exline 1955; Madsen 1957; Howell 1962) unconvinced. Madsen (1957) suggested a relationship with nereidiform polychaetes, while Durham (1971, 1974) maintained the possibility of relationship with holothurians, pending restudy. Until such reinvestigation the relationship of *Redoubtia* (both specimens USNM 83924 and 83925, and additional material) is problematic, and both previously illustrated specimens may represent major extinct groups.

- *"Selkirkia" gracilis* (Walcott 1911) and *"Selkirkia" fragilis* (Walcott 1911)

Walcott (1908, 1911) recognized three species of the tubicolous fossil *Selkirkia: S. fragilis, S. gracilis,* and *S. major,* the latter now referred to *S. columbia* on nomenclatural grounds (Conway Morris 1978). *Selkirkia columbia* is a tubicolous priapulid worm, whereas the other two species apparently are unrelated to the priapulids (Conway Morris 1977d). *"Selkirkia" fragilis* (Walcott 1911, pl. 19, fig. 8) was originally described from the *Ogygopsis* Shale, a prolific fossil locality adjacent to and apparently slightly older than the Burgess Shale (Walcott 1908; Fritz 1971; Collins et al. 1983). This species appears to be absent from the Burgess Shale, soft parts associated with the tube are not known, and its relationships are uncertain. *"Selkirkia" gracilis* (Walcott 1911, pl. 19, fig. 9), however, is relatively abundant in the Burgess Shale (about 37 specimens in the Phyllopod bed). However, almost without exception, specimens include only the gradually tapering tube, which unlike *S. columbia* usually terminates in a finely drawn apex. The walls were evidently very thin (Walcott 1911), presumably organic with minimal or nonexistent mineralization. One specimen appears to have associated soft parts, and although a cnidarian relationship is possible, the affinities of *"S." gracilis* are best regarded as problematic.

- *Tubulella flagellum* (Matthew 1899)

This tubiculous fossil was first described from the *Ogygopsis* Shale by Matthew (1899a), with a subsequent mention by Walcott (1908). The generic name proposed by Howell (1949) as *Urotheca* Matthew 1899 was already assigned to a reptile genus. *Tubulella* is widespread in the Stephen Formation, with examples documented from the Burgess Shale, as well as various localities on Mount Field, Mount Stephen, and Odaray Mountain (Collins et al. 1983). Examples are also known from other Cambrian rocks. *Tubulella pennsylvanica,* which is closely comparable with *T. flagellum,* occurs in the Lower Cambrian Kinzers Formation (Resser and Howell 1938; Campbell 1969; Conway Morris 1977d), and possibly the overlying Middle Cambrian unit, which, though placed by Campbell (1971) in the Kinzers Formation, is more probably a lateral equivalent of the Conestoga Formation (M. E. Kauffman, personal communication to S. C. M.). Records of *Tubulella* from the Cambrian of Newfoundland (Matthew 1899b) require confirmation.

The tube of *T. flagellum* was narrow, gently tapering from the apertural region, with an extremely high length–width ratio. As Matthew (1899a) noted, the tube is finely annulated, and he referred to its composition as "chitinous." However, while the tube appears to have been fairly robust with an enhanced preservation potential that may explain in part its fairly wide distribution, the precise composition is presently under investigation (Conway Morris and Pye, in preparation). Complete specimens, which may exceed 90 mm in length, are relatively uncommon in the Burgess Shale, and the majority consist of short sections of tube. Given the apparently quiet environment of deposition and the general lack of dismemberment of the biota during transport, it is possible that fragmentation arose by predation (see Conway Morris 1985a) or, less plausibly, by extensive bioturbation in the preslide environment.

The affinities of *Tubulella* are not clear. Suggestions (e.g., Matthew 1899a; Howell 1962) that it represents an annelid tube seem to be without firm foundation (Kozłowski 1967). However, comparisons might also be drawn with *Byronia,* a tubicolous genus that was originally described from the *Ogygopsis* Shale (Matthew 1899a) but does not occur in the Burgess Shale. *Byronia* has been compared with the cnidarian *Stephanoscyphus* (Kozłowski 1967, 1968; Glaessner 1971), a tube that houses the polyp generation of the scyphomedusan Coronatae (e.g., Werner 1973), and such comparisons might be extended to *Tubulella.* Although a relationship for *Tubulella* might also be sought among the pogonophores, it does not appear to have any close similarity with Early Phanerozoic examples such as *Paleolina, Sabellidites,* or *Saarina* (e.g., Sokolov 1972). Pending its restudy, the status of *Tubulella* is best regarded as problematic.

- *Wiwaxia corrugata* Matthew 1899

Much of the body of this lepidote metazoan (Fig. 15) was covered by flattened scalelike sclerites. A row of elongate spinose sclerites arising from a dorsolateral zone on either side appears to have served in defense, judging from the arrangement and shape of the sclerites and the occurrence of examples showing breakage. The ventral surface of *Wiwaxia* was apparently naked and, if muscular, may have served as a locomotory organ. Little of the internal anatomy is preserved, but located within the anterior region of the alimentary canal was a feeding apparatus consisting of two (rarely

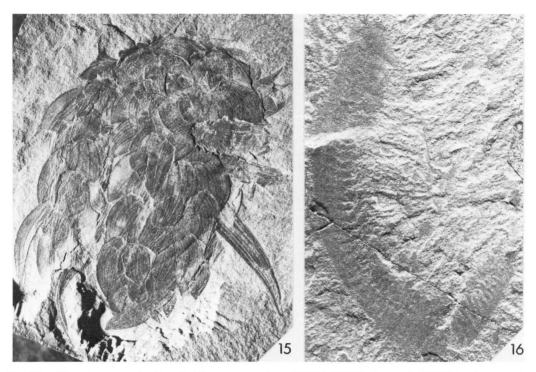

Fig. 15.   *Wiwaxia corrugata.* Dorsal-oblique compaction. USNM 199894, low-angle UV light from southwest. ×3.

Fig. 16.   *Worthenella cambria.* USNM 57643, low-angle UV light from northwest. ×2.5.

three) rows of posteriorly directed teeth (Conway Morris 1985a).

Wiwaxia has been allied with polychaetes (Walcott 1911; Jell 1981), but this seems unlikely. Wiwaxiids may have been derived from a turbellarian ancestor by the evolution of sclerite-secreting dorsal and lateral areas, through-gut, and denticulate feeding apparatus. Such an evolutionary sequence strongly resembles the proposed pathway by which mollusks arose from a turbellarian ancestor (e.g., Stasek 1972). *Wiwaxia* is not a mollusk, however, despite similarities such as the apparently locomotory foot and the radulalike feeding apparatus. The sclerites were not secreted along the margin of a continuous underlying mantle fold as in mollusks, but rather from dispersed and discrete nodes. Moreover, the mode of growth necessitated a molting cycle, because the sclerites were secreted at a fixed size. One specimen of *Wiwaxia* even appears to be preserved in the act of shedding its exuvium. The denticulate feeding apparatus recalls the molluscan radula both in shape and position, but in the absence of histologic information it is difficult to determine whether the resemblance is convergent.

It is possible that wiwaxiids and mollusks were derived from the same turbellarian stock, so that the feeding apparatus and radula are homologous,

evolving prior to the divergence of the groups. The available evidence, however, is not sufficient to distinguish between this possibility (Conway Morris 1985a) and the alternative notion that the general similarity between *Wiwaxia* and mollusks is convergent (Bengtson and Conway Morris 1984), representing similar adaptations to epifaunal life as grazers or deposit feeders (see also Chapter 16 by Dzik, this volume). Small, isolated sclerites from various Lower Cambrian sections, especially from the Tommotian, have been identified with reasonable assurance as those of wiwaxiids (Bengtson and Missarzhevsky 1981; Jell 1981; Bengtson and Conway Morris 1984). This stratigraphically earlier occurrence is of considerable interest for a variety of reasons. Using *Wiwaxia* from the Burgess Shale as a template, a hypothetical reconstruction of the arrangement of the now-dispersed sclerites may be attempted (Bengtson and Conway Morris 1984). Early wiwaxiid sclerities were evidently subject to the same constraints of growth as those of *Wiwaxia* itself, and these animals also presumably grew in a series of molts. The link between molting cycles and hormonal control in arthropods raises the possibility that comparable mechanisms operated in wiwaxiids. Further speculation suggests that hormonal changes may have fueled evolutionary changes in wiwaxiids by heterochrony, *Wiwaxia* representing a neotenic descendant of Early Cambrian forms. The size of the Tommotian wiwaxiid sclerites indicates an animal smaller than *Wiwaxia,* and unlike the Burgess Shale representatives they were mineralized with walls most probably composed of calcium carbonate. The loss of this relatively robust skeleton in favor of at most a very weakly mineralized composition by Middle Cambrian times led to the effective disappearance of wiwaxiids from the fossil record. This renders their total stratigraphic range highly uncertain, and adds to the complexity of our deciphering of the early metazoan fossil record.

• *Worthenella cambria* Walcott 1911

This species is known only from a single poorly preserved specimen (Fig. 16), and since Walcott's description (1911) it has received little attention (e.g., Bock 1962). While it awaits detailed redescription, preliminary study suggests that Walcott's observations (1911) will not require radical alteration. As noted by Walcott, the body is elongate and composed of numerous divisions, presumably representing segments. Details of the anterior are especially obscure, but it appears to have borne specialized appendages [palps and tentacles of Walcott (1911)]. The appendages of the rest of the body are also incompletely understood.

Walcott (1911) considered *Worthenella* to be a polychaetous annelid, and even suggested that associated organic debris on the surface of the shale represented detached scales (elytra). This appears highly unlikely, and there is little evidence to support Walcott's taxonomic assignment. *Worthenella* shows a certain similarity to uniramian arthropods such as the myriapods. However, it is best to regard this poorly known animal as *incertae sedis,* pending a complete reappraisal including elucidation of the structure of the appendages.

DISCUSSION

The Burgess Shale provides us with a window on marine life at the end of the initial stage in the Late Precambrian–Early Paleozoic adaptive radiation of the Metazoa (Sepkoski 1979). This radiation is considered to have taken place in a "vacuum" of essentially empty ecospace, with low levels of competition. The wide range of morphologies that evolved is presumably a reflection of this unique opportunity; most, if not all of the major modern phyla were established, as well as a large number (the problematica) that have subsequently become extinct. A mosaic of characters appeared in different combinations in different organisms. *Anomalocaris,* for example, has the kind of segmented limb characteristic of the arthropod phyla but combines a suite of characters

otherwise unlike any arthropod (Whittington and Briggs 1985). *Wiwaxia* has a feeding apparatus similar to the radula of mollusks, but its armor of spiny and scaly sclerites was probably molted, unlike any mollusk (Conway Morris 1985a). There is no obvious way to determine whether such individual attributes indicate relationship with the phyla for which they are characteristic, or whether they are the result of convergence. Schram (1981, 1982) has described a similar phenomenon in the adaptive radiation of the Carboniferous Malacostraca, which he terms "stochastic mosaicism." Cambrian problematica are not, of course, confined to the Burgess Shale, but occur, for example, in the Lower Cambrian Tommotian faunas (Bengtson 1977; Chapter 8 by Rozanov, this volume), but here the soft parts are lacking.

The analysis of the mode of life of problematica ultimately depends largely on a comparison with living animals—the living perspective of a uniformitarian approach. A hypothetical Cambrian observer would have found it impossible to predict which of the organisms around him had the greater potential for success (Conway Morris and Whittington 1979). Their success can only be measured retrospectively, with reference to the fossil record, although their low fossilization potential makes their range and abundance impossible to assess reliably. The small number of problematic taxa in the Lagerstätten of the Silurian of Wisconsin (Mikulic et al. 1985) and the Scottish Midland Valley (Scourfield 1937), and in the Carboniferous Mazon Creek biota of Illinois (Nitecki and Schram 1976; Nitecki and Solem 1973; Johnson and Richardson 1969), for example, are not obviously related to the Cambrian forms discussed in this paper. It appears that the diversity of problematica is greatest in the Cambrian and declines thereafter, the Cambrian problematica becoming extinct relatively soon. Once the Early Paleozoic "vacuum" was filled, the ecospace vacated by phyla that became extinct was colonized by radiation from within the ranks of surviving phyla. The opportunity for establishing new body plans was limited once the initial adaptive radiation was over.

The morphology of the problematica may appear to us to be bizarre and "experimental," but this is only by comparison with the familiar living phyla. Raup (1978) has explored the possible role of stochastic processes in the extinction of higher taxa and suggests, for example, that if the Carboniferous and Permian were repeated, the echinoids might become extinct in place of the blastoids. Similarly if the Early Paleozoic were rerun, the wiwaxiids might take the place of the mollusks (Conway Morris 1985a). The groups that we regard as problematica are only problematic from a modern perspective. All the major body plans may have started out with a similar chance of surviving and diversifying. The organisms reviewed here may have been simply unluckier than the phyla that remain to inhabit the modern oceans.

Acknowledgments

We are grateful to H. B. Whittington for advice and encouragement during the course of our work on the Burgess Shale biota, and for supplying the photographs used in Figs. 3, 4, and 11.

This paper is Cambridge Earth Sciences publication 577.

REFERENCES

Barham, E. G., Ayer, N. J., and Boyce, R. E. 1967. Macrobenthos of the San Diego Trough: photographic census and observations from bathyscaphe, *"Trieste." Deep Sea Res.* **14,** 773–784.

Bengtson, S. 1976. The structure of some Middle Cambrian conodonts, and the early evolution of conodont structure and function. *Lethaia* **9,** 185–206.

———. 1977. Aspects of problematic fossils in the early Palaeozoic. *Acta Univ. Upsal. Abstr. Uppsala Dissert. Fac. Sci.* **415,** 71 pp.

———. 1983. The early history of the Conodonta. *Fossils and Strata* **15,** 5–19.

——— and Conway Morris, S. 1984. A comparative study of Lower Cambrian *Halkieria* and Middle Cambrian *Wiwaxia. Lethaia* **17,** 307–329.

——— and Missarzhevsky, V. V. 1981. Coeloscleritophora, a major group of enigmatic Cambrian metazoans. *U.S. Geol. Surv. Open-File Rep.* **81-743,** 19–21.

Bock, W. 1962. Systematics of dichotomy and evolution. *Geol. Cent. Res. Ser.* **2,** 1–299.

Briggs, D. E. G. 1976. The arthropod *Branchiocaris* n. gen., Middle Cambrian, Burgess Shale, British Columbia. *Bull. Geol. Surv. Canada* **264,** 29 pp.

———. 1979. *Anomalocaris,* the largest known Cambrian arthropod. *Palaeontology* **22,** 631–664.

———. 1981. The arthropod *Odaraia alata* Walcott, Middle Cambrian, Burgess Shale, British Columbia. *Phil. Trans. R. Soc. Lond. (B.)* **291,** 541–584.

———. 1983. Affinities and early evolution of the Crustacea: the evidence of the Cambrian fossils. *In* Schram, F. R. (ed.). *Crustacean Phylogeny.* A. A. Balkema, Rotterdam, pp. 1–22.

———, Clarkson, E. N. K., and Aldridge, R. J. 1983. The conodont animal. *Lethaia* **16,** 1–14.

——— and Mount, J. D. 1982. The occurrence of the giant arthropod *Anomalocaris* in the Lower Cambrian of Southern California, and the overall distribution of the genus. *J. Paleont.* **56,** 1112–1118.

——— and Whittington, H. B. 1985. The mode of life of the Burgess Shale arthropods. *Trans. R. Soc. Edinb. (Earth Sci.)* **76,** 149–160.

Campbell, L. D. 1969. Stratigraphy and paleontology of the Kinzers Formation, Southeastern Pennsylvania. Unpubl. M.Sc. thesis, Franklin and Marshall College, Lancaster, Pa.

———. 1971. Occurrence of *"Ogygopsis* Shale" fauna in southeastern Pennsylvania. *J. Paleont.* **45,** 437–440.

Collins, D. H., Briggs, D., and Conway Morris, S. 1983. New Burgess Shale fossil sites reveal Middle Cambrian faunal complex. *Science* **222,** 163–167.

——— and Rudkin, D. M. 1981. *Priscansermarinus barnetti,* a probable lepadomorph barnacle from the Middle Cambrian Burgess Shale of British Columbia. *J. Paleont.* 55, 1006–1015.

Conway Morris, S. 1976a. A new Cambrian lophophorate from the Burgess Shale of British Columbia. *Palaeontology* **19,** 199–222.

———. 1976b. *Nectocaris pteryx,* a new organism from the Middle Cambrian Burgess Shale of British Columbia. *N. Jb. Geol. Paläont. Mh.* **1976,** 705–713.

———. 1977a. A redescription of the Middle Cambrian worm *Amiskwia sagittiformis* Walcott from the Burgess Shale of British Columbia. *Paläont. Z.* **51,** 271–287.

———. 1977b. A new metazoan from the Cambrian Burgess Shale of British Columbia. *Palaeontology* **20,** 623–640.

———. 1977c. A new entoproct-like organism from the Burgess Shale of British Columbia. *Palaeontology* **20,** 833–845.

———. 1977d. Fossil priapulid worms. *Spec. Pap. Palaeont.* **20,** 1–95.

———. 1978. *Selkirkia* Walcott, 1911 (Priapulida): proposed designation of a type-species under the plenary powers. Z.N.(S.) 2171. *Bull Zool. Nom.* **35,** 49–50.

———. 1979a. Burgess Shale. *In* Fairbridge, R. W. and Jablonski, D. (eds.). *The Encyclopedia of Paleontology.* Dowden, Hutchinson & Ross, Stroudsburg, Pa., pp. 153–160.

———. 1979b. The Burgess Shale (Middle Cambrian) fauna. *Ann. Rev. Ecol. Syst.* **10,** 327–349.

——— (ed.). 1982. *An Atlas of the Burgess Shale.* Palaeontological Association, London.

———. 1985a. The Middle Cambrian metazoan *Wiwaxia corrugata* from the Burgess Shale and *Ogygopsis* Shale, British Columbia, Canada. *Phil. Trans. R. Soc. Lond. (B.)* **307,** 507–586.

———. 1985b. Non-skeletalized lower invertebrate fossils: a review. *In* Conway Morris, S., George, J. D., Gibson, R., and Platt, H. M. (eds.). *The Origins and Relationships of Lower Invertebrates.* Clarendon Press, Oxford, pp. 343–359.

——— and Robison, R. A. 1982. The enigmatic medusoid *Peytoia* and a comparison of some Cambrian biotas. *J. Paleont.* **56,** 116–122.

———— and Whittington, H. B. 1979. The animals of the Burgess Shale. *Sci. Am.* **241,** 122–133.

Croneis, C. and McCormack, J. 1932. Fossil Holothuroidea. *J. Paleont.* **6,** 111–148.

Delle Cave, L. and Simonetta, A. M. 1975. Notes on the morphology and taxonomic postion of *Aysheaia* (Onychophora?) and of *Skania* (undetermined phylum). *Monit. Zool. Ital.* **9,** 67–81.

Durham, J. W. 1971. The fossil record and the origin of the Deuterostomata. *Proc. North Am. Paleont. Conv. Chicago 1969* **H,** 1104–1132.

————. 1974. Systematic position of *Eldonia ludwigi* Walcott. *J. Paleont.* **48,** 750–755.

Dzik, J. 1978. Larval development of hyolithids. *Lethaia* **11,** 293–299.

Emschermann, P. 1982. Les Kamptozoaires. Etat actuel de nos connaissances sur leur anatomie, leur dévelopement, leur biologie et leur position phylogénetique. *Bull. Soc. Zool. Fr.* **107,** 317–344.

Farmer, J. D., Valentine, J. W., and Cowen, R. 1973. Adaptive strategies leading to the ectoproct ground-plan. *Syst. Zool.* **22,** 233–239.

Fritz, W. H. 1971. Geological setting of the Burgess Shale. *Proc. North Am. Paleont. Conv. Chicago 1969* **I,** 1155–1170.

Frizzell, D. L. and Exline, H. 1955. Monograph of fossil holothurian sclerites. *Bull. Mo. School Mines* **89,** 1–204.

Glaessner, M. F. 1971. The genus *Conomedusites* Glaessner and Wade and the diversification of the Cnidaria. *Paläont. Z.* **45,** 7–17.

Häntzschel, W. 1975. Miscellanea (suppl. 1), trace fossils and problematica. *In* Teichert, C. (ed.). *Treatise on Invertebrate Paleontology,* part W. Geological Society of America and Univ. of Kansas Press, Lawrence, 269 pp.

Heezen, B. C. and Hollister, C. D. 1971. *The Face of the Deep.* Oxford Univ. Press, New York, 659 pp.

Herring, P. J. and Clarke, M. R. (eds.). 1971. *Deep Oceans.* Arthur Barker, London, 320 pp.

Hofmann, H. J. 1971. Precambrian fossils, pseudofossils, and problematica in Canada. *Geol. Surv. Canada Bull.* **189,** 146 pp.

Howell, B. F. 1949. New hydrozoan and brachiopod and new genus of worms from the Ordovician Schenectady Formation of New York. *Bull. Wagner Inst. Sci. Phila.* **24**(1), 1–10.

————. 1962. Worms. *In* Moore, R. C. (ed.). *Treatise on Invertebrate Paleontology,* part W, *Miscellanea.* Geological Society of America and Univ. of Kansas Press, Lawrence, pp. W144–W177.

Hutchinson, G. E. 1930. Restudy of some Burgess Shale fossils. *Proc. U.S. Natl. Mus.* **78**(11), 1–24.

Jell, P. A. 1981. *Thambetolepis delicata* gen. et sp. nov., an enigmatic fossil from the Early Cambrian of South Australia. *Alcheringa* **5,** 85–93.

Johnson, R. G. and Richardson, E. S. 1969. Pennsylvanian invertebrates of the Mazon Creek area, Illinois: the morphology and affinities of *Tullimonstrum. Fieldiana: Geol.* **12,** 119–149.

Korotkevich, V. S. 1967. Systematic position of *Amiskwia sagittiformis* from Middle Cambrian of Canada. *Paleont. J.* **4,** 115–118 [AGI transl. of *Paleont. Zh.*].

Kozłowski, R. 1967. Sur certains fossiles ordoviciens à test organique. *Acta Palaeont. Polon.* **12,** 99–132.

————. 1968. Nouvelles observations sur les conulaires. *Acta Palaeont. Polon.* **13,** 497–531.

Lemche, H., Hansen, B., Madsen, F. J., Tendal, O. S., and Wolff, T. 1976. Hadal life as analysed from photographs. *Vidensk. Meddr. Dansk Naturh. Foren.* **139,** 263–336.

Lohmann, H. 1922. *Oesia disjuncta* Walcott, eine Appendicularie aus dem Kambrium. *Mitt. Zool. Mus. Hamb.* **38,** 69–75.

————. 1933–1934 Appendiculariae. *In* Kukenthal, W. and Krumbach, T. (eds.). *Handbuch der Zoologie, Tunicata,* pp. 15–202.

Madsen, F. J. 1957. On Walcott's supposed Cambrian holothurians. *J. Paleont.* **31,** 281–282.

Marek, L. and Yochelson, E. L. 1976. Aspects of the biology of Hyolitha (Mollusca). *Lethaia* **9,** 65–82.

Margulis, L. and Schwartz, K. V. 1982. *Five Kingdoms.* Freeman, San Francisco, 338 pp.

Matthew, G. F. 1899a. Studies on Cambrian faunas. no. 3, Upper Cambrian fauna of Mount Stephen, British Columbia. The trilobites and worms. *Trans. R. Soc. Can. Ser. 2* **5,** 39–66.

————. 1899b. The Etcheminian fauna of Smith Sound, Newfoundland. *Trans. R. Soc. Can. Ser. 2* **5,** 97–123.

McKenzie, K. G. 1983. On the origin of Crustacea. *In* Lowry, J. K. (ed.). Papers from the Conference on the Biology and Evolution of Crustacea. *Aust. Mus. Mem.* **18,** 21–43.

Mikulic, D. G., Briggs, D. E. G., and Kluessendorf, J. 1985. A Silurian soft-bodied fauna. *Science* **228,** 715–717.

Nitecki, M. H. and Schram, F. R. 1976. *Etacystis communis,* a fossil of uncertain affinities from the Mazon Creek fauna (Pennsylvanian of Illinois). *J. Paleont.* **50,** 1157–1161.

———— and Solem, A. 1973. A problematic organism from the Mazon Creek (Pennsylvanian) of Illinois. *J. Paleont.* **47,** 903–907.

Okulitch, V. J. 1955. Archaeocyatha. *In* Moore, R. C. (ed.). *Treatise on Invertebrate Paleonotology,* part E, *Archaeocyatha and Porifera.* Geological Society of America and Univ. of Kansas Press, Lawrence, pp. E1–E20.

Owre, H. B. and Bayer, F. M. 1962. The systematic position of the Middle Cambrian fossil *Amiskwia. J. Paleont.* **36,** 1361–1363.

Raup, D. M. 1978. Approaches to the extinction problem. *J. Paleont.* **52**, 517–523.

Raymond, P. E. 1935. *Leanchoilia* and other mid-Cambrian Arthropoda. *Bull. Mus. Comp. Zool. Harv.* **76**(6), 205–230.

Resser, C. E. and Howell, B. F. 1938. Lower Cambrian *Olenellus* zone of the Appalachians. *Bull. Geol. Soc. Am.* **49**, 195–248.

Robison, R. A. 1985. Affinities of *Aysheaia* (Onychophora), with description of a new Cambrian species. *J. Paleont.* **59**, 226–235.

Rolfe, W. D. I. 1969. Phyllocarida. *In* Moore, R. C. (ed.). *Treatise on Invertebrate Paleontology, part R, Arthropoda 4.* Geological Society of America and Univ. of Kansas Press, Lawrence, pp. R296–R331.

———, Schram, F. R., Pacaud, G., Sotty, D., and Secretan, S. 1982. A remarkable Stephanian biota from Montceau-les-Mines, France. *J. Paleont.* **56**, 426–428.

Rowell, A. J. 1982. The monophyletic origin of the Brachiopoda. *Lethaia* **15**, 299–307.

Runnegar, B. 1980. Hyolitha: status of the phylum. *Lethaia* **13**, 21–25.

———, Pojeta, J., Morris, N. J., Taylor, J. D., Taylor, M. E., and McClung, G. 1975. Biology of the Hyolitha. *Lethaia* **8**, 181–191.

Schram, F. R. 1975. A Pennsylvanian lepadomorph barnacle from the Mazon Creek area, Illinois. *J. Paleont.* **49**, 928–930.

———. 1981. On the classification of the Eumalacostraca. *J. Crust. Biol.* **1**, 1–10.

———. 1982. The fossil record and evolution of Crustacea. *In* Abele, L. G. (ed.). *The Biology of Crustacea,* vol. 1. Academic Press, New York, pp. 94–147.

Scourfield, D. J. 1937. An anomalous fossil organism, possibly a new type of chordate, from the Upper Silurian of Lesmahagow, Lanarkshire—*Ainiktozoon loganense* gen. et sp. nov. *Proc. R. Soc. Lond. (B.)* **121**, 533–547.

Seilacher, A. 1962. Form und Funktion des Trilobiten—Daktylus. *Paläont. Z.* **36** (Hermann Schmit Festband), 218–227.

Sepkoski, J. J., Jr. 1979. A kinetic model of Phanerozoic taxonomic diversity. II. Early Phanerozoic families and multiple equilibria. *Paleobiology* **5**, 222–251.

Simonetta, A. and Delle Cave, L. 1978. Notes on new and strange Burgess Shale fossils (Middle Cambrian of British Columbia). *Mem. Soc. Tosc. Sci. Nat. Ser. A* **85**, 45–49.

Sokolov, B. S. 1972. Vendian and Early Cambrian Sabelliditida (Pogonophora) of the USSR. *Int. Geol. Congr.* **23** (1968, *Proc. Int. Paleont. Union*), 79–86.

Stanley, G. D. and Stürmer, W. 1983. The first fossil ctenophore from the Lower Devonian of West Germany. *Nature* **303**, 518–520.

Stasek, C. R. 1972. The molluscan framework. *In*

Florkin, M. and Scheer, B. T. (eds.). *Chemical Zoology,* vol 7. Academic Press, New York, pp. 1–44.

Størmer, L. 1944. On the relationships and phylogeny of fossil and Recent Arachnomorpha. A comparative study on Arachnida, Xiphosura, Eurypterida, Trilobita, and other fossil Arthropoda. *Skr. Norske Vidensk—Akad.* **5**, 1–158.

Stürmer, W. and Bergström, J. 1978. The arthropod *Cheloniellon* from the Devonian Hunsrück Shale. *Paläont. Z.* **52**, 57–81.

Szaniawski, H. 1982. Chaetognath grasping spines recognized among Cambrian conodonts. *J. Paleont.* **56**, 806–810.

———. 1983. Structure of protoconodont elements. *Fossils and Strata* **15**, 21–27.

Tarlo, L. B. 1960. The invertebrate origins of the vertebrates. *22nd Int. Geol. Congr.* 113–123.

Thayer, C. W. 1975. Morphologic adaptations of benthic invertebrates to soft substrata. *J. Mar. Res.* **33**, 177–189.

Thompson, I. and Jones, D. S. 1980. A possible onychophoran from the middle Pennsylvanian Mazon Creek beds of northern Illinois. *J. Paleont.* **54**, 588–596.

Valentine, J. W. 1975. Adaptive strategy and the origin of grades and ground-plans. *Am. Zool.* **15**, 391–404.

———. 1977. General patterns of metazoan evolution. *In* Hallam, A. (ed.). *Patterns of Evolution as illustrated by the Fossil Record.* Elsevier, Amsterdam, pp. 27–57.

Walcott, C. D. 1908. Mount Stephen rocks and fossils. *Can. Alp. J.* **1**, 232–248.

———. 1911. Middle Cambrian annelids. *Smithson. Misc. Coll.* **57**, 109–144.

———. 1912. Middle Cambrian Branchiopoda, Malacostraca, Trilobita and Merostomata. *Smithson. Misc. Coll.* **57**, 145–228.

———. 1918. Geological explorations in the Canadian Rockies. *In* Explorations and fieldwork of the Smithsonian Institution in 1917. *Smithson. Misc. Coll.* **63**, 3–20.

———. 1931. Addenda to descriptions of Burgess Shale fossils (with explanatory notes by C. E. Resser). *Smithson. Misc. Coll.* **85**, 1–46.

Werner, B. 1973. New investigations on systematics and evolution of the Class Scyphozoa and the phylum Cnidaria. *Publ. Seto Mar. Biol. Lab.* **20**, 35–61.

Whiteaves, J. F. 1892. Description of a new genus and species of phyllocarid crustacean from the Middle Cambrian of Mount Stephen, British Columbia. *Can. Rec. Sci.* **5**, 205–208.

Whittington, H. B. 1971. The Burgess Shale: history of research and preservation of fossils. *Proc. North Am. Paleont. Conv. Chicago 1969* **I**, 1170–1201.

———. 1975. The enigmatic animal *Opabinia regalis,* Middle Cambrian, Burgess Shale, British Columbia. *Phil. Trans. R. Soc. Lond. (B.)* **271,** 1–43.

————. 1978. The lobopod animal *Aysheaia pedunculata* Walcott, Middle Cambrian, Burgess Shale, British Columbia. *Phil. Trans. R. Soc. Lond. (B.)* **284,** 165–197.

————. 1980. The significance of the fauna of the Burgess Shale, Middle Cambrian. *Proc. Geol. Assoc. Lond.* **91,** 127–148.

————. 1981. Cambrian animals: their ancestors and descendants. *Proc. Linn. Soc. N.S.W.* **105,** 79–87.

————. 1982. The Burgess Shale fauna and the early evolution of metazoan animals. *In "Palaeontology, Essential of Historical Geology,"* Mucchi, Modena, Italy, pp. 11–24.

———— and Briggs, D. E. G. 1985. The largest Cambrian animal, *Anomalocaris,* Burgess Shale, British Columbia. *Phil. Trans. R. Soc. Lond. (B.),* **309,** 569–609.

Wills, L. J. 1963. *Cyprilepas holmi,* Wills 1962, a pedunculate cirripede from the Upper Silurian of Oesel, Esthonia. *Palaeontology* **6,** 161–165.

Yochelson, E. L. 1977. Agmata, a proposed extinct phylum of early Cambrian age. *J. Paleont.* **51,** 437–454.

REPOSITORIES

Repositories are abbreviated as follows in the text and figure explanations: MCZ, Museum of Comparative Zoology, Harvard; USNM, National Museum of Natural History, Smithsonian Institution, Washington D.C.; ROM, Royal Ontario Museum, Toronto.

14. THE ENIGMA OF GRAPTOLITE ANCESTRY: LESSON FROM A PHYLOGENETIC DEBATE

Adam Urbanek

HISTORY OF THE DEBATE

The debate on the affinities of graptolites provides a model of a phylogenetic controversy that may be profitably analyzed in terms of philosophy of science. The gradual emergence of phylogenetic concepts, from vague suggestions to a fully formulated paradigm, the coexistence of a number of such paradigms at a given time or predominance or even almost complete monopoly of one of them, the persistence and revival of old concepts in modified versions, the relative independence of validity of the general conclusions from their sometimes erroneous premises, the discovery of what may be called anomaly from the standpoint of a given paradigm—all these are surprisingly conspicuous features of the history of ideas on graptolite ancestry.

In such a context, the most recent concepts also seem to be temporary and subject to future changes or modifications. They cannot be isolated from earlier concepts from which they either inherited some common ideas or to which they arose in opposition. The conceptual anatomy of the current views on graptolite affinities becomes more understandable when their intellectual pedigree is exposed.

Primary pool of ideas

Early speculations concerning the graptolite affinities were naturally confined to the problem of their systematic position. Possessing an unusual combination of organizational features, graptolites were the subject of highly diverse opinions on their place in the system of living beings. There were even doubts whether they were organic at all.

Carl Linné, who first distinguished the class of natural objects called *Graptolithus,* considered them in the first edition of *Systema Naturae* (Linnaeus 1735) to be of similar nature as dendritic incrustations. Later he described under the same name fossil plants and graptolites. Despite Linné's intentions, the name *Graptolithus* was later used to denote truly organic remains. Brommel (1727) and Brongniart (1828) believed that graptolites were remains of marine plants. Their animal nature was first recognized by Walch (1771), who considered them to be minute cephalopods.

A coral, "polypide," or generally coelenterate nature of the graptolites was proposed around 1830. This idea, usually in its hydrozoan version, was supported by many distinguished paleontologists of the second half of the nineteenth century (Hall 1865; Allman, 1872; Nicholson 1872; Lapworth 1873, 1876). Although prevailing, this view never achieved a general acceptance. Salter (1866) assigned graptolites to Bryozoa, while others pointed to *Rhabdopleura* as an organism, in some aspects, the most similar to graptolites—especially Richter (1871), but also Allman (1872). Paradoxically, these two theses were not mutually exclusive because at that time *Rhabdopleura* was considered to be an aberrant bryozoan.

The early ideas on the systematic position of graptolites were best summarized by Karl von Zittel (1880). He saw no reason for comparing graptolites with plants, cephalopods, or foraminifers (as once suggested), and found little substantiation for relation with sea pens or horny corals in general. He considered only the relationship with hydrozoans and with *Rhabdopleura,* treated as an aberrant bryozoan, as sound hypotheses. However, the presence of a common canal and a wide communication of thecae, and chitinous periderm instead of calcified walls argue against the bryozoan affinities of graptolites. The resemblance to Hydrozoa is much closer. Zittel placed the sessile genera (then frequently separated from true graptolites, but now assigned to Graptolithina) within the hydroid suborder Companulariae, and the free-living, "genuine" graptolites within a new suborder, the Graptolithidae. All these he placed within the Hydroida.

In the second half of the nineteenth century, three main concepts of graptolite affinities emerged: (1) the coelenterate, in most cases hydrozoan; (2) the bryozoan; and (3) the pterobranch or, more exactly, the *Rhabdopleura* concept. Only the first was based on a detailed anatomic comparison (by Allman 1872). This was also the most widespread concept among paleontologists and zoologists, and it was elevated to the status of canonical knowledge by Zittel (1880).

These early concepts of graptolite affinities had common features. The historical bias focused the attention on true graptolites, the pelagic Grapto-loidea (= Rhabdophora of Allman). Sessile and primitive graptolites were little known, poorly understood, and hence usually excluded from phylogenetic considerations. This is why the "solid axis" (= nema or virgula) and the "common canal" were regarded as structures characteristic of all graptolites. On the other hand, these incompletely known graptolite structures were compared with imperfectly understood structures in living colonial groups. This caused such misconceptions as the comparison of the solid axis with the stolon of *Rhabdopleura.* In general, these early ideas were mere intuitions, although many of them proved later to be valuable and stimulating.

At the turn of the twentieth century Wiman (1895) proposed that graptolites were in many ways unique and belonged to a completely extinct branch not clearly related to any living animals. This somewhat "agnostic" idea was based on careful morphologic studies of graptolites isolated from the rocks by chemical methods. A number of paleontologists shared Wiman's agnostic opinion (Perner 1894; Ruedemann 1895; Frech 1897; Elles 1922).

The concepts elaborated in the late nineteenth century constitute the main pool of ideas for later authors. The revival of the classical concepts of graptolite affinities was based on the elimination of a number of old misconceptions concerning the structure of graptolites and on a better knowledge of their presumed living relatives. While old errors were eliminated, new ones crept into the logic of interpretations.

Schepotieff's pterobranch hypothesis

The rise of knowledge of living pterobranchs provided the basis for Schepotieff's (1905) penetrating considerations on their relationship with graptolites. He was convinced that the general features of graptolite organization, such as the presence of sicula and a solid axis in colonies of pelagic graptolites, and the microstructure of the skeleton, were incompatible with the coelenterate ancestry. In contrast, he held that graptolites, and especially uniserial monograptids, were closely related to *Rhabdopleura* and, consequently, should be placed in the same phylum with Pterobranchia. Later Schepotieff (1910) considered graptolites as a polyphyletic group, with the scandent graptoloids with the nema incorporated within the stipes (Axonophora) being closely related to *Rhabdopleura,* and the group with its nema free (Axonolipa) to *Cephalodiscus.*

Schepotieff's views were an odd blend of apt conclusions and surprisingly inadequate substantiations (Fig. 1). While emphasizing the importance of the peculiar initial portion of the colony, he suggested that the sicula may be compared with closed, pointed, and resting terminal parts of the creeping tubes of *Rhabdopleura.* Consequently, he regarded the sicula as representing the distal and not the proximal end of the graptolite colony. He was quite correct in pointing out that the "black stolon" of *Rhabdopleura* has no analogy in the animal kingdom, but altogether wrong in comparing this stolon with the nema or virgula of axonophorous graptoloids. Although Schepotieff was well aware that both pterobranchs and graptolites possess a peculiar structure of the skeleton, he emphasized instead the dark-pigmented substance in the middle layer of the periderm as the most important common feature of their fine structures. This dark substance, however, is probably an artifact. His main conclusion, almost prophetically correct, was that graptolites were not at all related to coelenterates but closely related to *Rhabdopleura* and pterobranchs in general.

The sources of Schepotieff's contradictory views are in his methodologic decisions. First of all, he believed *Monograptus* to represent the basic type of the Graptolithina and considered this late, specialized, and secondarily simplified form to be an ancestral form. Moreover, he was concerned that etching changes the microstructure of graptolites and based his observations on less reliable thin sections. In spite of these shortcomings, Schepotieff firmly introduced the idea of the pterobranch ancestry of graptolites. This idea proved to be highly heuristic.

The coelenterate paradigm

The coelenterate theory doubtlessly gained the greatest popularity, and the coelenterate analogies were most convincing for the majority of paleontologists. The organic and, presumably, chitinous nature of the skeleton in graptolites and in many Cnidaria, the polymorphism of colonial individuals, and the so-called common canal were most frequently quoted in this context.

The concept of the common canal as a continuous tubular cavity along the rhabdosome, comparable with the coenosarcal canal of the hydrocaulus in hydroid colonies, was for a long time a controversial issue. Wiman (1895) demonstrated that each theca in dendroid colonies budded directly from the preceding theca. The common canal is, therefore, a secondary structure, the sum of the proximal portions of the successive thecae, and not a primary structure from which thecae are emerging.

However, Bulman (1937, 1938) concluded that the budding individuals represented initially a coenosarcal structure—that is, a common colonial tissue—only later differentiated into individual zooids. A number of characters (e.g., the growth zone of irregular fuselli in the middle of the rhabdosome in some diplograptids) were interpreted by Bulman as evidence of vestiges of coenosarclike structures in Graptoloidea. The

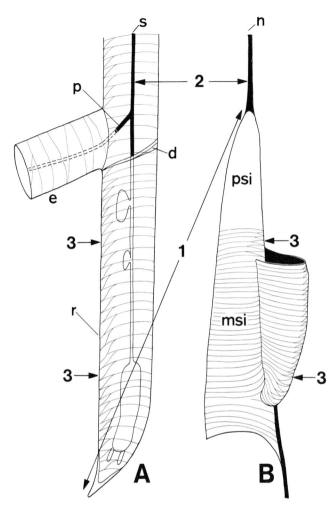

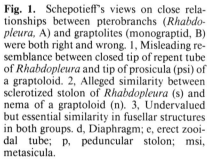

Fig. 1. Schepotieff's views on close relationships between pterobranchs (*Rhabdopleura,* A) and graptolites (monograptid, B) were both right and wrong. 1, Misleading resemblance between closed tip of repent tube of *Rhabdopleura* and tip of prosicula (psi) of a graptoloid. 2, Alleged similarity between sclerotized stolon of *Rhabdopleura* (s) and nema of a graptoloid (n). 3, Undervalued but essential similarity in fusellar structures in both groups. d, Diaphragm; e, erect zooidal tube; p, peduncular stolon; msi, metasicula.

presence of "something analogous to the coenosarc of Coelenterata" gave a new impetus to the old idea of coelenterate origin of graptolites.

The coelenterate affinity of graptolites explained convincingly such structures as different floats, vesicles, and similar rare parts of the graptoloid rhabdosomes. This is especially true for the enigmatic compound rhabdosomes, the synrhabdosomes of Ruedemann (1895), which focused the thinking of generations of paleontologists on the siphonophoran model, with its rare contemporary survivors (Berry 1978).

The polymorphism of dendroid colonies was also interpreted by comparison with coelenterates, especially hydrozoans. It became gradually apparent from the works of Wiman and Bulman that there were three types of thecae in the Dendroidea, namely hydro- and bithecae and tubular thecae housing the so-called budding individuals (later termed stolothecae). The chain of budding individuals was compared by Bulman with the coenosarc of the Hydrozoa, while bithecal individuals were usually considered to be nematophores or protective zooids. Hydrothecae (later renamed autothecae) housed main zooids, presumably comparable with gastrozooids or nourishing individuals of the polymorphic colonies of Hydrozoa. Allman (1872) thought that the hydrothecae of dendroids and the corresponding thecae of Graptoloidea lack the diaphragm, a basal constriction separating the thecal cavity from the common canal. Such diaphragms are always present in the hydrothecae of thecophorous hydrozoans. Allman concluded on these rather insignificant grounds that graptolite colonies were composed solely of individuals homologous with nematophores of plumularians, and that their hydranths were suppressed. Bulman (1938), however, doubted if a direct comparison between the polymorphism of Dendroidea and particular polymorphs of Hydrozoa is reliable, although he was convinced that such comparison has a better parallel among Coelenterata than among any other groups. Furthermore, such processes as the loss of bithecal individuals in the early evolution of Graptoloidea may also be explained by coelenterate models.

In the late 1930s, the coelenterate paradigm was widely favored. This was best expressed by Bulman (1938): "Detailed comparison of the graptolites with any living class has not provided fruitful results, but if they are to be placed in some existing phylum (which seems a reasonable supposition) it is perhaps as a separate class of Coelenterata that their assemblage of characters can most readily be accommodated."

Graptolites as nearest relatives of bryozoans

"The authors, while enjoying the ease of a steamer chair on a joint trip to Europe in 1922, and discussing somewhat idly at first the taxonomic position of the graptolites, discovered to their mutual great satisfaction that they had arrived independently at the conviction that the graptolites are an early and long extinct branch of the bryozoans." In such an unusual way Ulrich and Ruedemann (1931) announced their firm belief in the bryozoan affinities of graptolites and presented one of the best substantiations of this hypothesis. Their argument was as follows:

1. There is a correspondence between the primary zooecium of the bryozoan colony and the sicula of graptolites, and especially a full analogy between the protoecium and the ancestrula of Bryozoa and the prosicula and the metasicula of the Graptolithina, respectively.

2. The mode of budding from the sicula compares well with the mode of budding from the ancestrula.

3. The graptolite thecae and the bryozoan zooecia and zooids have a fundamental bilateral symmetry.

4. The similarity of the general habit and details of structure of many graptolites to those of the colonies of Cyclostomata and Cryptostomata implies a close relationship between these groups.

5. The presence of chitinous zooecia in Recent Ctenostomata is also a condition typical of graptolites.

6. The muscle scars of certain biserial graptoloids correspond to the muscular system of phylactolaemate bryozoans.

7. The presence of thecae on stipes in certain graptolites is comparable with the zooecia of the bryozoans.

From this evidence Ulrich and Ruedemann (1931) drew a general conclusion that although graptolites were closely related to bryozoans, they "represent a more primitive branch than the bryozoans proper" and are close to the ancestral lophophorates.

The reasoning of Ulrich and Ruedemann was strongly criticized by Kozłowski (1938, 1949).

The resemblances cited by Ulrich and Ruedemann are very general and, in fact, do not imply any definitive systematic position. Such resemblances as the habit of the colony or the shape of its thecae may be easily explained as convergence due to a similar mode of life of colonial organisms. Bilateral symmetry is a very general and common feature in a large number of metazoans. An apparently profound resemblance in the structure of early astogenetic stages of both groups reflects a common sequence of morphogenetic events associated with the attachment of free-living larvae and their later metamorphosis. The budding patterns are not specific enough to provide a safe basis for phylogenetic conclusions.

Certain resemblances used by Ulrich and Ruedemann to substantiate their thesis were based on obvious misunderstandings. The presence of presumed muscle scars on the inner surfaces of the thecae, also suggested by Haberfelner (1933), has never been proved on isolated specimens. When found on flattened material, these are most probably traces of thickened edges of the interthecal septa pressed through the walls of the rhabdosome. The problems of colonial polymorphism and chitinous nature of graptolite colony skeleton are discussed below. The idea of bryozoan affinities of graptolites never gained a wide acceptance, but nevertheless, bryozoans still offer useful models for biologic interpretation of graptolite structures (Kozłowski 1949) and life cycles (Urbanek 1983).

Kozłowski's classical hypothesis—a new version of the pterobranch paradigm

Kozłowski's (1938) discovery of the astonishingly well-preserved graptolites in Tremadocian (Ordovician) cherts of Poland led him to a number of important observations. Specimens sectioned with a microtome and acid-bleached, for the first time provided a clear picture of the structure of periderm, the evidence for the presence of an extensive stolon system in the sessile orders of graptolites, and a complete account of dendroid development (Kozłowski 1949).

These investigations show that the periderm of Dendroidea and other orders of Graptolithina consists of two components, the primary consisting of short transverse growth bands referred to as fuselli, and the secondary (or cortex) being laminated (Fig. 2). The specific "fusellar" microstructure of the primary component was considered by Kozłowski as the most characteristic feature of all graptolites as well as living and fossil Pterobranchia. It was this similarity that Kozłowski considered to be essential and indicative of a close phylogenetic relationship between both groups.

According to Kozłowski (1949, 1966a), the uniqueness of the fusellar structure lies in the form of the elementary units (fuselli) and their ar-

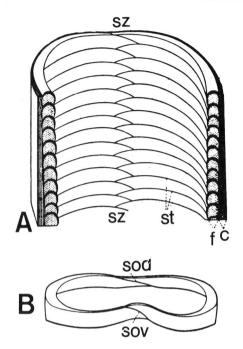

Fig. 2. Kozłowski's microstructure of dendroid graptolite.
A. Main component of thecal wall: cortical (c) and fusellar (f)
elements, and isolated fusellar growth bands forming zigzag suture (sz) and interfusellar sutures (st). B. Relationships between two adjacent fuselli with ventral and dorsal oblique
structures (sov, sod). C. Transparency of stolon system of
dendroid graptolite with triad pattern of buddings. a_2–a_4, b_2–
b_4, s_2–s_4, Successive auto-, bi-, and stolotheca, and their internal portions (ib_3–ib_5, is_3–is_5; autotheca marked by faint dotted
lines). Reproduced by permission from Kozłowski (1949).

rangement. The restricted distribution of such
structures among animals was also emphasized.
Kozłowski (1966a) stressed that "such a structure
is recorded in all graptolites, as well as in Recent
and fossil representatives of the Pterobranchia...
in both the Graptolithina and Pterobranchia, the
walls of theca have an identical and very specific
structure." Thus, the fusellar structure was regarded by Kozłowski as diagnostic among invertebrates, as the presence of feathers is diagnostic
among vertebrates. It constituted the basis of
Kozłowski's views on the close affinities between
graptolites and pterobranchs (Fig. 3).

In addition to this main criterion, Kozłowski
(1949) considered also the chemical nature of the
organic skeleton of graptolites and pterobranchs.
He regarded graptolites to be "chitinous." The
term chitin was used as a general term to denote
any organic, flexible skeletal structure, without
reference to its chemical composition. Later,
when the first biochemical data indicated the nonchitinous and probably proteinaceous nature of
pterobranch and graptolite skeletons, Kozłowski
(1966a) included them into his criteria of homology. Other criteria of the close relationship between graptolites and pterobranchs include the

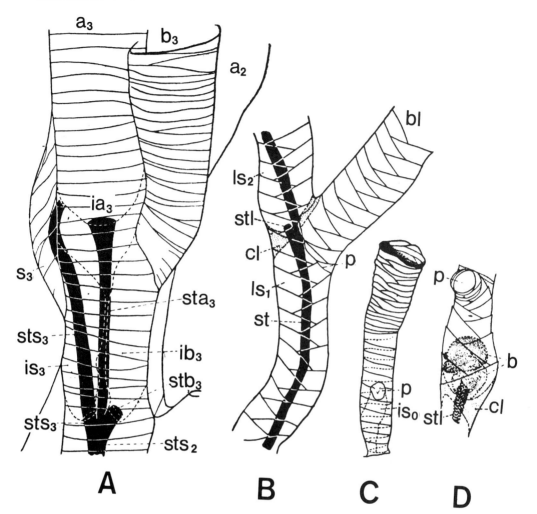

Fig. 3. Resemblance between dendroid graptolites (A) and rhabdopleurid pterobranchs (B) as seen in shape and arrangement of unique growth bands (fuselli), presence of internal stolon system and peculiar mode of budding of primary blastozooid in the dendroid sicula (C), and in all blastozooids of rhabdopleurid pterobranchs (D) involving resorption of wall and formation of pore (porus, p). a_2–a_3, Corresponding autothecae; b, blastozooid bud; b_3, corresponding bitheca; bl, lateral branch of zooidal tube; cl, diaphragm separating adjacent zooidal lodgings; is_0, ia_3, ib_3, is_3, internal portions of corresponding thecae; ls_1, ls_2, successive stolonal tubes comparable to graptolite stolothecae; s_3, corresponding stolotheca; st, stolon; stl, lateral or peduncular stolon; sta_3, stb_3, sts_3, autothecal, bithecal and stolothecal portions of the stolon system of a triad. Reproduced by permission from Kozłowski (1949).

presence of internal stolons and the peculiar budding pattern of primary zooids in rhabdopleurid pterobranchs and in the majority of graptolites (Fig. 3).

The stolon of *Rhabdopleura* ("black stolon") is a string of soft tissues, later coated by a strongly pigmented peridermal sheath and situated inside the thecal tube. Kozłowski repeatedly emphasized that such an internal stolon is a unique feature that pterobranchs share only with graptolites. Kozłowski provided firm evidence for the stolon system in Dendroidea and other sessile graptolites. Therefore, the absence of stolon in Graptoloidea may be considered a secondary phenomenon, probably due to skeletal sheath reduction.

In rhabdopleurid pterobranchs new buds are formed on naked portions of the young stolon, inside the creeping tube of the colony. The newly formed zooid produces an orifice in the upper wall of the tube. A similar opening (porus), used by the new zooid to emerge from the parental theca (or tube) and to secrete its individual lodging, appears on one of the walls of the sicula in sessile graptolites and also in primitive graptoloids (Fig. 3). Thus, all zooids in rhabdopleurides and all primary buds in the majority of Graptolithina share the same pattern of development—that is, "perforatory" budding (Kozłowski 1938).

Kozłowski drew two main conclusions: one concerning the morphogenesis of the skeleton and its way of secretion, and the other on the phylogenetic affinities and biologic organization of Graptolithina. He explained the formation of the graptolite periderm as a twofold process. On the one hand, this process was assumed to involve secretion of the fusellar component from the glands situated on the cephalic disc on the edge of the growing thecal tube, as in pterobranch zooids. On the other hand, there is no good analogy to the graptolite cortex in *Rhabdopleura* and *Cephalodiscus*. Because of the laminated nature of the cortex, Kozłowski suggested that it was secreted under the envelope of soft tissues covering the outer surface of the rhabdosome (extrathecal tissues). Thus, Kozłowski's model (1938, 1949) of secretion of the primary component of graptolite

skeletons was inspired by pterobranchs, but the model of secretion of the cortex was based on an analogy with ctenostome bryozoans.

The recognition of the phylogenetic relationship between the Pterobranchia and the Graptolithina provided a starting point for biologic interpretation of graptolite colonies. The Rhabdopleurida, rather than the Cephalodiscida, are considered by Kozłowski (1938, 1949) to be the closest relatives of the graptolites; they also supply a model for the polymorphism in dendroid colonies. The colonial skeleton of *Rhabdopleura* (Fig. 4) comprises only two kinds of tubes, the zooidal and stolonal tubes. The former are comparable to the graptolite autothecae; the latter correspond to the stolothecae that house the so-called budding individuals. The stolothecae were occupied by buds responsible for secretion of their walls as well as for gradual elongation of the stolon. At regular intervals the stolon within stoloth-

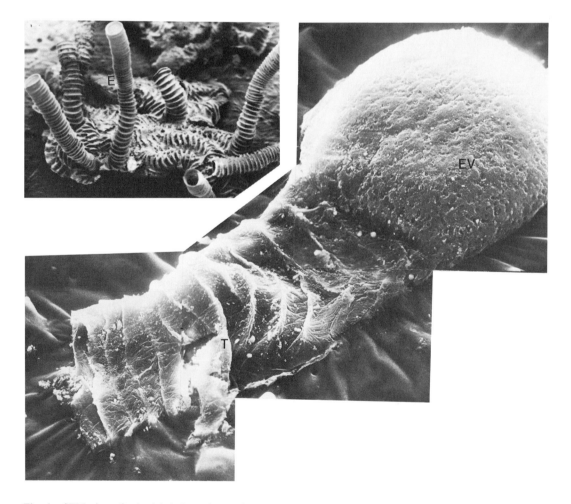

Fig. 4. SEM view of colonial skeleton (coenecium, tubarium) of *Rhabdopleura compacta* (Hincks).   A. Repent and erect (E) and tubes.   B. Early developmental stage of the colony, embryonal vesicle (EV), and initial portion of primary repent tube (T).

ecae gave rise to triads of daughter buds. Stolozooids were not a different kind of individuals but rather juvenile autozooids, because each autotheca is a direct continuation of the parental stolotheca. Thus, the trimorphism of graptolite thecae corresponds to the dimorphism of graptolite individuals. The bithecae are much smaller and have distinctly simpler apertures than the autothecae. Kozłowski pointed out that *Cephalodiscus* displays an extreme sexual dimorphism (it is weaker in *Rhabdopleura)*, and suggested that the graptolite autothecae might contain females and the bithecae, males.

This interpretation differs fundamentally from earlier comparisons with the highly differentiated polymorphism and the functional specialization of zooids in coelenterate colonies. The suppression of bithecae in planktic graptoloids has also found an entirely new explanation. While the males (bizooids) were gradually eliminated, the females (autozooids) were transformed into hermaphroditic zooids.

New light was shed also on the graptolite astogeny. Successive stolothecae produce indeed a continuous chain of thecal tubes on one side of the branch, but the resulting structure is different from the coelenterate common canal and the graptolite stipe differs from the hydrocaulus of hydroid colonies. In specialized graptoloids (e.g., *Monograptus*) the basal portions of the successive thecae make in fact a secondary structure comparable to the common canal. But direct budding proceeds even in such forms from the preceding zooid, but not from the common canal. In primitive graptolites (e.g., Dendroidea), in turn, the rhabdosome is composed entirely of thecae, and the colony cosists solely of zooids. Therefore, the concept of coenosarc cord within the common canal—the alleged evidence for coelenterate affinities—is unwarranted.

Reception of Kozłowski's ideas

Kozłowski's (1938) preliminary note was published just before World War II, while the greater part of impressions of Bulman's (1938) section on the Graptolithina, with an addendum considering Kozłowski's results, was destroyed during the war. But Bulman fully recognized the importance of Kozłowski's work. While Kozłowski with his materials and scripts shared the tragedy of Nazi-occupied Warsaw (see Kozłowski 1949; Kielan-Jaworowska and Urbanek 1978), Bulman (1942, 1945) restated the main points of Kozłowski's views and applied them to the interpretation of Caradoc dendroids.

Soon after the war, Kozłowski (1947) published an article on graptolite affinities where he summarized his main arguments. The article was followed by a monograph (Kozłowski 1949). Within a few years his views became generally accepted

by the scientific community. In addition to the authoritative opinion of Bulman (1949, 1955), an important role was also played by zoologic treatises (Dawydoff 1948; Waterlot 1952) and popular textbooks (Shrock and Twenhofel 1953; Beklemishev 1970). The crucial support, however, came from a series of discoveries and contributions made by Kozłowski himself. The discovery of fossil pterobranchs, first in the Paleocene and Cretaceous (Kozłowski 1949, 1956) and finally in the Ordovician (Kozłowski 1961, 1967, 1970), shattered the objections that Kozłowski derived graptolites, an ancient fossil group, from geologically much younger, or even exclusively Recent groups. Kozłowski (1963) demonstrated also a remarkable similarity in the early stages of development between the tuboid graptolite colony and the embryonic vesicle of *Rhabdopleura*. The discovery of Crustoidea, a new order of sessile graptolites with great similarity to rhabdopleuran pterobranchs, crowned the list of arguments supporting Kozłowski's hypothesis (Kozłowski 1962).

While accepting Kozłowski's ideas on the systematic position of graptolites, some authors expressed doubts concerning his concept of the skeleton secretion (Bulman 1955, 1790). As a matter of fact, Kozłowski never precisely defined the relation between the soft parts and the skeleton. A combination of the pterobranch (for the fuselli) and the bryozoan mode of secretion (for the cortex) appeared implausible. Beklemishev (1970) concluded that the secretion of the graptolite periderm followed entirely the pterobranch mode. He ascribed the formation of the cortical coating to secretionary activity of the zooids, which left their zooidal tubes and crept over the outer surface of the thecae, thus covering them with secondary layers of a peridermal substance. Beklemishev was also the first to emphasize that the formation of fusellar growth bands, so peculiar to both groups, cannot be attributed to such different parts of the body as the cephalic disc (in the Pterobranchia) and the epithelial membrane (in the Graptolithina). By rejecting Kozłowski's morphogenetic concepts but accepting his phylogenetic conclusions, Beklemishev (1951, 1970) anticipated the most recent developments.

On the benches of the opposition

Bohlin (1950) was the first to criticize Kozłowski's arguments. He argued that the resemblance between graptolites and pterobranchs "is altogether superficial" and that graptolites arose from unknown ancestors and became extinct without descendants. His agnosticism was inconsistent, however, because after considering the relation of soft parts to the skeleton, he suggested that "the entire wall of the rhabdosome was formed between two epithelia and may be comparable to the

Stützlamellae (Mesoglea) in the Hydrozoa." Following from this basic assumption, Bohlin concluded that there was no place for mesoderm in graptolites and consequently, they were diploblastic and probably a specialized group of coelenterates. At the same time Bohlin emphasized that the high degree of morphogenetic integration within graptolite colonies implies the secretion of both cortical and fusellar layers under the coating of a soft tissue. This contrasts to the pterobranch mode of secretion and is incompatible with close kinship between pterobranchs and graptolites.

The coelenterate concept of graptolite affinities soon received more support. Decker (1956; see also Decker and Gold 1957; Decker and Hassinger 1958) believed that he found evidence for the presence of bithecae also in graptoloids and for the occurrence of a new kind of thecae, called gonothecae, both in the Dendroidea and in the Graptoloidea. Decker concluded that all graptolites had separate sexes, as indicated by the consistent co-occurrence of male bithecae and female, barrel shaped gonothecae. Autothecae were interpreted as feeding zooids, while nematothecae were considered to be an obvious indication of the presence of stinging cells—an important coelenterate characteristic. Decker claimed also to have found remains of the soft parts of autothecal, bithecal, and gonothecal zooids, frequently showing "a single row of extensible tentacles," again a distinct difference from the plumelike lophophore of *Rhabdopleura*. Decker and his coauthors regretted that Kozłowski "made the terrible mistake of placing graptolites with *Rhabdopleura* under Hemichordata" (Decker 1956; Decker and Gold 1957; Decker and Hassinger 1958).

Hyman (1959) also expressed strong criticism of Kozłowski's views. First of all, she saw no point in his argument that the graptolites like the pterobranchs secrete chitinous tubes that enclose the soft body, because the pterobranch coenoecium was definitely not chitinous. She neglected the fusellar structure and emphasized that the oldest known fossil pterobranchs were "far removed from graptolites in geological time." Hyman found Kozłowski's interpretation of graptolite polymorphism "not very convincing," but she was much impressed by Decker's observations. Only one argument advanced by Kozłowski survived her devastating criticism—the presence of a stolon in *Rhabdopleura* and in sessile graptolite orders—but stolons occur commonly throughout the animal kingdom. Therefore, the hemichordate affinities of graptolites were rejected by Hyman (1959) "as insufficiently grounded."

Kozłowski (1966a) answered the criticisms with a considerable delay. He pointed out Bohlin's insufficient knowledge of the anatomy of graptolites as well as coelenterates, for structures suggested by Bohlin's diagrams are unknown in the Coelenterata; for example, the periderm of the thecate Hydroida is never covered by ectoderm. Kozłowski questioned the validity of Decker's observations and interpretations as imaginary and shaped by a preconceived idea of the coelenterate nature of graptolites. In fact, Decker interpreted apertural spines in monograptids as nematothecae, thus proving that his understanding of graptolite anatomy was very imperfect. Both Bohlin and Hyman were criticized by Kozłowski for neglecting the presence of the internal stolon with a sheath, and of the fusellar structure. Kozłowski demonstrated that fuselli are randomly arranged in the graptolite *Mastigograptus*—essentially a pterobranch condition—and concluded: "So it may be stated that the tubes of the tubarium in the Rhabdopleuridea and the tubes of the coenoecium in the Cephalodiscoidea have in general a structure identical to the thecae of the Graptolithina and that a structure of such type does not occur in any other animal, either Recent or fossil" (Kozłowski 1966a). The specificity of the fusellar structure, the discovery of the Crustoidea with some transitional features between Pterobranchs and graptolites, and the proteinaceous nature of the periderm in both the groups (see below) were the main arguments for a close phylogenetic relationship between the Graptolithina and the Pterobranchia.

Paleobiochemistry and the systematic position of graptolites

The organic, flexible material of the graptolite skeleton usually was described as "chitin" because of analogy to the exoskeleton of many invertebrates, but there was little actual evidence to consider the graptolite material to be chitin. Wiman (1901), who undertook the first elementary analysis of graptolite periderm, thought that its nitrogen–carbon ratio was suggestive of its nonchitinous nature. Kraft (1926), however, insisted on the chitinous composition of the graptolite skeleton.

Therefore, when Rudall (1955) established that the zooidal tubes of *Rhabdopleura* contain no chitin, Hyman (1959) took this as an argument against Kozłowski's hypothesis on the close relationship between rhabdopleurid pterobranchs and graptolites. She documented that chitin occurred less frequently among the lophophorate phyla than it had been believed (Hyman 1958), and argued that the graptolite and the pterobranch skeletons differ fundamentally in chemical composition, even though she was aware that Kozłowski and his fellow graptolitologists used the term chitin in a rather vague sense.

New light was shed on this issue by the studies of Foucart (1964; see also Foucart et al. 1965; Foucart and Jeuniaux 1966) on the chemistry of the graptolite periderm and the skeletal tissues of extant pterobranchs. Hydrolysates from two Silurian monograptids and an Ordovician diplograptid revealed high amounts of glycine, serine, and alanine, which suggested that the graptolite skeleton was scleroproteinaceous in nature. The evidence that neither graptolites nor pterobranchs contained any trace of chitin devastated the old chitin fallacy and provided a strong support for the close phylogenetic relationship between the Graptolithina and the Pterobranchia.

Graptolite periderm under TEM

Around 1970, Kozłowski's classical hypothesis on graptolite affinities and the secretion of their skeleton seemed well established. However, the advances of electron microscopy demanded testing the hypothesis at the ultrastructural level. Early attempts (Wetzel 1958; Kraatz 1964, 1968; Berry and Takagi 1970, 1971; Rickards et al. 1971) gave rather confusing results. It was not until the joint studies by Urbanek and Towe (Towe and Urbanek 1972, 1974; Urbanek and Towe 1974, 1975) that reliable ultrastructural data on the graptolite skeleton were supplied (Fig. 5). They showed that although the fusellar component and the cortex could be easily recognized under TEM, these were relatively differentiated and organized structural systems. The fusellar tissue consists of non-banded, fairly delicate fibrils (average diameter 600 Å) that usually are wavy, branching, or irregularly anastomosing to produce a meshy or spongy appearance, without any traces of preserved ground substance. In contrast, the cortical tissue consists of coarser (average diameter 1500 Å) and straight fibrils; they are arranged parallel to one another and usually interconnected by numerous, more-or-less uniformly spaced rods. The cortical fibrils are packed into layers separated by sheets, and oriented at different angles in adjacent layers.

The appearance and spatial interrelations of the cortical fibrils appeared strongly suggestive of their collagenlike nature (Towe and Urbanek 1972). They resemble strikingly the "plywood" pattern of such extant collagen materials as the annelid cuticle and the vertebrate ocular tissues. Frequently observed passages of a single fusellar fibril into a fibril of the so-called outer lamella within the same fusellus suggested that fusellar fibrils also represent the collagen class of proteins, whereas distinct differences between fusellar and cortical fibrils could be ascribed to the remarkable polymorphism of collagen depending on the milieu.

The ultrastructural studies thus pointed to the collagen-like nature of the standard graptolite periderm (Urbanek and Towe 1974, 1975), and even the strongly modified reticular skeleton of retiolitids proved to be a somewhat aberrant form of collagen (Towe and Urbanek 1974; Urbanek and Towe 1975). These results have revealed distinct differences between the fibrous components of the fusellar tissue in pterobranchs and graptolites. Especially striking was the contrast between these observations and the data obtained by Dilly (1971) for *Rhabdopleura*.

While the graptolites' fusellar tissue consists of interconnected, wavy fibrils producing a spongy pattern (Fig. 6), both the rhabdopleurid (Dilly 1971) and cephalodiscid pterobranchs (Urbanek 1976a) have their zooidal tubes made of thin, straight fibrils, loosely dispersed in ample matrix (Fig. 7). Dilly (1971) distinguished three types of fibrils embedded in *Rhabdopleura* tube. Two of them, each with internal helical electron-dense lines surrounded by a sheath of less electron-dense material, were classified as keratin-like, while the nature of the remaining long, thicker and structureless fibrils was obscure. The keratinous nature of the fibrous material in *Rhabdopleura* was questioned by Bairati (1972) and doubted by Urbanek (1976a); the appearance, the pattern and the preliminary biochemical results all seemed incompatible with the concept of the collagenous nature of *Rhabdopleura* fibrils.

New wave of the phylogenetic agnosticism

The substantial differences between the fusellar tissues of the Pterobranchia and the Graptolithina were summarized by Urbanek (1976a) and provided the basis for his critical evaluation of the pterobranch concept of graptolite ancestry.

Urbanek (1976a) argued that full homology of a tissue should include its structural specificity—for example, a similarity of ultrastrucutral units, resemblance of the overall pattern, and consistency in chemical composition. Therefore, he doubted if there was a true homology between the fusellar tissue of graptolites and pterobranchs.

Whereas Kozłowski (1938, 1949) suggested a dualistic model of secretion of the peridermal tissues, Urbanek (1976a, b) postulated secretion by a single secretory organ or portion of the body. TEM observations provided evidence that the fusellar and the cortical tissues rather frequently occur as intercalations within each other, and even a single fusellus may contain some cortical fabric enclosed within its pellicle (Urbanek and Towe 1974, 1975). Such paradoxes could hardly be explained within the framework of Kozłowski's classical hypothesis of secretion, because they suggest easy and smooth shifts in secretion of the graptolite skeletal "tissues." The ultrastructural data thus indicated that the whole graptolite periderm was secreted in the same manner by the same part of the body. Urbanek (1976a) believed the secretion of the periderm within the perithecal membrane (i.e., a body fold surrounding thecal wall from the outside and inside) to be compatible with all the ultrastructural data (Fig. 8B). The surface of the periderm seemed featureless and rather

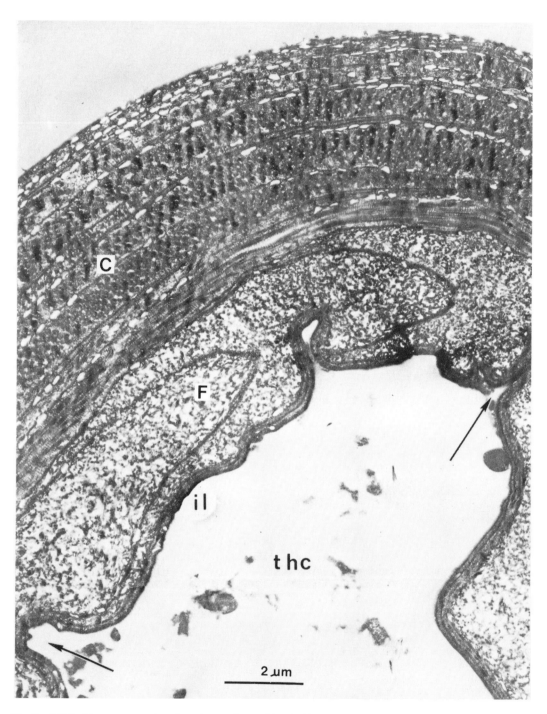

Fig. 5. TEM of transverse section through autotheca of dendroid *Dictyonema* sp. Note distinct difference between cortical (C) and fusellar (F) components of periderm, and multilayered inner lining (il) delimiting thecal cavity (thc) with characteristic, probably compressional, infoldings (arrows). Reproduced by permission of Smithsonian Institution Press from Urbanek and Towe (1974, pl. 1).

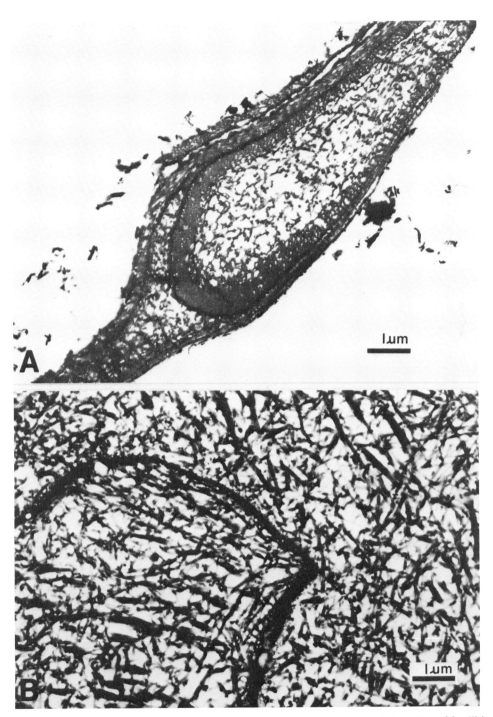

Fig. 6. TEM of fusellar tissue of Graptolithina. A. Longitudinal section through boundary of fuselli in mono-graptid *Pristiograptus dubius* (Suess); compare with Fig. 7A. B. Fusellar boundary and fabric in dendroid *Dictyonema* sp. exhibiting characteristic meshwork of fibrils. Reproduced by permission of Smithsonian Institution Press from Urbanek and Towe (1974, pl. 14, fig. 3; 1975, pl. 15, fig. 2).

Fig. 7.   Transverse section of zooidal tube of *Rhabdopleura compacta* (Hincks).   A. Overlap of fuselli and loose fibrous material within ground substance.   B. Enlarged thick fibril with enigmatic helical dense line.

smooth on a few unpublished SEM micrographs made by Urbanek prior to his TEM studies. Therefore, secretion under a continuous envelope of soft tissues appeared as a plausible interpretation.

Kirk (1972, 1974) had earlier considered the problem of the extrathecal tissue, and because of difficulties in anatomic and functional association of the cephalic disc with the extrathecal tissue required to account for the deposition of the cortical layers, doubted if the graptolite fuselli could have been secreted by a cephalic disc. She suggested, like Bohlin (1950), that both the fusellar and the cortical tissues were secreted by an epithelium.

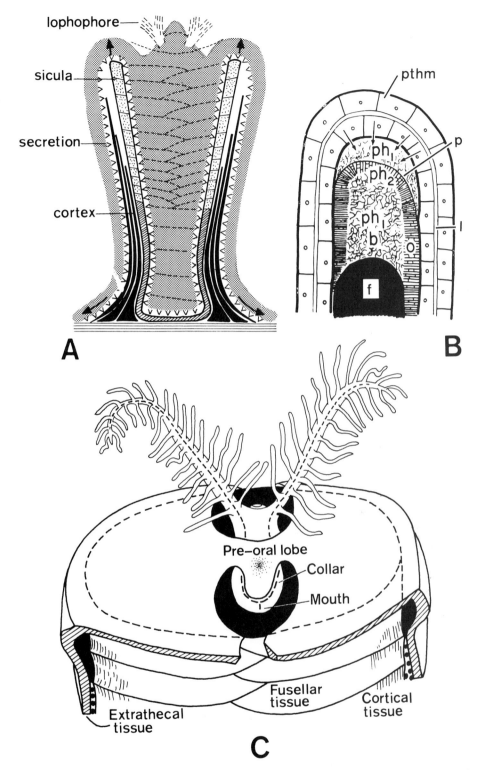

Fig. 8. Relationships of soft parts to the skeleton in graptolites proposed by Kirk (A), Urbanek (B), and Rickards (C). A. Sicula of sessile graptolite with continuous secretory epithelium forming epithelial fold (dotted) that covers skeleton (solid black) from outside. B. Wall of graptolite theca, made of fuselli (f) and surrounded by double-layered perithecal membrane (pthm), which envelops skeleton and secretes particular components of the fuselli (b, base, o, outer lamella, and p, outer pellicle) in consecutive secretionary phases (ph₁, ph₂). C. Fusellar and cortical component of skeleton with peculiar differentiation of preoral lobe into peripheral portion (secreting fuselli) and outer fold, forming cortex. Reproduced by courtesy of the author (Kirk 1972) by permission from Urbanek (1976a, fig. 6B), and by permission of Cambridge University Press from Rickards (1975, text-fig. 76).

Kirk (1972) envisaged a double-layered epithelial evagination, with one layer secreting the fuselli and the other secreting the cortex (Fig. 8A). In turn, Rickards (1975) rejected the mantle evagination concept and suggested that graptolite zooids had a specially modified bifid and lobate cephalic disc, extending into the extrathecal tissue on the lateral slopes of the thecae (Fig. 8C). Such a modification could, at least theoretically, allow the disc to reach the growing edge of the theca, and the cortex to be secreted by the extrathecal tissue. However, in order to explain development of the nema and its derivatives, Rickards was forced to accept an additional source of the extrathecal tissue—a tissue spreading from the prosicula and filling in the hollow nema to be later extruded from its tip. This complicated model is hardly compatible with the pterobranch morphology, although Rickards (1975) considered pterobranchs as the nearest relatives of graptolites.

The observations on the pattern of the fibrous component and the considerations on the models of secretion in the Pterobranchia and the Graptolithina led Urbanek (1976a) to agnosticism. If the pterobranchs and the graptolites were indeed closely related, a much closer resemblance in ultrastructure and morphogenesis of their skeletons should be expected. Consequently, Urbanek concluded that the systematic position of graptolites within Bilateria remains an unresolved problem.

The cortical bandages sensation

It was, however, the formation of cortex that became the proverbial bone of contention. Andres (1976, 1977) demonstrated with light microscope that the periderm surface in some graptoloids is densely covered with ribbonlike elements. These "Verdickungsbänder" or "Cortex-Bänder" usually have scarplike edges, are rather irregularly arranged, and are thinner than the fuselli of the underlying theca. The entire cortex consists of such bands, which could be deposited by the cephalic discs of zooids creeping over the surface of the rhabdosome. A similar process has been observed in *Cephalodiscus,* and hence the "cortex formation does not imply a principal difference between graptolites and pterobranchs" (Andres 1977). Almost at the same time, Crowther and Rickards (1977) drew similar conclusions from SEM studies. The bandage structure is probably universal in graptolites, and a cortical bandage is the fundamental unit of the cortex. Each bandage is a bundle of cortical fibrils parallel to its long axis and usually enclosed by an outer membrane (pellicle). Like fuselli, the cortical bandages were also secreted by the cephalic discs of graptolite zooids "which crept out of the thecal tubes in order to deposit . . . each bandage" (Crowther and Rickards 1977).

This discovery provided much support for the pterobranch paradigm of graptolites (Fig. 9). SEM studies supplied further evidence for close relationship between the fusellar and the cortical tissues, and documented ringlike annulations in cortical fibrils, suggestive of their collagenous nature. Moreover, a number of observations made by Andres (1976, 1977, 1980) indicated a close resemblance between the fusellar structures in certain sessile graptolites (*Mastigograptus* and *Micrograptus*) and fossil rhabdopleurids *(Rhabdopleurites).* Some graptolites have randomly arranged fuselli, without a regular zigzag suture, while others tend to have an orderly superposition of fuselli and an incipient zigzag suture (Andres 1977). Transitional groups, such as the Crustoidea, have a number of features in common with pterobranchs (Andres 1980). Consequently, attempts were made to demonstrate that the entire structural diversity observed in the Graptolithina (the skeletal network of retiolitids, the apertural spines, the nema, and the heavy secondary thickening of the periderm) is compatible with the morphogenetic potential of the Pterobranchia (Andres 1980; Crowther 1978, 1980).

Structural differences between the pterobranch and graptolite skeletons

The uniformity in molecular composition of the pterobranch and the graptolite skeletons contrasts with their considerable structural differences. The graptolite skeleton is heteromorphic, with the cortex and the fusellar component, or fusellum (cf. Kühne 1955), having their fibrous material organized differently. The pterobranch skeleton, however, is structurally homomorphic. Moreover, the ultrastructure of the fuselli is strikingly different between pterobranchs and graptolites.

The fine structure of the periderm of encrusting graptolites of the order Crustoidea appeared crucial in order to explain the origin of these surprising differences, because the crustoids seem to represent a morphologic type that is intermediate between the Rhabdopleurida and the Dendroidea. Andres (1980) proposed a model of transition from the pterobranch structure to the graptolite structure. The model assumes that the cortex is originally deposited in form of narrow layers of fusellar material separated by sheets that are only later covered by layers of the regular cortical fabric. This interpretation implies that phylogenetically primitive fusellar fabric occurs in crustoids as a component of the early layers of cortex, thus resulting in a recapitulatory effect and representing a *sui generis* missing link in the evolution of secretion (Fig. 10).

Urbanek and Mierzejewski (1984), however, suggest that early layers of crustoid cortex are produced as an extension of the lateral limbs of the fuselli [dependent mode of cortex formation as defined by Urbanek (1976a)]. In fact, the early

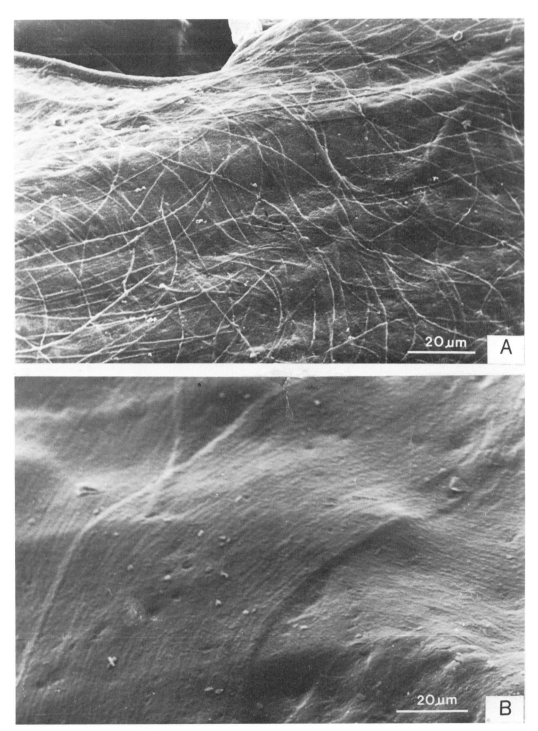

Fig. 9. SEM view of cortical bandages in Ordovician diplograptids. A. *Orthograptus gracilis* (Roem.) with numerous bandages around aperture of theca. B. *Climacograptus brevis mutabilis* Strachan with enlarged bandages showing faint traces of course cortical fibrils.

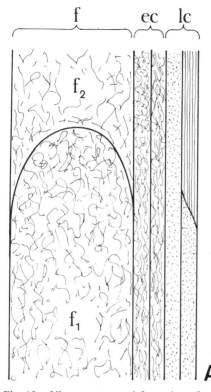
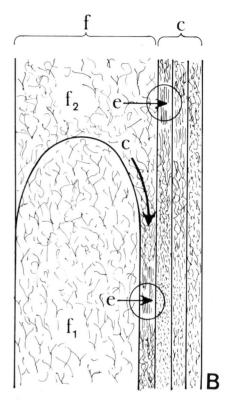

Fig. 10. Ultrastructure and formation of cortex in Crustoidea as interpreted by Andres (A) and Urbanek and Mierzejewski (B). Fusellar or primary component of periderm (the fusellum, f) made up of a number of successive fusellar growth bands (f_1, f_2). In A overlain by early cortex (ec) of fibrils of a fusellar type later replaced by layers of regular cortical fabric (late cortex, lc), and thus producing a recapitulatory effect. In B successive layers of cortex (c) appear over the fusellum (f) by accumulation of lateral fusellar limbs (f_2), producing strong overlap and gradual "corticization" (c and heavy arrow) of primarily fusellar content of limb; changes into a different matrix either of tightly packed fibrils (paracortex) or of material lacking individualized fibrils (pseudocortex). Occasionally a more regular parallel arrangement of fibrils resembles the genuine cortex (the eucortex, e). These different approaches suggest Crustoidea as a transient link in the cortical fibrillogenesis.

layers show a gradual "corticization" of the primarily fusellar content of the limb. These early layers become eventually overlain with layers deposited later at the surface and independent of the fuselli (Fig. 11). Thus, the crustoid cortex is composed primarily of multiple sheets. The scanty material inbetween is either organized into a tightly packed meshwork of fibrous material (paracortex) or almost homogeneous and devoid of any fibrous elements (pseudocortex). More distinct and parallel fibrils, resembling the genuine cortical fabric (eucortex) of dendroids, appear rarely.

These observations allow for at least two important conclusions. First, it becomes clear that the term tissue is untenable with respect to the fusellar and the cortical components of the graptolite skeleton, and may be used only in a purely descriptive sense. In fact, these components are little more than different patterns of the physical organization of collagen or a collagen-like material, which grade into each other.

Second, the phylogenetic changes affected first the primary component, and transformed the straight, parallel fibrils characteristic of pterobranchs into wavy, interconnected fibrils found in the fuselli of graptolites. This radical change in the appearance and the pattern of fibrils could be accompanied by only minor changes at the molecular level (amino acid composition, primary structure), but it resulted in a remarkable difference in the fine structure of the primary component of the pterobranch and the graptolite periderm.

The crustoids, however, have one remarkable peculiarity, namely, the fibrils found in some graptoblasts (Mierzejewski 1984). Graptoblasts are tiny bodies with a flat lower and a convex upper wall, intimately related to crustoid colonies. Their enigmatic function has been recently compared with that of hibernacula, or dormant zooecia, in ctenostome bryozoans (Urbanek 1983). Morphologically, they represent the arrested and closed terminal portions of stolothe-

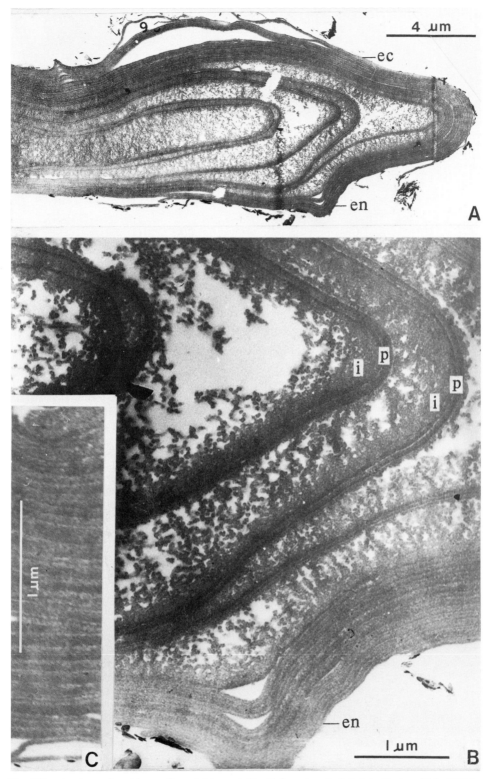

Fig. 11. Ultrastructure of periderm of crustoid graptolite *Bulmanicrusta latialata* Kozłowski in TEM section through thickened margin of apertural apparatus. A. Margin with accumulation of sheets producing outer (ec, ectocortex) and inner cortical deposits (en, endocortex). B. Enlargement of A showing reduced fuselli with an incipient outer lamella (i) and an outer pellicle (p). C. Substructure of endocortex. Reproduced by permission of Lethaia Foundation from Urbanek and Mierzejewski (1984, fig. 5).

cae. In the majority of graptoblasts the upper wall is made of a standard fusellar tissue like that in the crustoid thecae. In some graptoblasts, however, it consists exclusively of aberrant straight fibrils, strongly resembling the fibrils of *Cephalodiscus* but entirely different from the fusellar fibrils of graptolites.

Amino acid composition of pterobranch coenoecia

The chemical nature of the pterobranch colonial skeleton (coenoecium) became a crucial issue at this point of the debate on graptolite affinities. The available data (Allman 1869; Sars 1874; Andersson 1907; Schepotieff 1907; Dilly 1971; Bairati 1972; Urbanek 1976a) were inconclusive, although Rudall (1955) and Sundara-Rajulu et al. (1982) ruled out the presence of chitin in the pterobranch coenoecium.

Theoretically, the fibrils of the pterobranch skeleton could be closely related biochemically to collagen, for example, representing an evolutionary precursor of the scleroprotein. In this case, the gap between pterobranchs and graptolites might have been bridged by gradual changes in the primary structure, associated also with changes in other properties of the fibrils. But it could be also conjectured that the structural differences between pterobranchs and graptolites are also associated with some biochemical dissimilarities.

The evidence had to come from biochemical analysis of the fresh extracellular skeleton of *Cephalodiscus* and *Rhabdopleura*. Adequate material was collected by Urbanek and Zieliński (1981) and P. N. Dilly, and the amino acid analysis shows that coenoecia of *Cephalodiscus* and *Rhabdopleura* contain considerable quantities of collagenous material with relatively high hydroxyproline and low hydroxylysine levels (Armstrong et al. 1984). The pterobranch collagen differs from the standard vertebrate collagen, but it is remarkably similar in amino acid composition to the ox-bone collagen.

These results are in many ways surprising. Although it is still unknown which particular part of the coenoecium contains collagen, it is most probably associated with the fibrous components of the periderm. However, the fibrous material in the skeletons of *Rhabdopleura* and *Cephalodiscus* lacks the characteristic ultrastructural features of collagen, some of which have been recognized in the graptolite skeletal tissues. This paradox may be at least partially explained as a secondary effect of differences in the molecular orientation of tropocollagen subunits within a single fibril (e.g., a random assembly of such units in pterobranchs and a more orderly arrangement in graptolites). A number of the ultrastructural features of pterobranch fibrils [e.g., the presence of internal helical

structures described by Dilly (1971)] remains, however, enigmatic.

The presence of collagen in coenoecia of Recent pterobranchs and the collagen-like nature of the skeletal material of graptolites remove the main stumbling block to interpretation of several striking resemblances between both the groups as true homologies. Armstrong et al. (1984) concluded that, in conjunction with other data, the essential similarity of the chemical composition of the skeleton was strongly indicative of a close phylogenetic relationship between the Pterobranchia and the Graptolithina.

ANALYSIS OF THE PHYLOGENETIC DEBATE

Popper (1976, 1979) suggested that all scientific discussions start with a problem (P1) to which a tentative solution, or tentative theory (TT1), is offered; this theory is later criticized in an attempt at error elimination (EE1), resulting in a new problem (P2), according to the scheme:

$$P1 \rightarrow TT1 \rightarrow EE1 \rightarrow P2 \rightarrow TT2 \rightarrow$$
$$EE2 \rightarrow P3 \ldots$$

Each problem is always formulated within the context of a problem situation (PS), which reflects the state of the knowledge.

This Popperian scheme may be applied to interpret the phylogenetic debate on the systematic position of graptolites. The basic problem was formulated early and remained essentially the same throughout the debate. What are the nearest relatives (if any?) of the graptolites, or what is the systematic position of graptolites in the animal kingdom? The formulation of the problem, however, was related to the changing problem situations. Consequently, the solution to the problem, the tentative theory, had in each case a somewhat different meaning. New problem situations were emerging owing to the growth of knowledge, especially the influx of new data on graptolites and their potential relatives, or owing to radical changes in methodology of systematic biology. Ten successive problem situations can be distinguished in the debate on graptolite affinities.

During the time span 1860–1895, the primary pool of ideas was formulated, and old, semiintuitive concepts were partially tested (PS1). A new situation (PS2) was created by Wiman (1895) and his attempt to interpret the budding and astogeny in the Dendroidea. Thus, the ignorance concerning the organization of sessile graptolites was eliminated and the primary pool of ideas was supplemented with a new viewpoint—the agnostic paradigm. Schepotieff's (1905) studies on pterobranchs resulted in a more complete definition of the pterobranch paradigm (PS3).

Bulman's (1932, 1938) treatise on graptolites, Ulrich and Ruedemann's (1931) modernization of the classical bryozoan concept, and the preliminary note by Kozłowski (1938) with a new version of the pterobranch paradigm, produced a unique situation (PS4) characterized by a controversy between three main and mutually exclusive concepts of graptolite ancestry.

Owing to Kozłowski's monograph (1949), a new problem situation developed (PS5): the domination of the pterobanch paradigm—best reflected by Bulman (1955). The criticism of Kozłowski's views by Bohlin (1950) and Hyman (1959), especially the chitin–protein confusion, somewhat diminished the impact of Kozłowski's work and resulted in a hesitant position of some paleontologists and zoologists (PS6). The biochemical inconsistency was soon resolved by Foucart (1964; Foucart et al. 1965; Foucart and Jeuniaux 1966), whose results permitted Kozłowski (1966a) to defend and develop his theses (PS7).

However, an attempt to test Kozłowski's theory by TEM studies on the skeletal fabric in pterobranchs and graptolites, brought about the discovery of an *ultrastructural anomaly* (Urbanek and Towe 1974, 1975; Urbanek 1976a). The validity of Foucart's paleobiochemical results was also questioned, as well as Dilly's (1971) interpretation of the nature of *Rhabdopleura* fibrils. Inconsistencies in Kozłowski's *morphogenetic* thesis were soon exposed, leading to Urbanek's (1976a) uniform membrane model of secretion. This is why the logic of Urbanek's conclusions was similar to Wiman's "agnostic concept," for it undermined also the validity of Kozłowski's *phylogenetic thesis* (PS8).

Independent studies on the periderm surface micromorphology by Andres (1976, 1977) and Crowther and Rickards (1977) soon revealed the presence of ribbonlike structural elements in the cortex ("cortical bandages"), implying an essentially pterobranch mode of secretion. This is why the membrane model of secretion of the graptolite periderm, as suggested by Kozłowski (1938, 1949) and supported by Urbanek (1976a), was rejected and the pterobranch affinities of graptolites were strongly defended. These studies provided a basis for a partial solution of the basic problem and created a new problem situation (PS9). Certain issues remained unresolved, however; among them were the nature of the structural differences in the fibrous material of the graptolite and the pterobranch periderms, and the origin of such specialized rhabdosomal structures as the basal disc, the nema, or the virgula, and the virgular derivatives.

Therefore, the most recent studies were focused on the structural differences between pterobranchs and graptolites—in particular, on such primitive graptolites as the crustoids (Andres

1980; Urbanek 1983) and on the chemical nature of the pterobranch skeleton (Armstrong et al. 1984). The results created again a new problem situation (PS10), because the collagenous nature of the pterobranch skeleton allows for its homology with the collagenous skeleton of graptolites. Recent SEM research also provided firm evidence for the essential uniformity of the mode of secretion in both groups, which strongly supports their close phylogenetic relationships.

The current problem situation implies a certain modification of the pterobranch paradigm in order to accommodate new data, in particular those obtained by TEM and SEM studies. As compared with Kozłowski's classical version (1949, 1966a), this *modified pterobranch paradigm* involves alteration of his *morphogenetic thesis,* assuming an essentially uniform pterobranchlike mode of secretion. It fully accepts his *phylogenetic thesis* on close affinities between graptolites and pterobranchs *because* of the uniformity of the morphogenetic patterns and the molecular uniformity and microstructural similarity of the skeletal materials, and *in spite of* the differences in the ultrastructural organization of the fabric.

THE PRESENT STATE OF THE KNOWLEDGE

The present state of the debate on systematic position of the Graptolithina can be conveniently described at conventionally distinguished levels of structural organization. For this discussion three levels are recognized: *molecular–submicroscopic, microscopic,* and *macroscopic.* Different problems appear to be crucial at each particular level of organization, for example, (1) the composition of the skeletal material and the nature of its unit elements at the molecular–submicroscopic level, (2) the unit elements of secretion and their arrangement at the microscopic level, and (3) the initial stages of colony development, the polymorphism of zooids, and their budding patterns at the macroscopic level.

Molecular–submicroscopic level

At the molecular level, the composition of the skeletal material seems essentially the same in graptolites and in pterobranchs. Recent data strongly indicate biochemical degradation of the fossilized graptolite periderm, and no amino acids were found in hydrolysates of the graptolite remains (Urbanek and Towe 1974; Crowther 1980). Earlier results suggesting high amounts of amino acids in the periderm hydrolysates (Foucart 1964; Foucart et al. 1965) appear questionable. Therefore, the only reliable informaton comes from TEM observations that provided data suggestive of the collagenlike nature of the fibrous material of the graptolite periderm (Towe and Urbanek 1972; Urbanek and Towe 1974, 1975). They also demonstrated mutual substitution and transition of one kind of fibrils into the other, for example, within the microfusellar tissue (Urbanek 1976b). Moreover, even the strongly modified reticular skeleton of retiolitid graptolites is composed of aberrant collagen-like fibrils, strikingly similar to obliquely banded collagen fibrils (Towe and Urbanek 1974; Urbanek and Towe 1975), but with a considerable share of noncollagenous, possibly nonproteinaceous (carbohydrate?) material, probably an equivalent of the ground substance in extant collagen systems. These basic TEM data were later supported by SEM observations that demonstrated a characteristic annulation of fibrils comparable with the periodic banding of collagen (Crowther and Rickards 1977; Crowther 1980).

Thus, the hemichordates seem to have remained faithful to collagen as the skeletal material throughout their history, although they made use of the remarkable polymorphism of collagen, producing different types of fibrils and different patterns of their arrangement. This essential uniformity at the molecular level yields to considerable variability at the submicroscopic level of organization. In the Graptolithina, tropocollagen units assemble into fibrils significantly different from those of the Pterobranchia. They differ not only in size and appearance but also in the pattern of their arrangement (Figs. 6 and 7).

While the pterobranchs have their fusellar fibrils loosely dispersed and straight, the graptolites have them interconnected and wavy. This transformation of the pattern may be regarded as a unique ultrastructural character of all graptolites, an autapomorphy, developed by elaboration of the mechanism of fibrillogenesis (a_2 in Fig. 13).

The difference in fibril appearance and size also supports a discontinuity between the Pterobranchia and the Graptolithina in ultrastructural organization of the skeleton. This discontinuity is only partly bridged by the recent discovery of transient fibrils in the fusellar covering (blastotheca) of graptoblasts (Mierzejewski 1984, in press). These transient fibrils have essentially pterobranch organization (straight, with a lucent core) but are close to the graptolite fusellar fibrils in size. Their presence in graptoblasts may be explained as biochemical regression to a more primitive stage of fibrillogenesis, owing to the inhibition of growth (Urbanek 1983), which disturbs also the processes of fibrillogenesis. This phenomenon is also indicative of relative lability between the newly acquired graptolite and the phylogenetically old pterobranch systems of fibrillogenesis (Fig. 12A).

The gap between the fusellar fibrils of the pterobranch and the graptolite periderms is partly filled also in another way. Dilly (1975) found a reticulum of coarse, silver-staining fibers in the repent part of the tubarium of *Rhabdopleura*. This tissue surrounds dormant buds and forms a matrix in close proximity of the black stolon. It is very similar to the fusellar fabric of graptolites and "may perhaps suggest an affinity between the graptolites and the pterobranchs" (Dilly 1975). Having examined thin sections, Urbanek (1976a) came to the conclusion that this resemblance was superficial and due to superposition of very coarse ribbon-shaped but entirely separated fibrils. However, most recent discoveries suggest that the pterobranchs can produce densely crowded thick fibrils, which might represent a first step in the evolution of fibrillogenesis toward the graptolite fusellar tissue (Fig. 12B).

Another character developed within the Graptolithina is the differentiation of cortical material. In pterobranchs, the secondary walls consist of the same material as the fuselli, that is, of multiple sheets or growth bands with fusellar contents. The Graptolithina, however, show a trend toward increasing differentiation of the cortical material. In the Crustoidea, the cortex is made up either of multiple sheets with little or no material between them (pseudocortex), or with tightly packed wavy fibrils of fusellar aspect (paracortex). It only rarely attains the level of eucortex, with typical cortical fibrils straight and rather coarse, and arranged parallel to one another. This implies that the origin of differentiated cortex within the Graptolithina was somewhat later than the transformation of the fibrous component in the fusellum (Urbanek and Mierzejewski 1984). It may be classified as another unique feature of the Graptolithina, an autapomorphy (see a_3 in Fig. 13).

The virgular fabric is a novelty of minor significance but with a great morphogenetic potential for a variety of rodlike structures in the Graptoloidea (e.g., the nema, or the virgula, and the retiolitid skeleton). It may be regarded as a partial autapomorphic character (a_4 in Fig. 13), which appeared only late in the graptoloid evolution. As shown by Urbanek and Towe (1975), the Didymograptina still had the nema of standard cortical fibrils.

The sequence of character-states in phylogeny of collagenous derivatives at the molecular-submicroscopic levels is shown in Fig. 13. It involves the appearance of capability to synthesize collagen by the cells of early metazoans or their immediate ancestors (a), a symplesiomorphy; the use of collagenous secretion for extracellular skeletal constructions made of numerous growth bands (a_1), a novelty attained by the common ancestor of pterobranchs and graptolites and therefore an apomorphy; the transformation of straight and loose fusellar fibrils of pterobranchs into an interconnected meshwork of fusellar fi-

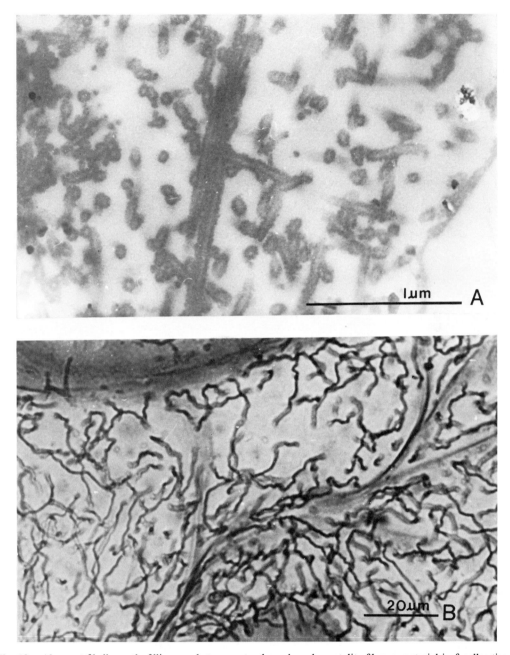

Fig. 12. Aberrant fibrils partly filling gap between pterobranch and graptolite fibrous material in fusellar tissue. A. TEM view of coarse but straight fibrils with lucent core in fusellar envelope of Ordovician graptoblast. B. Coarse silver-staining fibrils from repent tube in *Rhabdopleura compacta* (Hincks).

brils in graptolites (a₂), a primary autapomorphy; the appearance of distinct differences between the fusellar and the cortical fibrils, characteristic of some crustoids and all postcrustoid graptolites (a₃), and thus a secondary autapomorphy; and finally the appearance of virgular fibrils in the bulk of Graptoloidea (a₄), a partial autapomorphy.

In this way, the fusellar and the cortical "tissues" were produced, as well as more specialized structural systems, such as the microfusellar "tissue" or fibrils in the retiolitid skeleton and in the virgula. The skeletal "tissues" of graptolites (Fig. 14) are therefore nothing more than extracellular collagenous systems of fibrils representing differ-

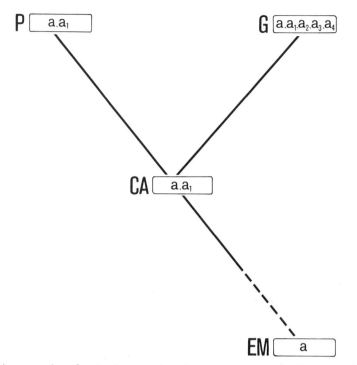

Fig. 13. Cladistic interpretation of molecular evolution of collagen and its derivatives within Pterobranchia and the Graptolithina. Traced from early metazoans (EM) through a common ancestor (CA) of pterobranchs and graptolites; collagenous derivatives exhibit a certain character-state transformation sequence: $a-a_1 \ldots -a_4$. After early metazoans or their immediate ancestors acquired an ability to synthesize collagen (a, symplesiomorphy or a shared primitive character inherited from a remote common ancestor), the cephalodiscid like common ancestors (CA) of pterobranchs (P) and graptolites (G) invented a novelty, namely extracellular skeletal constructions (coenoecia) revealing a fusellar structure caused by a specific mode of secretion (a_1, synapomorphy or a shared derived character). While pterobranchs (P) remain at this level of differentiation, using such secretions for constructing a number of different lodgings, graptolites (G) were great experimenters with collagenous material (a_2, a_3, a_4). They transformed the loose fibrous fusellar material into a tight meshwork (a_2, autapomorphy or a character unique for all the graptolites) to produce later a distinct cortical fabric differing from that in the fuselli (a_3, also autapomorphy, a unique feature common for almost all the Graptolithina). Finally, within advanced Graptoloidea, a specialized polymorph of collagen-like material, the virgular fabric, was evolved and used for constructing rodlike specialized structures in the rhabdosome (a_4, a partially autapomorphic character).

ent types of physical organization of tropocollagen units, sometimes with a share of interfibrillar material of obscure nature.

Recent studies revealed considerable quantities of collagenous material in *Rhabdopleura* and *Cephalodiscus* (Armstrong et al. 1984). It is more likely that this material occurs in the fibrils and not in the ground substance. Thus, there exists a complete uniformity at the molecular level in the skeletons of pterobranchs and graptolites. This is important in at least two aspects. First, the molecular unity complements the criteria of homology of hemichordate skeletal tissues. Second, it implies the phylogenetic significance of the collagenous nature of skeletal elements of the Pterobranchia and the Graptolithina (Fig. 13).

The presence of collagen per se is of minor significance. It is inherited from a very remote common ancestor, most probably from early metazo-

ans, as indicated by the ample presence of collagen (so-called spongine) in the mesogleal skeleton of sponges and its universal morphogenetic role as the binding material of the metazoan body. Therefore, the presence of collagen is here classified as a symplesiomorphy. Somewhat more specific is the use of collagen in extracellular structures—most probably an invention of the common ancestors of the Pterobranchia and the Graptolithina. These collagenous extracellular constructions (coenoecium or tubarium) may be regarded as a synapomorphy, and a highly successful adaptation. While both pterobranchs and graptolites exhibit a remarkable uniformity in the chemical composition of their skeleton, their skeletal biochemistry sharply contrasts to two other large groups of colonial animals—the bryozoans and the coelenterates—which have skeletons composed of chitin and various, often noncolla-

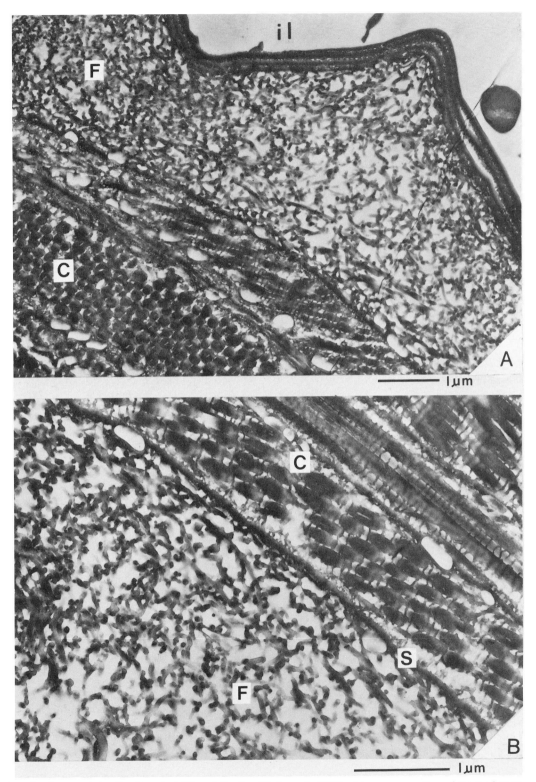

Fig. 14. Structural details of cortical (C) and fusellar (F) tissue in TEM of transverse section of theca of *Dictyonema*. il, Inner lining of thecal cavity; S, sheet fabric. Reproduced by permission of Smithsonian Institution Press from Urbanek and Towe (1974, pl. 2).

genous, proteins (see Ryland 1970; G. Chapman 1974).

Microscopic level

Under light microscope, the skeletal material of the Graptolithina is resolved into two basic components: fuselli and cortical bandages (Figs. 6 and 9). They are secretionary units of the fusellum and the cortex, respectively; their formation is due to the secretionary activity of the cephalic disc of the zooids. The fuselli find their equivalents in the growth bands of the pterobranch skeleton; a comparison of the cortical bandages with the units of secretion in the secondary deposits of pterobranchs is less obvious.

Each fusellus is a spindle shaped strip, tapering at both ends to produce an oblique suture where the ends contact other fuselli or join together forming an annular segment of the theca (Fig. 14). Owing to superposition, the fuselli form zigzag sutures, usually regularly bilaterally symmetric, with each fusellus producing only half a ring. Kozłowski (1949, 1966a) considered this structure of peridermal derivatives in the Graptolithina to be unique and highly specific, shared only with pterobranchs. However, there are a number of differences in shape, arrangement, and overlap of the growth bands between both the groups.

In rhabdopleurid pterobranchs, erect zooidal tubes are formed by superposition of annular fusiform growth bands, with their oblique sutures arranged randomly, whereas the graptolite thecae are composed of semiannular fuselli with two regular zigzag sutures (Fig. 15). There are also differences in the mode of superposition and overlap of adjacent fuselli between rhabdopleurid pterobranchs and graptolites (Urbanek 1976a). Although they imply a difference in the mode of deposition of fusellar bands, most recent studies suggest their minor phylogenetic significance. Almost the entire spectrum of variation in fuselli, including both the rhabdopleurid and the graptolite types, can be found in a single theca of a crustoid graptolite (Urbanek and Mierzejewski 1984). Kozłowski (1966a) and Andres (1977) described sessile graptolites where the random arrangement of fuselli, characteristic of *Rhabdopleura,* changes in a single theca into an orderly arrangement with two zigzag sutures, and the annular bands are replaced by semiannular ones. The end members— that is, the fusellar systems of Recent *Rhabdopleura* and, on the other hand, of the majority of graptolites—are bridged by some fossil rhabdopleuroids and such aberrant graptolites as *Mastigograptus* or *Micrograptus.*

Individual fuselli consist of a body of fusellar fabric and an outer pellicle, or an envelope made of sheet fabric (*Dictyonema* type of fuselli). In some cases, an outer lamella underlining the outer pellicle is present (*Acanthograptus* type of fuselli).

In contrast to the fusellar body made of a meshwork of fibrils, the outer lamella is composed of numerous, parallel fibrils thus resembling the cortical tissue but enclosed within the fusellar envelope. Rarely, a peculiar variety of the fusellar "tissue," the microfusellar "tissue," occurs; it has a strongly reduced fusellar body and a high share of cortical-like fabric, represented by a thick outer lamella (Urbanek 1976b). The accumulation of such strongly reduced microfuselli may lead to formation of a specific form of cortex, thus producing a link between two apparently contrasting types of skeletal material: the fusellar "tissue" and the cortical "tissue" (Fig. 16).

The growth bands lack an outer lamella in pterobranchs and crustoid graptolites; hence, the *Dictyonema* type of fuselli may be a primitive one. The sessile graptolites have both the types of fuselli, while the Graptoloidea appear to have only the more advanced, the *Acanthograptus,* type of fuselli. Although accumulations of narrow, fusellar stripes strongly resembling the microfuselli occur in tuboid graptolites, genuine microfuselli occur only in graptoloids and seem to represent a specialized, rather than primitive, feature of secretion. By analogy to pterobranchs, the secretion of the graptolite fuselli could be related to a cephalic disc of the graptolite zooids. The observations made by Dilly on tube secretion by living zooids of *Rhabdopleura* and *Cephalodiscus* are still unpublished, but it is fairly safe to assume that the formation of the fuselli is primarily related to the glandular zone on the cephalic disc and to the painting movements of this lobate segment of the body. The role of similar glands scattered over the body or situated on the tentacles is not clear. The width of the fuselli is in some way related to the width of the cephalic disc or at least of its glandular zone. Therefore, the narrowness of the microfuselli may be due to reduction of the cephalic disc in the terminal phase of its secretionary activity. A seasonal reduction in size of the zooid and the growth bands was also observed in *Rhabdopleura* (Stebbing 1970).

Bilateral symmetry is strongly emphasized in the organization of the Graptolithina, at both the microscopic and the macroscopic levels. Few examples of asymmetry in the structure of the thecae are obviously secondary, and their gradual development from a bilaterally symmetric ancestor could be traced in some lineages (Urbanek 1966, 1970).

The bilateral symmetry of graptolites is unique in that it sets in very early and remains throughout the astogeny from the microscopic level to the macroscopic pattern of the stipe and often the rhabdosome. While the prosicula has a monoaxial–radial symmetry, the metasicula reveals a pronounced bilateral symmetry. This is expressed in the arrangement of the fuselli that produce consistently the dorsal and the ventral zigzag suture.

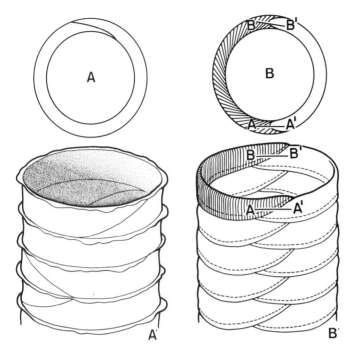

Fig. 15. Arrangement and shape of fusellar growth bands in *Rhabdopleura* (A, A′), in graptolites (B, B′) in transverse section (A, B) and side view (A′, B′). A.-A′ and B–B′, Plane of sectioning the thecal wall in graptolite. Note unilateral overlap of adjacent fuselli in rhabdopleurid and essentially bilateral overlap in the graptolites. Reproduced by permission from Urbanek (1976a, figs. 2 and 3).

However, at least in some sessile graptolites *(Mastigograptus, Kozłowskitubus),* the early portion of the metasicula has randomly arranged fuselli, without regular zigzag sutures. Such an irregular arrangement of the fuselli indicates a somewhat undefined symmetry of the young metasiculozooid. Nevertheless, the blastozooid thecae of the majority of graptolites reveal a perfect bilateral symmetry.

Phylogenetic significance of the pronounced bilateral symmetry of graptolites was frequently discussed. A number of hydrozoans display a remarkable, though secondary, bilateral symmetry, and therefore such a symmetry per se does not exclude coelenterate affinities. However, the bilateral symmetry of the Graptolithina is doubtlessly primary and not secondary as in coelenterates. On the other hand, its primary nature does not sufficiently substantiate the bryozoan affinities of graptolites. The only safe conclusion from the presence of such symmetry is that the Graptolithina should be placed within the Bilateria (= Bilateralia).

Under light microscope, the graptolite cortex appears distinctly laminated. Under TEM, however, these layers are resolved into bunches of relatively thick, unbranched fibrils arranged parallel to each other and enveloped by a bounding sheet fabric. In many graptoloids the surface of the cortex is covered with ribbonlike cortical bandages (Fig. 9). This indicates that the entire cortex is built of such units, superimposed and crowded one over another to produce a tightly packed, stratified structure. Cortical bandages are widely distributed in different groups of the Graptolo-

idea, while the surface of the rhabdosomes proved to be featureless and rather smooth in the majority of the examined sessile graptolites. This might argue against the universal nature of cortical bandages within the Graptolithina and for their secondary and specialized character (Urbanek et al. 1980).

However, Andres (1980) and Chapman and Rickards (1982) demonstrated the presence of sinuous, ribbonlike units with variously oriented fibrils on the surface of the periderm in a few dendroids. Liberally interpreted, these structures might be considered as cortical bandages. But there is always the danger that they merely represent "dilapidation" of the periderm, exposing a number of underlying layers, without providing unequivocal information on the shape of secretionary units (cf. Urbanek 1978).

The universality of cortical bandages as units of secretion of cortical deposits is in fact corroborated by ultramicrotome sections, which frequently display tapering layers (Urbanek and Towe 1974), suggesting a lens shaped transverse section of each individual layer and possibly a ribbonlike outline at the surface. Crowther (1980) proposed to distinguish two types of cortical bandages: the scarp-edged and the tapered-edge bandages. The scarp-edged cortical bandages are relatively clearly visible under SEM or even light microscope, while the tapered-edge bandages do not show up in surface micromorphology. The majority of sessile graptolites seem to possess the tapered-edge units of cortex secretion, while the scarp-edged bandages are present only in the advanced Graptoloidea. The latter units are most

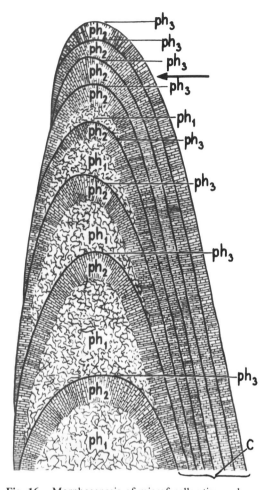

Fig. 16. Morphogenesis of microfusellar tissue showing gradual passage of microfuselli into an accumulation of cortical layers (c). ph_1–ph_3, successive phases of fibrillogenesis forming fusellar components in each microfusellus of a meshwork of fibrils (ph_1); formation of outer lamella of ordered fibrils resembling cortical fabric (ph_2), and envelope of sheet fabric (ph_3). Above arrow the microfuselli are reduced and entirely deprived of fusellar content, thus producing cortical material. Reproduced by permission from Urbanek (1976b, fig. 3).

clearly expressed in some diplograptids, which played, therefore, an important role in the recognition of the presence of cortical bandaging; but diplograptids do not constitute the norm for the surface micromorphology of the Graptolithina. The dichograptids and most of the studied sessile graptolites do not show bandaging, although their cortical deposits may be eventually resolved into bandagelike units.

This is also the case with pterobranchs. Except for the longitudinal increments on the surface of the thecal tubes of *Cephalodiscus levinsoni* described by Harmer (1905), which need to be reexamined, there is no firm evidence for the ribbon-

like units of secretion in this group. The examples quoted in the literature may represent secondary deposits over the primary fusellar layer, but they lack the characteristics of a cortical bandage. This is confirmed by the unpublished studies by Dilly, Urbanek, and Mierzejewski on *Cephalodiscus*, its periderm being either smooth or rough and patchy, but without bandaging. There are, of course, secondary deposits in *Cephalodiscus*, but they consist either of numerous growth bands that do not differ in nature from the fusellar growth bands (and therefore hardly deserve to be named "bandages" and still less "cortical bandages"), or they are built of multiple sheets, sometimes in the form of fairly thick intercalations or envelopes (Fig. 17).

The graptolite cortex and the fusellum have their fibrous material organized differently. Exceptions from this basic pattern are due to specialization involving replacement of the primary components of the periderm by the secondary fabric (e.g., retiolitids; Towe and Urbanek 1974). Such a differentiation is unknown in the Pterobranchia, whose secondary deposits consist either of multiple layers produced by sheets or pellicles with little or no material in between (Urbanek 1976a), or of the same material as the fuselli (Urbanek 1976a; Andres 1980). Thus, there is a remarkable difference between pterobranchs and graptolites in the fine structure of their periderm.

The ultrastructure of the fusellar fabric in crustoids does not differ much from that in the other graptolite orders. The distinct difference between the fibril arrangements in the growth bands (fuselli) of pterobranchs and crustoids leaves no doubts that the fundamental graptolite fibril pattern was acquired prior to the appearance of the Crustoidea, that is, at the precrustoid stage.

The mechanisms of cortical fibrillogenesis may have been attained at the crustoid stage of graptolite phylogeny but only incompletely. The bulk of the crustoid cortex consists of multiple sheets with little or ill-defined intersheet material. The ancestral stage of cortex formation in the Graptolithina may be tentatively visualized rather as a multiple deposition of sheets and the cortical fibrillogenesis *in statu nascendi* (Fig. 11). This may indicate that pseudocortex and paracortex represent the primitive structural patterns of secondary deposits in the Graptolithina, in analogy to the multiple sheets deposited by cephalodiscid pterobranchs. The presence of multiple sheets may, then, be considered as the essential component of secondary deposits in both the groups, but evaluation of this similarity poses a number of difficulties.

Taking the presence of a sheet (membrane) as a character-state, shared by the pterobranch and the crustoid secondary deposits, one could conclude that the crustoid cortex, made up predominantly of multiple sheets, is primitive. [This understand-

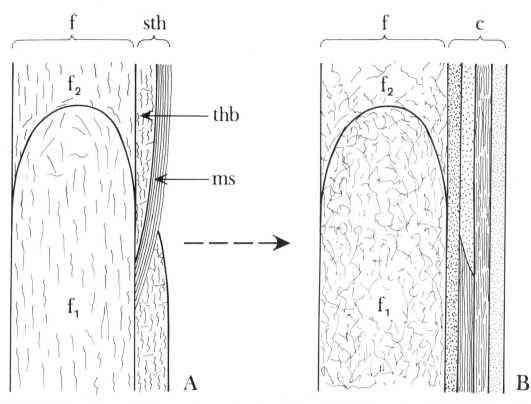

Fig. 17. Major features of primary and secondary deposits in colonial skeletons of cephalodiscids (A), and in graptolites (B). Compare a distinct difference in shape and arrangement of fibrils within primary component of their skeletons (f, fusellum) and within the secondary thickening of cephalodiscid pterobranchs (sth) composed of thickening bands (thb), multiple sheets (ms), and cortex of graptolites (c) of cortical fibrils. f_1–f_2, Successive fuselli. Arrow indicates direction of phylogenetic changes. Reproduced by permission of Lethaia Foundation from Urbanek and Mierzejewski (1984, fig. 2).

ing of primitiveness differs from the one proposed by Andres (1980) on the basis of a recapitulatory effect.] An evaluation of the thus-defined primitiveness of the crustoid cortex must be tentative, because the cortex is relatively simple as a structural system, comprising only a few subunits.

The specificity of the fusellar structure, which was so strongly emphasized by Kozłowski (1938, 1949, 1966a), survives the scrutiny of the modern analysis. The fusellar growth bands of pterobranchs and graptolites are essentially similar in their shape and arrangement, and imply an analogous mode of secretion in both groups. They show no close analogy to the incremental growth lines or growth bands in coelenterates or bryozoans. The frequently quoted example of the peridermal tube in the polypoid generation of the scyphomedusan order Coronatae (e.g., *Stephanoscyphus*) or related fossil forms *(Byronia)* offers only a remote resemblance. *Stephanoscyphus* has a slim conical tube made up of transverse annular bands with elaborated sculpture of fine longitudinal striae. Individual annuli develop by discontinuous secretion by an epithelial secretory band

situated below the head part (capitulum) of the polyp's body, while the longitudinal striation results from secretion of the peridermal substance by individual gland cells (Werner 1976; D. M. Chapman and B. Werner 1972). *Stephanoscyphus* and related forms, some of them colonial, may at first glance seem to be a "giant *Rhabdopleura*," as the annuli are not always complete and sometimes display an oblique suture. However, there is convincing evidence that the substance involved is chitin (D. M. Chapman 1966).

In spite of their microscopic similarity, the pterobranch fuselli are radically different from the graptolite fuselli in the ultrastructural organization of their fibrous content; but this ultrastructural anomaly may be now explained as a result of collagen polymorphism. Nonetheless, this difference is one of the best criteria for the pterobranch–graptolite boundary in phylogeny and classification.

The cortical bandages may be regarded as homologous to the fusellar growth bands, based on essentially the same mode of secretion by the same part of the zooid's body. The differences in their ultrastructure, size, and, in part, shape (lack of tapering ends) appear to be secondary. The smaller size of the cortical bandages compared to the underlying fuselli may be ascribed to a decrease in size of the cephalic disc in ontogeny of the zooids. The presence of coarse and straight cortical fibrils, instead of wavy and thinner fusellar ones, represents only a different phase of the physical organization of collagen. (Urbanek and Towe 1974).

Therefore, there is no reason to contrast the primary (fusellar) and the secondary (cortical) peridermal structures of graptolites as two different skeletal tissues. Most probably, they are both due to a uniform morphogenetic mechanism and represent different patterns of the physical organization of the same substance—collagen.

Early astogeny and the primary zooids

Early developmental stages of graptolite colonies can be best interpreted by comparison with early stages of the colony formation in *Rhabdopleura*. They also significantly resemble the initial stages of bryozoan astogeny, while coelenterates offer only a remote analogy.

All graptolite colonies start with a primary zooid placed within a peculiar theca called sicula. Sessile graptolites have discophorous siculae (cf. Kozłowski 1971); free-living graptolites have nematophorous siculae. No doubt the former represent the ancestral morphoecologic type of organization, while the latter are secondary and specialized. Discophorous siculae are known in a number of dendroids and in a few tuboids. The siculae of the Camaroidea are hitherto unknown, while the single finding of a sicula assigned to the Crustoidea is uncertain (Kozłowski 1971).

The discophorous sicula has a basal disc, a means of attachment to the substrate. This defines its general adaptive type as opposed to the free-living mode of life of the nematophorous sicula (Fig. 18). In the latter, the basal disc is replaced secondarily by a threadlike nema, whose function remains largely enigmatic. The nematophorous sicula is morphoecologically specialized and therefore beyond the scope of the present review.

The sicula is always composed of prosicula and metasicula. The prosicula of discophorous graptolites is bottle shaped or vesicular as in the Tuboidea, or cylindrical as in the sessile Dendroidea, while it is conical in free-living graptolites. It is always thin walled and apparently structureless under light microscope. In contrast, the metasicula is tubular and made of superimposed growth bands, fuselli, thus resembling in this respect all the other thecae of the graptolite colony. The difference between the prosicula and the metasicula is so great that Kozłowski (1949, 1966a) postulated metamorphosis of the prosicular zooid, probably a sexually produced larval individual

(oozooid), before the beginning of metasicular growth.

The discophorous sicula of some tuboids, in particular *Epigraptus* (= *Idiotubus;* see Kozłowski 1971), seems to represent an especially primitive condition, strongly resembling *Rhabdopleura*. Its vesicular prosicula has a large flat attachment surface; the tubular metasicula consists of the initial repent part and the terminal erect part (Fig. 18B). The absence of a helical line and sclerotized internal stolon inside the prosicula resembles the early astogenetic stages of *Rhabdopleura*.

A somewhat more advanced stage is represented by the sicula of *Kozłowskitubus* [= *Dendrotubus?* of Kozłowski (1961, 1971)], with a bottle shaped prosicula and an entirely erect metasicula (Fig. 18D). The ascending, cylindrical part of the prosicula displays faint traces of a helical line, while the sclerotized initial portion of the stolon (prostolon) may be seen inside the prosicula.

A still more advanced stage is represented by the sessile dendroids (as opposed to the free-living Anisograptidae). The prosicula is cylindrical, with a small basal disc at the bottom. It has distinct helical lines throughout its length, and the internal part of the sicular stolotheca is made of a membranous wall with a strongly sclerotized stolon (Fig. 18E).

In all these cases, the budding of the first blastozooid happens by a unique mechanism. The initial bud produces a characteristic foramen (porus) by resorption and/or perforation of the prosicular wall (Figs. 18 and 19), whereas the metasicula emerges directly from the circular aperture of the prosicula. After producing a porus in the wall of the prosicula, the first blastozooid secretes the wall of an outer portion of the sicular stolotheca, to form later the first diad (as in the Tuboidea) or triad (as in the Dendroidea). Somewhat unusual is the initial budding in *Epigraptus* (= *Idiotubus*), for the first blastozooid is a bitheca that coils around the prosicula.

Early descriptions of the initial part of *Rhabdopleura normani* colonies (Schepotieff 1907) provided little ground for a closer comparison with the early astogenetic stages of the graptolite colony. Although Schepotieff found in *Rhabdopleura* an "embryonal vesicle," an ovoid chamber comparable to the sicula and particularly the prosicula of the Graptolithina, this vesicle was always empty and surrounded with a peculiar structure ("embryonal ring") representing a repent zooidal tube with stolon inside. The latter structure has no analog in the graptolites.

However, Stebbing (1970) described the early astogeny in *Rhabdopleura compacta,* which has a direct bearing on the pterobranch–graptolite relationship. As the planuloid, ciliated, and lecithotropic larvae of *R. compacta* settle after a short

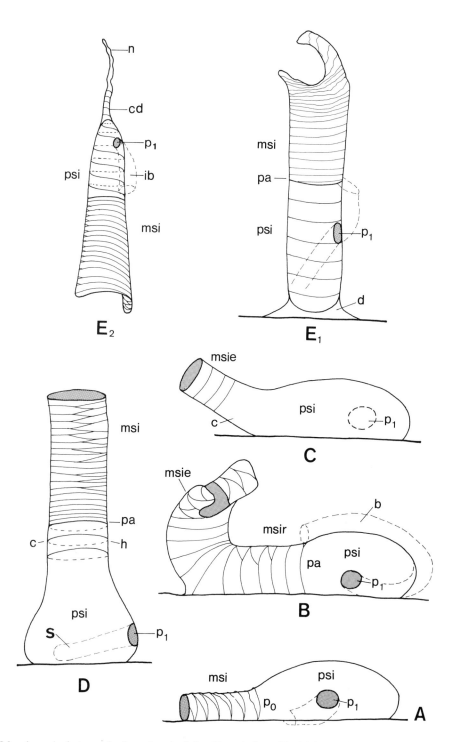

Fig. 18. Morphoecologic type of colony, largely defined by relation of its initial portion to substrate. A. Repent embryonal vesicle of *Rhabdopleura*. B. Sicula of *Epigraptus* (= *Idiotubus*), a tuboid graptolite with prosicula entirely adnate and metasicula partly erect. C. Prosicula of presumed crustoid with its apertural portion (c) and initial part of metasicula erect. D. *Kozłowskitubus*, an advanced tuboid with prosicula attached to substrate only by its broad base; and first traces of the helical line (h) appearing on erect collum of prosicula (c). E_1 and E_2. Siculae of sessile (E_1) and free-living (E_2) dendroid; in E_2 small area of attachment, basal disc (d) later in E_2 transformed into different structure, the nema (n). These forms demonstrate a gradual transition from the extreme discophorous type to the primitive nematophorous type. b, Sicular bitheca; c, collum or neck; cd, cauda or a taillike extension of prosicula; d, basal disc, h, helical line, ib, initial bud; msi, metasicula; msie, erect portion of metasicula; msir, repent portion of metasicula; p_1, porus of foramen for the first blastozooid; pa, prosicular aperture; psi, prosicula; S, prostolon. Highly modified from Stebbing (1970), Kozłowski (1966a), and Hutt (1974).

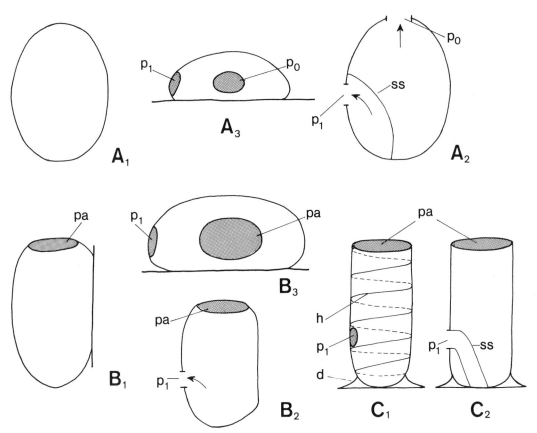

Fig. 19. Early developmental stages in *Rhabdopleura* (A_1–A_3), in primitive tuboid (B_1–B_3), and in advanced dendroids (C_1–C_2). In rhabdopleurids settled larva encapsulates itself in complete "embryonal vesicle" (A_1) to produce later two foramens; one for a metamorphosed oozooid (p_0) and the other for the first blastozooid (p_1). (A_2, A_3) Separated from the rest of the cavity of the embryonal vesicle by a single septum (A_2, ss). In graptolites the opening for metamorphosed zooid is preformed (pa in B_1; prosicular aperture), while porus foramen for first blastozooid is newly formed (p_1 in B_2 and B_3). The appearance of the helical line (h) of the internal portion of the sicular stolotheca (ss in C_2) and of the basal disc (d) in dendroids result in their prosicula being a definitely more specialized structure than the embryonal vesicle of *Rhabdopleura*.

period of free swimming, they undergo metamorphosis. First of all, the larva encapsulates itself in a completely sealed vesicle made of a coenoecial substance. Later, its body differentiates into tagmata, cephalic disc, arm buds, and metasome with a rudiment of peduncle. After completing the metamorphosis from an amorphous mass of tissue into a juvenile oozooid, the individual, still sealed inside the embryonal vesicle, breaks its wall ("perforatory" budding!) and starts to secrete the first growth bands of the primary zooidal tube (Fig. 19A). The initial portion of this tube is repent and composed of semiannular growth bands, producing a zigzag suture characteristic of all creeping tubes in rhabdopleurids, but its distal part is erect and made of annular growth bands (Fig. 18A). This tube is comparable to the metasicula, whereas the "embryonal vesicle" resembles the prosicula of discophorous graptolites. Thus, the entire coenoecial structure formed by the pri-

mary zooid of the *Rhabdopleura* colony (i.e., oozooid) may be compared with the sicula of the Graptolithina.

Further events involve production of successive buds. According to Stebbing (1970), the first blastozooid appear as a small bud at the base of the peduncle of the primary zooid. Later, the young blastozooid secretes a septum separating a small chamber inside the embryonal vesicle, comparable with the sicular stolotheca recognized in some sessile graptolites. Then, not unlike the oozooid, the first blastozooid breaks the wall of the vesicle and starts to secrete its own zooidal tube. According to Dilly (1973), however, the metamorphosed larva forms two separate body stalks, which develop on either side of the metasome. One of them represents the peduncle of the primary zooid; the other is the rudiment of the blastozooid bud, which appears independently of the contractile stalk of the oozooid. More advanced stages of the early astogeny were not ana-lyzed, although Stebbing (1970) mentioned that both the oozooid and the first blastozooid pro-duced subsequently new buds on their peduncles, which implies that the oozooid is capable of pro-ducing more than one daughter individual. The interpretation of further astogenetic events is un-clear. Is the first blastozooid a "blastozooïde in-achevé" (leading zooid) of the first creeping tube, comparable with a budding individual, or does it behave in a different way? And is the stolon a di-rect derivative of the peduncle of the primary zooid, or is it formed in a different way?

In spite of these and other questions, there is an essential similarity between the early astogeny in sessile graptolites and in *Rhabdopleura*. It is ex-pressed in the general morphology of the initial portion of the colony and in the sequence of non-fusellar ("structureless") and fusellar portions in the theca of the primary zooid. The single septum within the embryonal vesicle of *Rhabdopleura,* which separates the chamber of the first blasto-zooid, may be compared with the tabular walls of the inner portion of the sicular stolotheca in den-droid graptolites.

There are also some noteworthy differences. First of all, every bud in *Rhabdopleura,* including the metamorphosed primary individual oozooid, emerges by means of resorption or perforation of the thecal wall. In sessile graptolites, in contrast, the metasiculozooid emerges using a preformed opening (prosicular aperture), and only the initial bud (first blastozooid) produces a porus in the wall of the prosicula. Subsequent blastozooids emerge to the exterior through a free opening in the growing theca, following the pattern of aper-tural budding. In the evolution of graptoloids there is a progressive delay in the formation of the pore, which moves from the prosicula into a more distal portion of the metasicula; in monograptids it is even replaced by an apertural notch, or em-bayment, for passing the initial bud without re-sorption (Fig. 19A–C). The formation of the pro-sicular aperture by incomplete encapsulation of the larva may thus be regarded as a first step in this direction.

The second major difference in astogeny be-tween rhabdopleurids and graptolites lies in the presence of a helical line within the wall of the prosicula in most graptolites. The helical line proved to be an internal suture (Urbanek and Towe 1975), and the prosicula appears to be made of a spirally coiled band that may sometimes dis-integrate (Kozłowski 1971). The helical line seems to be a new phylogenetic acquisition of the Graptolithina. It is lacking in the embryonal ves-icle of *Rhabdopleura* and in the prosicula of some tuboid graptolites (Kozłowski 1971; Andres 1977). Andres compared the prosicular spiral band to the quasi-helical arrangement of fusellar sutures in the proximal portions of the metasicula

in some sessile graptolites *(Kozłowskitubus, Mas-tigograptus, Micrograptus).* This would imply a similarity in morphogenesis, with extension of spiral elements into the prosicular stage. Although ultrastructural studies revealed absence of essen-tial differences in the fabric of the prosicula and metasicula (Urbanek and Towe 1974), thus sup-pressing the difference recognized earlier under light microscope, the exceptional nature of Andres' (1977) suggestions requires a better justification.

The striking similarities between the embryonal vesicle of *Rhabdopleura* and the prosicula of some tuboids should be verified at the level of ul-trastructure. Although there are some data on the wall of the prosicula in *Kozłowskitubus* (Urbanek 1963), ultrastructural data on *Rhabdopleura* are wanting.

The present knowledge of the life cycle of *Rhab-dopleura* (Stebbing 1970) offers an insight into the presumed life story of sessile graptolite colonies, which complements the earlier speculations (Kozłowski 1949, 1971). The sessile tuboids had larvae with only a limited swimming ability, which probably formed prosicula only after hav-ing settled. The dendroids achieved a longer swimming period and secreted prosicular wall prior to settling, as their young prosiculae lack any traces of attachment. The settling of the lar-vae was followed by metamorphosis transforming the primary siculozooid (prosiculozooid) into the adult form (metasiculozooid). The appearance of the growth bands (fuselli) suggests a differentia-tion of the main body segments (tagmata) as a re-sult of the metamorphosis. The rest of the asto-geny reduces to repeated budding (blastogeny).

The similarities in early astogeny between grap-tolites and bryozoans are only very general, al-though the early stages of bryozoan colonies (es-pecially in Cyclostomata) offer an instructive analogy. The metamorphosis of the attached larva (preancestrula), occupying the bottle shaped part of the primary zooecium, into the ancestrula forming the tubular portion of the primary zooe-cium (Zimmer and Woollacott 1977) provides an analogy to the processes in the Graptolithina (his-togenetic and morphogenetic differentiation in the sicula, etc.). Kozłowski (1949) was correct in using such analogies without ascribing to them any phylogenetic significance. Still more remote are the similarities to hydrozoan colonies.

Polymorphism

Polymorphism, which is here defined as a discon-tinuous variation in morphology and function of the zooids within a single colony (Boardman and Cheetham 1969), is a striking features of the ma-jority of sessile graptolites. The Graptoloidea are monomorphic and have colonies composed of the thecae and zooids of only one kind; but the Dendroidea, Crustoidea, Tuboidea, and Camaroidea all display (with a few exceptions) a pronounced polymorphism.

The basic pattern, which is almost universal in the sessile orders of the Graptolithina, involves differentiation of the thecae into three categories: autothecae, bithecae, and stolothecae. The autothecae are the largest and frequently have an apertural apparatus or another elaboration of the aperture. In encrusting forms they are bottle or flask shaped, but in erect colonies they are conical, more-or-less expanded, and isolated at the aperture. The bithecae and the stolothecae are much narrower, tubular, and devoid of any apertural modifications. The stolothecae carry inside a section of stolon that divides at a certain point within the parental stolotheca; the stolon produces short branches leading to a bitheca, an autotheca, and also a daughter stolotheca that contains further extension of the stolon. The bithecae do not contain a stolon and usually are adnate to the adjacent autothecae.

Kozłowski (1938, 1949, 1966a) offered explanation of the graptolite polymorphism, based on interpretation of the autothecae as fully developed females, the bithecae as reduced males, and the stolothecae as secreted by immature autothecal zooids. A model for his interpretation is represented by some species of Cephalodiscus (e.g., C. sibogae), with their males reduced in size and bearing an atrophied lophophore.

Disappearance of the bithecae in some sessile genera or families (e.g., Cysticamaridae in camaroids, some Anisograptidae in dendroids), and especially in the Graptoloidea, probably was preceded by a transition to hermaphroditic autozooids. This change was in some way related to a radical transformation of the adaptive type (from the sessile benthic to the planktic pelagic mode of life). And again, the labile sexual conditions among the zooids in the pterobranch colonies may provide us with a model of this process.

Kozłowski's interpretation of the graptolite thecal polymorphism has been largely accepted (see Kirk 1973, for an alternative view). But it is still insufficient for understanding of the totality of polymorphism in graptolite colonies—for instance, the secondary polymorphism of autozooids in sessile graptolites that combine their basic sexual and nutritive functions with an additional specialization. While the pterobranchs exhibit only an incipient stage of sexual dimorphism, the bryozoans provide the most instructive models for comparison with the Graptolithina.

The basic bryozoan autozooid is a hermaphroditic organism, but in numerous instances the male and the female functions are delegated to morphologically distinct zooids (Silén 1977). When compared with such bryozoans, the graptolite autozooids should be termed gynozooids, and the bizooids should be named androzooids. Such an

interpretation, however, would imply that both the polymorphs in graptolite colonies were capable of feeding themselves like all autozooids of the Bryozoa. There is no doubt that the autothecal individuals in graptolite colonies possessed a complete feeding apparatus, but the feeding ability of the bithecal zooids is less certain. Their feeding apparatus was surely incomplete, but it could still have been effective enough to nourish them, or otherwise they could have been provided for by the autothecal zooids. Should the latter assumption be true, they ought to be compared, at least physiologically, with heterozooids rather than autozooids of bryozoan colonies. To make things worse, some bryozoan androzooids possess fairly well-developed tentacles that are devoid of cilia and therefore ineffective as a feeding device. In some dendroids the bithecae open into the cavity of the adjacent autothecae; they may represent individuals that were no longer self-supporting and therefore "lost interest" in communication with the external environment. This may be an extreme case, for the bizooids probably had a wide range of variation, including different degrees of the reduction of their feeding apparatus.

The secondary specialization of the autothecal zooids is expressed by such autozooidal polymorphs as the microthecae and conothecae in tuboids, the occluded autothecae in camaroids and some tuboids, and to some degree the graptoblasts of crustoids. All these secondary polymorphs feature functional specializations of the female zooids and their thecae, or, as in the case of graptoblasts, of their juvenile, premature growth stages. This, again, resembles the Bryozoa where the primary specialization of function into the male and the female opens quite different prospects for further differentiation. The morphogenetic potential of the male zooids is rather limited, but the female functions open ample possibilities for further specialization. This is well demonstrated by the occluded autothecae of camaroid and tuboid graptolites, which were convincingly compared by Kozłowski (1949, 1971) to the gonozooids (ovicells) of cyclostomatous bryozoans. Before the occlusion, that is, the sealing of the aperture by a diaphragm, the autothecae were evidently occupied by active zooids that later degenerated, probably supplying nutrients to the numerous embryos that developed inside and left traces of their presence in the foamy mass of a membranous material filling the cavity of the occluded autothecae. Kozłowski ingeniously interpreted the numerous oval bodies leaving their traces in this delicate material as polyembrions—the uniquely cyclostome mode of reproduction.

The tuboid conothecae (Bulman and Rickards 1966; Whittington and Rickards 1968) were distinctly more specialized than the occluded autothecae and represented swollen bodies with a narrow orifice (aperture) on a short neck, not unlike the female gonozooids in the Cyclostomata. They were formed at the termination of stolonal branches in a diad, with the accompanying theca being always a bitheca. Thus, they were the autothecae converted into some sort of incubatoria, and their function, as inferred from the structural analogies, probably involved reproduction and brooding (Kozłowski 1971). The conothecae occurring at irregular intervals along the stipes were identified in *Reticulograptus, Discograptus,* and *Epigraptus.*

Nothing but conjectures could be given on the function of the microthecae, another autothecal polymorph found in the tuboid graptolite *Tubidendrum.* They are common, irregularly spaced on the stipes, swollen in the middle, and with constricted apertures. Like the autothecae of some tuboids, the microthecae show a characteristic spiral coiling of the neck. To the best of my knowledge, no attempts have been made to define the function of the microthecae. They exhibit some resemblance to the nanozooids of cyclostome bryozoans, which are considered to have a cleaning function (Silén 1977). It is unclear, however, if this resemblance is relevant for understanding of the function of the microthecae.

The nature of the autothecal dimorphism in *Galeograptus,* a tuboid encrusting graptolite, is obscure. The autothecae on the proximal portions of sheaves (stipes) possess elaborate apertural modifications and are called umbellate thecae, while those placed more distally have only moderately expressed flanges (Bulman and Rickards 1966). The change from one type of the aperture to the other is abrupt; therefore, the umbrella shaped apertural modifications may be interpreted as an instance of autothecal astogenetic variation. Urbanek (1973) suggested that the morphogenetic gradients may be expressed along the sheaves of encrusting tuboids, and in fact such gradients may result in an abrupt morphologic change (threshold effect).

The function of the graptoblasts, enigmatic ovoid bodies associated with the parental stolothecae in crustoid colonies (Kozłowski 1962), was for a long time obscure, although their connection with dormancy and asexual reproduction was repeatedly suggested. Kozłowski (1949) considered the graptoblasts as sui generis cysts formed inside the autothecae, while Urbanek (1983) proposed that they were terminal portions of the stolothecae, arrested in growth and specialized for dormancy. Urbanek pointed out that the graptoblast zooids never functioned actively as feeding zooids (they lack the column and the apertural apparatus); instead of a short autothecal stolon, they possessed an elongated portion thereof, without any traces of branching or a node. Ultrastructural studies (Urbanek and Rickards 1974; Urbanek et al., in press) supplied more data on the outer fusellar, and the inner heavy, "crassal," components of the graptoblast wall (blastotheca and blastocrypt, respectively). Urbanek (1983) homologized the graptoblasts with the closed terminal portions of creeping tubes in *Rhabdopleura,* and their zooids with the resting zooids in such terminal chambers, capable of hibernation. On the other hand, he indicated a striking functional analogy between the graptoblasts and the hibernacula, that is, thick-walled dormant zooecia formed on the branches or stolons of some freshwater and brackish ctenostome bryozoans. The graptoblasts were dormant thecae capable of surviving a period of adverse environmental conditions. After germination, they probably produced small propagules that were ejected through a narrow opening and formed a new colony.

Kozłowski never attempted to identify the sexual nature of the sicular zooid. Was it hermaphroditic already in the sessile orders of the Graptolithina, or was it female, male, or sterile? The answer is impossible unless adequate living models are available. To my knowledge, there is no information on the sexual nature of the primary zooid in *Rhabdopleura.* The ancestrula in cyclostomate bryozoans is an agonadic, sterile individual (Hyman 1959). Was it the case also with the Graptolithina? It might well be, but the peculiar association of a bitheca with the sicula in primitive tuboids and the unusual behavior of the first bitheca in some Anisograptidae, escaping the suppression and associated with the sicula, provide an evidence against the sexual sterility of the siculozooid. These phenomena may be best interpreted by regarding the siculozooid as a hermaphrodite. In such a case, the associated sicular bitheca would be an adventitious male.

Stolon system

Stolons are slender outgrowths serving for spreading the colony, budding of new zooids, and communication among individuals in a single cormus. The stolon system plays an important role in shaping the architecture and the integration of the colony, but stolons in different animal phyla or even lower-rank taxa may exhibit a variety of structures that differ in their nature and origin. Therefore, the mere presence of a stolon has little or no phylogenetic meaning, except if accompanied by a number of essential similarities implying homology with a certain stoloniferous group.

The sessile orders of graptolites are characterized not only by the presence of a well-developed stolon system but also by its unique features bearing much similarity to the stolon of *Rhabdopleura* (Kozłowski 1938, 1949). As in pterobranchs, the stolon system in graptolites is internal, that is, enclosed within thecal or zooidal tubes. The stolon is provided with a thin, sclerotized sheath and is therefore comparable to the pectocaulus, or "black stolon," of *Rhabdopleura.* The latter is a sclerotized portion of the stolon, with a thread of soft tissues surrounded by a strongly pigmented peridermal encasement.

Sclerotized portions of the stolon (pectocaulus) and young stolons without a peridermal sheath (gymnocaulus) were studied by Schepotieff (1907), and their ultrastructure was more recently analyzed by Dilly (1975). Young stolons consist of the outer epidermis (ectoderm) enclosing a pair of cavities lined by the peritoneum (mesoderm) and separated from each other by the median mesentery. All of them are continuous with the epidermis, the peritoneum, and the coelom cavities of the zooid trunk. New buds are formed on this active young portion of the soft stolon, with the youngest bud nearest to the terminal zooid (Fig. 20). Older portions of the stolon secrete a peridermal sheath that turns black owing to sclerotization. When incorporated into the sheath, living parts of the stolon alter their structure and lose their coelomic cavities composed of an outer vacuolated tissue with groups of granules and an inner, syncytial core. The central lumen is frequently filled with a solid material of the same nature as in the sheath (Schepotieff 1907). Recent observations by Dilly (1975) indicate the presence of blood vessels and nerve fibers in the living contents of the black stolon in *Rhabdopleura.* Thus, the stolon ensures a true organic continuity between all the zooids of the colony. A number of short side branches of the stolon, the peduncular stolons, reach to the base of the peduncles of the zooids. The black stolon is no more capable of producing new buds, but it may give rise to new stolonal branches.

Fossil rhabdopleurid pterobranchs have quite significant peculiarities compared to the stolon system in extant *Rhabdopleura.* The stolon sheath exhibits constrictions and swellings. The ample branching may produce a number of stolons within a single stolonal tube. Moreoever, the cross section of the stolon changed from circular in Paleozoic forms, to somewhat flattened in Late Mesozoic and Early Cenozoic forms, to strongly flattened and embedded in the lower wall of the stolotheca in Recent species (Mierzejewski, in press). Peduncular stolons may be missing in some Ordovician genera (*Rhabdopleuroides* and *Rhabdopleurites*); when present, they are devoid of internal constrictions or diaphragms that occur in post-Paleozoic forms. Middle Cambrian forms, recently discovered, show no traces of stolons or their derivatives (Bengtson and Urbanek, in press), which suggests that their stolon was entirely soft and devoid of skeletal sheath.

The Graptolithina show a variety of features related to their stolon system. In the majority of sessile groups, the stolon is fairly sclerotized, provided with a peridermal envelope capable of preservation, and extending from the base of the prosicula to the growing tips of the branches.

Even within the Tuboidea, the degree of sclerotization may vary from the heavily sclerotized stolon system of *Dendrotubus* (Kozłowski 1963), to the nonsclerotized stolothecal portions of the stolon associated with the bithecal and autothecal stolons provided with a peridermal sheath in *Reticulograptus* (Bulman and Rickards 1966), to *Epigraptus* (= *Idiotubus*) with an entirely nonsclerotized stolon system (Kozłowski 1971). The latter condition, which is most probably primitive, must be distinguished from the secondary reduction of the peridermal sheath of the stolon as observed in anisograptids (e.g., in *Graptolodendrum;* Kozłowski 1966b). The peridermal sheath underwent complete reduction in free-living graptoloids.

The internal lumen of the stolon may have constrictions or annular thickenings (diaphragms). They are most commonly associated with the thecal bases or the stolonal nodes, and frequently produce a vesicular inflation of the stolon. Although recognized only in certain genera, such diaphragms seem to occur in the majority of sessile graptolites. Similar diaphragms, frequently multiple, occur in extant and in some fossil rhabdopleurids, usually at the end of the penduncular stolon. A rather unique structure of the stolon has been described in an aberrant encrusting graptolite (Mierzejewski 1982, 1985, in press). The stolons are provided with annular and helical internal thickenings, resembling to a great extent the sieve tubes of the vascular plants. The annular thickenings have an analog in *Rhabdopleura,* but the helical ones are so far unique and difficult to evaluate (primitive feature or specialization?).

In the majority of dendroids and tuboids, the stolon is smooth and uniformly wide, as in *Rhabdopleura.* However, in crustoids and some camaroids, the stolon has a beaded appearance, with swollen and constricted portions, not unlike some Paleozoic rhabdopleurids. Thus, the constant width of the stolon and the smoothness of its surface seem to have appeared independently in both the groups. The stolon system is internal and placed inside the tubular stolothecae both in graptolites and in pterobranchs. It differs, in this respect, from the coelenterate and the bryozoan stolons. However, it is noteworthy that among the Camaroidea, well-defined stolothecae occur only in the genus *Bithecocamara;* in the other genera, the stolons are directly embedded in the extracamaral tissue (undefined common tissue of the colony), which forms rather indistinct walls surrounding the stolon and representing the vestiges of the original stolothecae.

The stolothecal portions of the stolons usually are adnate to the inner surface of the wall in the sessile orders; in places, they are anchored to it with a loose mass of the tissue. In the Crustoidea, the stolons are embedded in the lower wall of the stolothecae as in some rhabdopleurids; but they

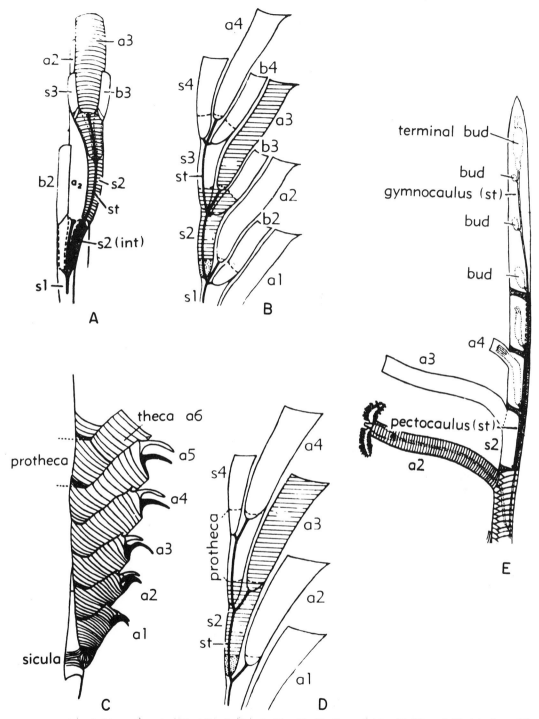

Fig. 20. Bulman's ideas of mode of budding in Dendroidea (A, B), Graptoloidea (C, D), and *Rhabdopleura* (E). Comparable portions of thecae (a, autothecae; b, bithecae; s, stolothecae) shaded or stippled. Note the essentially monopodial budding in *Rhabdopleura* and sympodial in graptolites. From Bulman (1955, fig. 8).

are circular in transverse section and thus less specialized than in some Mesozoic and Recent rhabdopleurids.

The graptolite stolons show a number of intrinsic resemblances to the stolon system in *Rhabdopleura,* but they also display a wide array of structural differentiation. This may substantiate the conclusion that the stolon systems in both the groups evolved to a large extent independently. Perhaps the immediate ancestors of graptolites had the "naked stolon" (gymnocaulus) (i.e., a nonsclerotized stolon), while the stolonal sheath was a later acquisition, attained independently in the pterobranch and the graptolite lines of descent.

Some bryozoans (Stolonifera) have creeping and branching stolons provided with a sclerotized sheath and superficially resembling the pterobranch and the graptolite stolons. Some of them, the "true stolons," are compound structures composed of many segments separated by a short of septa (diaphragms) from one another. These stolonal compartments are commonly considered to be equivalents of the zooids that lost the ability to form their own polypide and are placed end to end to produce a long series. This view was developed by a number of authors including Silén (1942); it has been modified by Jebram (1973), who argued that in most cases an individual stolonal segment does not represent complete arrested zooids (kenozooids), but only portions of an originally unitary autozooid bud. Such a compartmentalization of the stolon is unique to bryozoan colonies, without any analogy in the stolon system of Pterobranchia and the Graptolithina. Nevertheless, the communication via the stolon is ensured by the presence of another unique structure, the stolonal funiculus. It is a mesenchymatous band passing through pores in the diaphragms and interconnecting all the zooids. The funiculus plays an important role in the transport of substances toward the stolon's free end, which enables the sharing of resources in bryozoan colonies (Bobin 1977).

Therefore, the "true stolons" of the Stolonifera have a different origin and morphology from the stolons in the Pterobranchia and the Graptolithina. They develop on the basis of the "cystid-polypide" duality, so characteristic of the bryozoan morphogenesis but unknown in the Graptolithina. The stolonization of bryozoan kenozooids differs also from the elongation of the peduncle (or stolonization of the peduncle), as observed in the terminal leading zooid of *Rhabdopleura.*

Some representatives of another group of Ctenostomata, the Carnosa, form "pseudostolons," that is, slender outgrowths from the base of the autozooids. They consist of continuous threads of soft tissues enclosed in the sclerotized cuticle. Such "pseudostolons" resemble much more closely the stolons of *Rhabdopleura* and graptolites, than the "true stolons" of Stolonifera, but they are always "external" instead of being surrounded by the zooecial tube.

Budding patterns

The expansion of the *Rhabdopleura* colony and, consequently, the budding of new zooids as well as the elongation of the stolon, is intimately related to the permanently undeveloped terminal or "leading" zooid. This zooid is believed to produce a continually growing tip of a given branch, and the distal extension of its peduncle is considered to be responsible for the elongation of the soft stolon on which new zooidal buds appear. This widespread opinion is based on the observations of Lankester (1884), accepted later by Bulman (1955, 1970) and Dawydoff (1948). Lankester recognized such undeveloped individuals at the open ends of actively growing creeping tubes of *Rhabdopleura* (Fig. 21).

However, the entire problem of the advancement of the stolon in *Rhabdopleura* seems more complicated in the light of Schepotieff's (1907) observations, more recently quoted by Hyman (1959), but otherwise almost neglected in current literature. Schepotieff observed terminal zooids enclosed within sealed and pointed tubes (probably a hibernating phase) and provided, on their ventral side, with a short stolon projecting forward (cf. Schepotieff 1907, pl. 18, fig. 19). This indicates that the stolon is not a derivative of the elongated peduncle of the terminal zooid but that it is formed anew from the ventral side of the terminal zooid of a given branch (Fig. 21). Unfortunately, empirical observations to resolve this obvious contradiction are lacking.

Having accepted Lankester's observations, Bulman (1955, 1970) pointed out differences in the mode of colony growth in Rhabdopleura and the Graptolithina. For if those observations are correct, extant rhabdopleurids exhibit the *monopodial* mode of budding (Fig. 20). In contrast, every juvenile autotheca becomes at its turn the temporary leading zooid and produces a pair of lateral buds in graptolites, and in dendroids in particular (Wiman 1895). The details of the budding patterns differ, but the mode of graptolite budding is invariably *sympodial.*

At the first glance, these differences in the cormogenesis appear to be very substantial. However, even granting the uncertain conclusion on the monopodial mode of colony growth in *Rhabdopleura,* one should note the intermediate stages in development of the sympodial pattern of budding within the Graptolithina. In the sessile orders of graptolites, the lateral buds are always produced after a longer period of growth of the juvenile autozooid ("stolozooid," "budding individual"), which results in a considerable elonga-

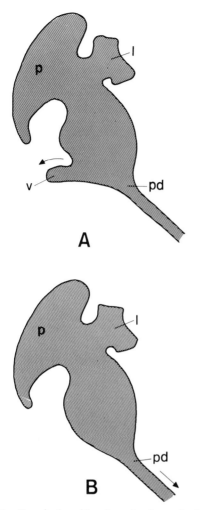

Fig. 21. Terminal zooid and mode of growth of stolon (Gymnocaulus) as suggested by Schepotieff (A) and Lankester (B). A. Elongation of stolon from a ventral outgrowth on metasome of terminal zooid (v, arrow). Stolon's advancement through elongation of terminal zooid peduncle (pd, arrow). l, Lophophore; p, preoral lobe.

tion of the stolon. The growing juvenile, or temporarily arrested individual, provides an analogy to the terminal undeveloped zooid of *Rhabdopleura*. The budding of a lateral zooid (or zooids) from this elongated portion of the stolothecal stolon resembles the formation of new buds in the terminal chamber of the tubarium in *Rhabdopleura*. The differences are those of degree. The main difference is that in graptolite colonies, the temporary leading zooid transforms into a definitive, mature autozooid, which delegates the function of further extension of the colony to a later zooid in a daughter stolotheca. This transformation may even be belated if a bifurcation diad is produced, with the sequence parental stolotheca–daughter stolotheca, and so forth, which results in

a prolongation of the morphogenetic arrest of growing thecae.

Graptoloids, however, represent a more advanced stage of sympodial budding, with successive thecae interconnected by rather short portions of the common soft tissue, corresponding to the naked soft stolon of sessile graptolites. Therefore, the characteristic lag in transformation of the juvenile bud into the fully grown zooid is lacking, or at least drastically shortened. Each zooid is generated by the preceding one almost immediately, skipping a longer period of an underdeveloped bud.

Compared with some Paleozoic rhabdopleurids, the Graptolithina exhibit remarkably orderly patterns of stolon branching and thecal budding. Kozłowski (1949) distinguished two patterns of thecal budding among sessile graptolites. In one case, the thecae are produced in triads, each composed of an autotheca, a bitheca, and a stolotheca, and preceded by trifurcation of the stolon itself (Fig. 3). In the other case, the thecae appear in various diads due to bifurcation of the stolon. Kozłowski (1949, 1962) regarded the budding pattern as a key feature in the evolution of sessile graptolites and an important diagnostic character of their orders. In the Dendroidea and the Crustoidea the stolon trifurcates at each node and the thecae appear in triads; in the Tuboidea and the Camaroidea the stolon bifurcates, producing at each node a diad. A number of sessile genera with dendroid habit associated with diad budding were consequently placed within the Tuboidea. According to Kozłowski, the two main evolutionary trends, that is, the Crustoidea–Dendroidea–Graptoloidea and the Tuboidea–Camaroidea, reflect a fundamental schism in the graptolite mode of budding. However, Kozłowski did not define the pattern of budding characteristic of the common ancestor of all the Graptolithina.

For purely formal reasons, diads can be considered as geometrically simpler than triads. Moreover, every formation of a peduncular stolon on the main stolon and every bifurcation producing two secondary stolons can be equated with a diad. Therefore, the essential budding pattern in rhabdopleurids is diad formation. It was later conveyed without much modification to the Tuboidea–Camaroidea. After the appearance of the zooid dimorphism and thecal trimorphism, which was the main event at the pterobranch–graptolite boundary, diad formation still remained the budding pattern. However, a variable composition of the diads resulted in uneven distribution of thecal types along the stipes, as observed in some tuboids. This apparent adaptive imperfection could be improved by a more rigid morphogenetic control of diad composition, for example, by ensuring the regular sequence of diads and equalizing the numbers of the autothecal and bithecae within the colony. But the graptolites explored another possibility involving consistent grouping of all three types of thecae in the form of a triad. Presumably, the appearance of triad budding was a later event, following the primary diad budding and the elaboration of dimorphism. In fact, some aberrant encrusting graptolites (Mierzejewski 1985, in press) display a remarkably irregular pattern of budding, with not only diads but also triads being incidentally formed. The latter are unusual, as their middle stolon is stolothecal instead of autothecal as is common in the Dendroidea.

This simple scheme may be questioned in the view of Andres' (1977) observations on other aberrant graptolites—*Mastigograptus* and *Micrograptus.* They display a regular triad budding, although their thecae are not clearly differentiated into autothecae and bithecae. If this conclusion is correct, a triad budding pattern was established, at least in this lineage, prior to the elaboration of thecal dimorphism.

The phylogenetic significance of such an irregular budding pattern, combining within a single colony the features of all sessile orders of the Graptolithina, is difficult to evaluate, but it may serve as a useful model of possible transient stages between the diad and the triad budding pattern, whether or not actually realized in the course of evolution.

CONCLUSIONS

1. The bulk of evidence suggests a close relationship between the Pterobranchia and the Graptolithina. However, the latter achieved a much higher level of colony organization. This is demonstrated by the highly orderly morphogenesis of the graptolite colony, and also by the appearance of a number of special colonial structures such as the nema and the nematularia. Their origin and function still are, to a great extent, obscure.

2. Morphologic differences between the Pterobranchia and the sessile Graptolithina (i.e., between the Rhabdopleurida and some Tuboidea and Crustoidea) are less pronouned than the differences between the sessile and the planktic orders within the Graptolithina. In this context, Beklemishev's concept (1970) of the Rhabdopleuroidea and the Cephalodiscoidea as orders of the class Graptolithoidea deserves to be reconsidered, because it eliminates the actualistic bias involved in the recognition of the extant Pterobranchia and the extinct Graptolithina as higher taxa of the same rank.

3. Cambrian graptolite faunas are too poorly known to allow for a reconstruction of the early phylogeny of the group. There still is no evidence to identify the course of evolution and the phylogenetic relationships between the sessile orders of Graptolithina.

4. The discovery of ultrastructural anomaly,

that is, a distinct difference between the Pterobranchia and the Graptolithina in the fusellar component of their skeleton—once the source of misleading conclusions—became the basis of a valuable research program. This program was instrumental in reaching a better understanding of the graptolite affinities. The ultrastructural differences are by no means trivial. Perhaps they represent the best diagnostic feature of the graptolite grade of rhabdosome construction.

5. The graptolites have ceased to be a problematic group, but they still pose quite a few problems. So the graptolite workers will never have a chance to rest on their laurels.

Acknowledgments

This article is a result of many discussions with D. Andres, P. Crowther, P. N. Dilly, V. Jaanusson, P. Mierzejewski, and R. B. Rickards. I am especially grateful to my wife Irina Bagaeva for reading the manuscript, to Mrs. E. Gutkowska-Leszak for line drawings, and to Mrs. J. Kobuszewska for bibliographic data. I am grateful to P. N. Dilly, P. Crowther, and P. Mierzejewski for their courtesy in providing the illustrations for Figs. 4, 7, and 12.

REFERENCES

Allman, G. J. 1869. On *Rhabdopleura,* a new form of Polyzoa, from deep-sea dredging in Shetland. *Q. J. Microsc. Sci.* **9,** 57–63.

———. 1872. On the morphology and affinities of graptolites. *Ann. Mag. Nat. Hist.* 4(9), 364–380.

Andersson, K. A. 1907. Die Pterobranchier der Schwedischen Südpolar Expedition 1901–1903. *Ergebn. Schwed. Südpolar Exped.* **5,** 1–122, Stockholm.

Andres, D. 1976. Graptolithen aus ordovizischen Geschieben und die Frühe Stammesgeschichte der Graptolithen. Ph.d. dissert. Freie Universität, Berlin, 75 pp.

———. 1977. Graptolithen aus ordovizischen Geschieben und die frühe Stammesgeschichte der Graptolithen. *Paläont. Z.* **51,** 52–93.

———. 1980. Feinstrukturen und Verwandtschaftsbeziehungen der Graptolithen. *Paläont. Z.* **54,** 129–170.

Armstrong, W. G., Dilly, P. N., and Urbanek, A. 1984. Collagen in the pretobranch coenecium and the problem of graptolite affinities. *Lethaia* **17,** 145–152.

Bairati, A. 1972. Comparative ultrastructure of ectoderm-derived filaments. *Boll. Zool.* **39,** 283–308.

Beklemishev, V. N. 1951. K postroenyu sistemy zhivotnykh. Vtorichnorotye (Deuterostomia), ikh proizkhozhdenye i sostav. *Usp. Sovr. Biol.* **32,** 256–270.

———. 1970 (Russian ed. 1951, 1964). *Princi-*

ples of Comparative Anatomy of Invertebrates, vol. 1, 490 pp.; vol. 2, 529 pp. Oliver and Boyd, Edinburgh.

Bengtson, S. and Urbanek, A. In press. *Rhabdotubus,* a Middle Cambrian rhabdopleurid hemichordate. *Lethaia.*

Berry, W. B. N. 1978. A contribution toward understanding the relative integration of graptolite colonies. *Acta Palaeont. Polon.* **23,** 449–462.

——— and Takagi, R. S. 1970. Electron microscope investigation of *Orthograptus quadrimucronatus* from the Maquoketa Formation (late Ordovician) in Iowa. *J. Paleont.* 44, 117–124.

——— and ———. 1971. Electron microscope study of a *Diplograptus* species. *Lethaia* **4,** 1–13.

Boardman, R. S. and Cheetham, A. H. 1969. Skeletal growth, intracolony variation, and evolution in Bryozoa—a review. *J. Paleont.* **43,** 205–233.

Bobin, G. 1977. Interzooecial communications and the funicular system. *In* Woollacott, R. M. and Zimmer, R. I. (eds.). *The Biology of Bryozoans,* Academic Press, New York, pp. 325–341.

Bohlin, B. 1950. The affinities of graptolites. *Bull. Geol. Inst. Univ. Uppsala* **34,** 107–113.

Brommel, M. Von 1727. Lithographiae Suecanae. *Acta Lit. Sueciae Uppsaliae* **1–2,** Uppsala.

Brongniart, A. 1828. *Histoire des Végétaux Fossiles.* Paris.

Bulman, O. M. B. 1932. On the graptolites prepared by Holm. *Ark. Zool.* **24A**(8), 1–46.

———. 1937. Carl Wiman's work on the structure of the Graptoloidea. *Bull. Geol. Inst. Univ. Uppsala* **27,** 10–18.

———. 1938. Graptolithina. *In* Schindewolf, O. H. (ed.). *Handbuch der Paläozoologie,* 2D. Borntraeger, Berlin, pp. 1–92.

———. 1942. The structure of the dendroid graptolites. *Geol. Mag.* **79,** 284–290.

———. 1945. A monograph of the Caradoc (Balclatchie) graptolites from limestones in Laggan Burn, Ayrshire. *Palaeontogr. Soc. Monogr.* **1,** 1–42.

———. 1949. The anatomy and classification of the graptolites. *13th Int. Cong. Zool. Paris,* pp. 529–535.

———. 1955. Graptolithina with sections on Enteropneusta and Pterobranchia. *In* Moore, R. C. (ed.). *Treatise on Invertebrate Paleontology,* part V. Geological Society of America and Univ. of Kansas Press, Lawrence, XVII + 101 pp.

———. 1970. Graptolithina with sections on Enteropneusta and Pterobranchia. *In* Teichert, C. (ed.). *Treatise on Invertebrate Paleontology,* part V, 2nd ed. Geological Society of America and Univ. of Kansas Press, Lawrence, XXXII + 163 pp.

——— and Rickards, R. B. 1966. A revision of Wiman's dendroid and tuboid graptolites. *Bull. Geol. Inst. Univ. Uppsala* **43,** 1–72.

Chapman, A. J. and Rickards, R. B. 1982. Peridermal (cortical) ultrastructure in *Dictyonema* f. *rhinanthiforme* Bulman, and the significance of its bitheca. *Paläont. Z.* **56,** 217–227.

Chapman, D. M. 1966. Evolution of the scyphistoma. *Symp. Zool. Soc. Lond.* **16,** 51–75.

——— and Werner, B. 1972. Structure of a solitary and a colonial species of *Stephanoscyphus* (Scyphozoa, Coronatae) with observations on periderm repair. *Helgoländer Wiss. Meeresunters.* **23,** 393–421.

Chapman, G. 1974. The skeletal system. *In* Muscatine, L. and Lenhoff, H. M. (eds.). *Coelenterate Biology. Reviews and New Perspectives.* Academic Press, London.

Crowther, P. R. 1978. The nature and mode of life of the graptolite zooid with reference to secretion of cortex. *Acta Palaeont. Polon.* **23,** 473–479.

———. 1980. The fine structure of graptolite periderm. *Spec. Pap. Palaeont.* **26,** 1–119.

——— and Rickards, R. B. 1977. Cortical bandages and the graptolite zooid. *Geol. et Palaeont.* **11,** 9–46.

Dawydoff, C. 1948. Embranchement des Stomochordés. *In* Grassé, P.-P. (ed.). *Traité de Zoologie,* vol. 11. Masson, Paris, pp. 367–489.

Decker, C. E. 1956. Place of graptolites in animal kingdom. *Bull. Am. Assoc. Petrol. Geol.* **40,** 1699–1704.

——— and Gold, I. B. 1957. Bithecae, gonothecae, and nematothecae in Graptoloidea. *J. Paleont.* **31,** 1154–1158.

——— and Hassinger, N. 1958. What higher magnification is doing for the study of graptolites. *J. Paleont.* **32,** 697–700.

Dilly, P. N. 1971. Keratin-like fibres in the hemichordate *Rhabdopleura compacta. Z. Zellforsch. Mikrosk. Anat.* **117,** 502–515.

———. 1973. The larva of *Rhabdopleura compacta* (Hemichordata). *Mar. Biol.* **18,** 69–86.

———. 1975. The dormant buds of *Rhabdopleura compacta* (Hemichordata). *Cell Tissue Res.* **159,** 387–397.

Elles, G. L. 1922. The graptolite faunas of the British Isles. A study in evolution. *Proc. Geol. Assoc.* **33,** 168–200.

Foucart, M. F. 1964. Paléobiochimie et position systématique des graptolites. *Mem. Licence Univ. Liège,* 120 pp.

———, Bricteux-Gregoire, S., Jeuniaux, C., and Florkin, M. 1965. Fossil proteins of graptolites. *Life Sci.* **4,** 467–471.

——— and Jeuniaux, C. 1966. Paléobiochimie et position systématique des graptolites. *Ann. Soc. Zool. Belg.* **95**(2), 39–45.

Frech, F. 1897. Graptolithiden. *Lethaea Geognostica* **1,** 544–684.

Haberfelner, E. 1933. Muscle-scars of Monograptidae. *Am. J. Sci.* **25**, 298–302.

Hall, J. 1865. Graptolites of the Quebec Group. *Geol. Surv. Can. Organic Rem.* **2**, 1–151.

Harmer, S. F. 1905. The Pterobranchia of the Siboga-Expedition with an account of other species. *In* Uitkomsten op zoologisch, botanisch, oceanographisch en geologisch gebied verzameld in Nederlandsch Oost-Indien 1899–1900 aan boord *H. M. Siboga*. **12**(26), 1–132.

Hutt, J. E. 1974. The development of *Clonograptus tenellus* and *Adelograptus hunnebergensis*. *Lethaia* 7, 79–92.

Hyman, L. H. 1958. The occurrence of chitin in the lophophorate phyla. *Biol. Bull.* **114**, 106–112.

———. 1959. *The Invertebrates.* V. *Smaller Coelomate Groups.* McGraw-Hill, New York, 783 pp.

Jebram, T. 1973. Zooid individuality and the brooding organs (Bryozoa). Additional aspects in the discussion about the phylogeny of the Bryozoa. *Z. Morph. Tiere* **75**, 255–258.

Kielan-Jaworowska, Z. and Urbanek, A. 1978. Dedication. Roman Kozłowski (1889–1977). *Acta Palaeont. Polon.* **23**, 415–425.

Kirk, N. H. 1972. Some thoughts on the construction of the rhabdosome in the Graptolithina, with special reference to extrathecal tissue and its bearing on the theory of automobility. *Univ. Coll. Wales, Aberystwyth, Dept. Geol. Publ.* **1**, 1–21.

———. 1973. Some thoughts on the construction and functioning of the rhabdosome in the Retiolitidae. *Univ. Coll. Wales, Aberystwyth, Dept. Geol. Publ.* **3**, 1–26.

———. 1974. More thoughts on the construction of the rhabdosome in the Dendroidea, in the light of the ultrastructure of the Dendroidea and *Mastigograptus. Univ. Coll. Wales, Aberystwyth, Dept. Geol. Publ.* **6**, 1–11.

Kozłowski, R. 1938. Informations préliminaires sur les Graptolithes du Tremadoc de la Pologne et sur leur portée théorique. *Ann. Mus. Zool. Polon.* **13**, 183–196.

———. 1947. Les affinités des Graptolithes. *Biol. Rev.* **22**, 93–108.

———. 1949. Les Graptolithes et quelques nouveaux groupes d'animaux du Tremadoc de la Pologne. *Palaeont. Polon.* **3**, 1–235.

———. 1956. Sur *Rhabdopleura* du Danien de Pologne. *Acta Palaeont. Polon.* **1**, 1–21.

———. 1961. Decouverte d'un Rhabdopleuride (Pterobranchia) Ordovicien. *Acta Palaeont. Polon.* **6**, 3–16.

———. 1962. Crustoidea, nouveau groupe de graptolites. *Acta Palaeont. Polon.* **7**, 3–52.

———. 1963. Le développement d'un graptolite tuboïde. *Acta Palaeont. Polon.* **8**, 103–134.

———. 1966a. On the structure and relationships of graptolites. *J. Paleont.* **40**, 489–501.

———. 1966b. *Graptolodendrum mutabile* n. gen., n. sp.—an aberrant dendroid graptolite. *Acta Palaeont. Polon.* **11**, 3–14.

———. 1967. Sur certains fossiles Ordoviciens à test organique. *Acta Palaeont. Polon.* **12**, 92–132.

———. 1970. Nouvelles observations sur les Rhabdopleurides (Pterobranches) Ordoviciens. *Acta Palaeont. Polon.* **15**, 3–17.

———. 1971. Early development stages and the mode of life of Graptolites. *Acta Palaeont. Polon.* **16**, 313–343.

Kraatz, R. 1964. Untersuchungen über die Wandstrukturen der Graptolithen (mit Hilfe des Elektronenmikroskops). *Z. Dtsch. Geol. Ges.* **114**, 699–702.

———. 1968. Elektronenmikroskopische Beobachtungen an *Monograptus* Rhabdosomen. *Aufschluss* **12**, 357–361.

Kraft, P. 1926. Ontogenetische Entwicklung und die Biologie von *Diplograptus* and *Monograptus. Paläont. Z.* **7**, 207–249.

Kühne, W. G. 1955. Unterludlow-Graptolithen aus Berliner Geschieben. *N. Jb. Geol. Paläont. Abh.* **100**, 350–401.

Lankester, E. R. 1884. A contribution to the knowledge of *Rhabdopleura. Q. J. Microsc. Sci.* **24**, 622–647.

Lapworth, C. 1873. Notes on the British graptolites and their allies. I. On an improved classification of the Rhabdophora. *Geol. Mag.* **10**, 500–504; 555–560.

———. 1876. On Scottish Monograptidae. *Geol. Mag.* **13**, 308–552.

Linnaeus, C. 1735. *Systema Naturae.* 1st ed. Lugduni Batavorum.

Mierzejewski, P. 1982. The nature of pre-Devonian tracheid-like tubes. *Lethaia* **15**, 148.

———. 1984. *Cephalodiscus*-type fibrils in the graptoblast fusellar tissue. *Acta Palaeont. Polon.* **29**, 157–160.

———. 1985. New aberrant sessile graptolites from glacial boulders. *Acta Palaeont. Polon.* **30**, in press.

———. In press. Ultrastructure, taxonomy, and affinities of some Ordovician and Silurian organic microfossils. *Palaeont. Polon.* **46**.

Nicholson, H. A. 1872. *Monograph of the British Graptolitidae.* Blackwood, Edinburgh, 133 pp.

Perner, J. 1894. Etudes sur les Graptolites de Bohême. 1ère partie: structure microscopique des genres *Monograptus* et *Retiolites*. Prague, 14 pp.

Popper, K. 1976. *Unended Quest. An Intellectual Autobiography.* Fontana-Collins, Glasgow, 255 pp.

———. 1979. *Objective Knowledge. An Evolutionary Approach.* Clarendon Press, Oxford, 395 pp.

Richter, R. 1871. Aus dem Thüringischen Schiefergebirge. *Z. Dtsch. Geol. Ges.* **23**, 231–256.

Rickards, R. B. 1975. Palaeoecology of the Graptolithina, an extinct class of the phylum Hemichordata. *Biol. Rev.* **50**, 397–436.

———, Hyde, P. J., and Krinsley, D. H. 1971. Periderm ultrastructure of a species of *Monograptus* (phylum Hemichordata). *Proc. R. Soc. Lond. (B.)* **178**, 347–356.

Rudall, K. M. 1955. Distribution of collagen and chitin. *Symp. Soc. Exp. Biol.* **9**, 49–71.

Ruedemann, R. 1895. Development and mode of growth of *Diplograptus* McCoy. *Rep. St. Geol. N. Y.* **14**, 217–249.

Ryland, J. S. 1970. *Bryozoans.* Hutchinson, London, 175 pp.

Salter, J. W. 1866. *In* Ramsay. The Geology of North Wales. *Mem. Geol. Surv. Great Britain* **3**.

Sars, G. O. 1874. On *Rhabdopleura mirabilis. Q. J. Microsc. Sci.* **14**, 23–45.

Schepotieff, A. 1905. Über die Stellung der Graptolithen im zoologischen System. *N. Jb. Mineral.* **2**, 79–98.

———. 1907. Die Pterobranchier. Anatomische und histologische Untersuchungen über *Rhabdopleura normani* Allman und *Cephalodiscus dodecalophus* M'Int. 1, 2 Knospungprozess und Gehäuse von *Rhabdopleura. Zool. Jb. Abt. Anat. Ontog.* **24**, 193–238.

———. 1910. Die Pterobranchier des Indischen Ozeans. *Zool. Jb. Abt. Syst.* **28**, 429–448.

Schrock, R. R. and Twenhofel, W. H. 1953. *Principles of Invertebrate Paleontology.* 2nd ed. McGraw-Hill, New York, 816 pp.

Silén, L. 1942. Origin and development of the cheilo-ctenostomatous stem of the Bryozoa. *Zool. Bidr. Uppsala* **22**, 1–59.

———. 1977. Polymorphism. *In* Woollacott, R. M. and Zimmer, R. I. (eds.). *Biology of Bryozoans,* Academic Press, new York, pp. 183–231.

Stebbing, A. R. D. 1970. Aspects of the reproduction and the life-cycle of *Rhabdopleura compacta* (Hemichordata). *Mar. Biol.* **5**, 205–212.

Sundara-Rajulu, G., Jeuniaux, C., Poulicek, M., and Voss-Foucart, M. F. 1982. Comparative value of chitosan test and enzymatic method for chitin detection. *In* Hirano, S. and Tokura, S. (eds.). *Chitin and Chitosan.* Sapporo, Japan, pp. 1–4.

Towe, K. M. and Urbanek, A. 1972. Collagen-like structures in Ordovician graptolite periderm. *Nature* **237**, 443–445.

——— and ———. 1974. Fossil organic material: a unique fibril ultrastructure in Silurian graptolites. *Proc. 8th Int. Congr. Electron Microscopy Canberra* **2**, 694–695.

Ulrich, E. O. and Ruedemann, R. 1931. Are the graptolites bryozoans? *Bull. Geol. Soc. Am.* **42**, 589–604.

Urbanek, A. 1963. On generation and regeneration of cladia in some Upper Silurian monograptids. *Acta Palaeont. Polon.* **8**, 135–254.

———. 1966. On the morphology and evolution of Cucullograptinae (Monograptidae, Graptolithina). *Acta Palaeont. Polon.* **11**, 291–544.

———. 1970. Neocucullograptinae n. subfam. (Graptolithina)—their evolutionary and stratigraphic bearing. *Acta Palaeont. Polon.* **15**, 163–388.

———. 1973. Organization and evolution of graptolite colonies. *In* Boardman, R. S., Cheetham, A. H., and Oliver, W. A., Jr. (eds.). *Animal Colonies. Development and Function through Time.* Dowden, Hutchinson and Ross, Stroudsburg, Pa., pp. 441–514.

———. 1976a. The problem of graptolite affinities in the light of ultrastructural studies on peridermal derivatives in pterobranchs. *Acta Palaeont. Polon.* **21**, 3–36.

———. 1976b. Ultrastructure of microfuselli and the evolution of graptolite skeletal tissues. *Acta Palaeont. Polon.* **21**, 315–331.

———. 1978. Significance of ultrastructural studies for graptolite research. *Acta Palaeont. Polon.* **23**, 595–629.

———. 1983. The significance of graptoblasts in the life cycle of crustoid graptolites. *Acta Palaeont. Polon.* **28**, 313–326.

———, Mierzejewska, G., and Mierzejewski, P. 1980. Scanning electron microscopy of sessile graptolites. *Acta Palaeont. Polon.* **25**, 197–212.

——— and Mierzejewski, P. 1984. The ultrastructure of the Crustoidea and the evolution of graptolite skeletal tissues. *Lethaia* **17**, 73–91.

———, ———, and Rickards, R. B. In press. New observations on the fine structure of graptoblasts. *Lethaia.*

———, and Rickards, R. B. 1974. The ultrastructure of some retiolitid graptolites and graptoblasts. *Spec. Pap. Palaeont.* **13**, 176–188.

——— and Towe, K. M. 1974. Ultrastructural studies on graptolites, 1: the periderm and its derivatives in the Dendroidea and in *Mastigograptus. Smithson. Contr. Paleobiol.* **20**, 1–48.

——— and ———. 1975. Ultrastructural studies on graptolites in the Graptoloidea. *Smithson. Contr. Paleobiol.* **22**, 1–48.

——— and Zieliński, K. 1981. Preliminary report on *Cephalodiscus* (Pterobranchia) from Admiralty Bay, King George Island, South Shetland Islands (West Antarctica). *Bull. Acad. Pol. Sci., Ser. Sci. Biol.* **29**, 257–262.

Walch, J. E. I. 1771. *Naturgeschichte der Versteinerungen zur Erläuterung der Knorr'schen Sammlung von Merkwürdigkeiten der Natur,* suppl. 3. Nürnberg.

Waterlot, G. 1952. Classe des Graptolites. *In* Piveteau, J. (ed.). *Traité de Paléontologie,* vol. 3. Masson, Paris, pp. 968–997.

Werner, B. 1967. *Stephanoscyphus* Allman (Scyphozoa, Coronatae), ein rezenter Vertreter der Conulata? *Paläont. Z.* **41**, 137–153.

Wetzel, W. 1958. Graptolites and their possible relations, compared by electron microscope. *N. Jb. Geol. Paläont. Mh.* **1958**, 307–312.

Whittington, H. B. and Rickards, R. B. 1968. New tuboid graptolite from the Ordovician of Ontario. *J. Paleont.* **42**, 61–69.

Wiman, C. 1895. Über die Graptolithen. *Bull. Geol. Inst. Univ. Uppsala* **2**, 239–316.

————. 1901. Über die Borkholmerschicht in Mittelbaltischen Silurgebiet. *Bull. Geol. Inst. Univ. Uppsala* **5**, 149–222.

Zimmer, R. I. and Woollacott, R. M. 1977. Metamorphosis, ancestrulae and coloniality in bryozoan life cycles. *In* Woollacott, R. M. and Zimmer, R. I. (eds.). *Biology of Bryozoans.* Academic Press, New York, pp. 233–271.

Zittel, K. A. von. 1880. *Handbuch der Paläontologie. Paläozoologie. l. Protozoa, Coelenterata, Echinodermata und Molluscoidea.* Oldenburg, München, 765 pp.

15. CONODONTS

RICHARD J. ALDRIDGE AND DEREK E. G. BRIGGS

Conodonts are chiefly represented in the fossil record by their scattered phosphatic skeletal elements, which are commonly abundant in marine strata of Cambrian to Triassic age. Although they may be found by careful scanning of bedding surfaces using a lens or a microscope, they are usually recovered by disaggregating or dissolving rock samples by mechanical or chemical methods. Conodont elements usually occur as discrete entities, but it is evident that each conodont animal bore a skeleton comprising several elements, often of considerable morphologic variety. The complete skeletons of some individuals have been found resting more-or-less undisturbed on bedding surfaces, giving direct evidence of the multielement structures of some species. Until recently, the soft parts of the conodont animal were totally unknown, but specimens from the Carboniferous of Scotland, the first of which was found by E. N. K. Clarkson and described by Briggs et al. (1983), have provided the best evidence to date of the nature of this enigmatic organism and of the position and function of its set of skeletal elements.

Conodonts were first recorded and named by Pander (1856). Although he only recovered isolated elements, it appears that he intended the name he applied (Conodonten) to refer to the group of organism from which these fossil remains were derived. Many subsequent authors, however, have used the term "conodont" for the skeletal elements alone, a practice endorsed in the first edition of the *Treatise on Invertebrate Paleontology* (Hass 1962). This usage has resulted in a need for terms like "conodont-bearing animal" (e.g., Hass 1962) or "conodontophorid" (e.g., Barnes and Fåhraeus 1975) for reference to the organism or to the taxonomic group. The ambiguity generated by these two different uses of the name "conodont" has yet to be fully resolved, although there now seems to be a body of opinion favoring Pander's original usage (e.g., Jeppsson 1982; Fåhraeus 1983). This approach has received some support in the revised edition of the *Treatise on Invertebrate Paleontology* (Robison 1981), where conodonts are treated as organisms and assigned to a separate phylum, the Conodonta. Hence the term "element," or "conodont element," is applied to the individual mineralized skeletal parts, and the unqualified term "conodont" refers to the complete animal. We follow this course in this

paper, but until consistent usage is achieved, we advocate continuing use of "conodont animal" when any possibility of ambiguity or misunderstanding arises.

CONODONT ELEMENTS

Conodont elements are formed of calcium phosphate; detailed chemical analyses of some Devonian specimens give a composition close to the carbonate apatite mineral francolite (Pietzner et al. 1968). Conical phosphatic elements first appear in the fossil record near the Precambrian–Cambrian boundary, and a variety of forms has been recorded from Cambrian strata (e.g., Müller 1959). Bengtson (1976) used histologic characters of these elements to discriminate three groups of Cambrian fossils: protoconodonts, paraconodonts, and euconodonts (Fig. 1). Protoconodont elements show three structural layers, with a thick, laminated middle layer bounded by thin inner and outer lamellae; only the middle layer appears to have been mineralized, and the apatite lamellae are arranged in a manner indicating accretion at the inner surface and the base of the element (Bengtson 1976, 1983b; Szaniawski 1982, 1983). Paraconodont elements display growth lamellae accreted at the inner and outer surfaces, but not continuous around the tip; each lamella is thick and composed of very fine crystallites with more organic matter than in euconodont elements (Müller and Nogami 1971; Bengtson 1983b). Euconodont, or true conodont, elements consist of a hyaline crown and a basal body, which both grew out from a nucleus. The lamellae of the crown are continuous around the upper surface; they are usually 0.2–1.2 μm thick (Barnes et al. 1973) and made up of apatite crystallites that are aligned with their C-axes parallel to the direction of growth (Lindström and Ziegler 1971). The basal body is also lamellar, although the lamellae are not always continuous around the lower surface (Müller and Nogami 1971); the mineral composition is broadly similar to that of the crown, but the crystallites are smaller and more organic matter is commonly included. Although most early conodont elements possess a basal body, it is absent from many Upper Paleozoic genera, in which its function may have been fulfilled by unmineralized tissue.

From morphologic, histologic, and strati-

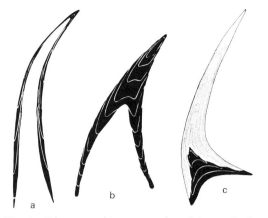

Fig. 1. Diagrammatic representation of the apatite lamellar structure of a protoconodont element (A), a paraconodont element (B), and a euconodont element (C). After Bengtson (1976), with permission.

CONODONT APPARATUSES

Close associations of aligned conodont elements on bedding planes were first described in the mid-1930s (Schmidt 1934; Scott 1934), and through the succeeding years several additional examples have been reported (e.g., Scott 1942; Du Bois 1943; Rhodes 1952; Schmidt and Müller 1964; Mashkova 1972; Avcin 1974; Norby 1976). The vast majority of records are from Carboniferous

graphic data, Bengtson (1976, 1983b) proposed a model of early conodont evolution in which protoconodonts evolved through paraconodonts to euconodonts by successively deeper retention of the elements in the secreting epithelium during growth (Fig. 2). Current evidence points to a development of euconodonts from paraconodonts in the Late Cambrian and is consistent with a link between protoconodonts and paraconodonts, although this relationship is less strongly affirmed (Bengtson 1983b). However, the possibility of a link between protoconodonts and euconodonts achieves greater significance in the light of the strong case put forward by Szaniawski (1982, 1983) for homology between protoconodont elements and the grasping spines of modern chaetognaths. The hypothetical sequence of links from chaetognaths, through protoconodonts and paraconodonts, to euconodonts has led Bengtson (1983b) to postulate that euconodont animals are a branch of the chaetognaths that had developed pharyngeal denticulation.

The development of euconodont elements was followed by considerable diversification of morphology. Conodont elements from latest Cambrian to latest Triassic strata may be broadly grouped into three shape categories: coniform, ramiform, and pectiniform (Fig. 3). The various forms within these groups and the terminology applied to them were fully described by Sweet (1981). In some Cambrian and Ordovician elements the crown is entirely hyaline, but in advanced conodonts the hyaline structure is interrupted by patches of opaque white matter (Fig. 4), which is finely crystalline and porous (Pietzner et al. 1968). The mode of formation and function of this white matter are not clearly understood (Lindström and Ziegler 1981, p. 49), but its distribution through different elements varies widely and is commonly a useful taxonomic character.

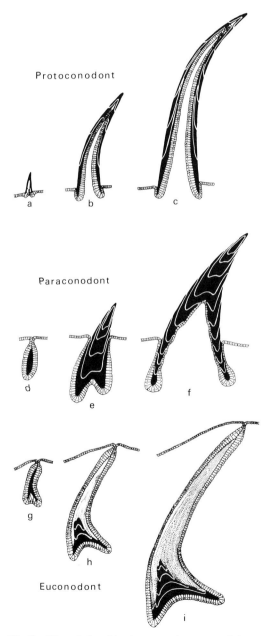

Protoconodont

Paraconodont

Euconodont

Fig. 2. The relationships between elements and the secreting epithelium during growth, as reconstructed by Bengtson (1976).

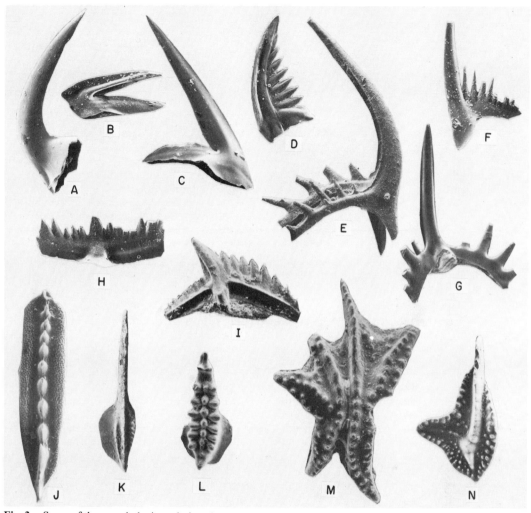

Fig. 3. Some of the morphologic variation shown by euconodont elements. A–D. Coniform elements, all lateral views. ×60. E–G. Ramiform elements. All ×40. E and F in lateral view, G in posterior view. H–N. Pectiniform elements. H and I lateral views, ×40; J upper view, ×80; K–N upper views, ×40. All illustrations are scanning electron micrographs.

shales, and some of the best specimens have yet to be described. Some of these natural assemblages are undoubtedly of coprolitic or similar origin and the nature of others is equivocal, but many clearly represent the complete skeletons of individual animals that rested undisturbed in the sediment after the organism died and decayed (Fig. 5). These assemblages provided the first direct evidence that conodonts possessed a skeleton comprising several elements, but they are so rare that they only give information about the skeletal structure of a very few species. Additional knowledge may be gained from clusters of elements found in acid-insoluble residues from dissolved limestone samples. In these clusters, two or more elements are fused together, commonly with specimens paired or in alignment (Fig. 6). They have been found more widely through the stratigraphic

sequence than natural assemblages (e.g., Lange 1968; Austin and Rhodes 1969; Pollock 1969; Behnken 1975; Landing 1977; Ramovš 1978, Nowlan 1979; Repetski 1980; Smith 1984), and while some may again be of coprolitic origin, most represent the partial skeletal apparatuses of single animals. The nature of the fusion has been interpreted as pathologic (e.g., Rexroad and Nicoll 1964; Pollock 1969) or diagenetic (e.g., Landing 1977; Nowlan 1979). As the occurrence of clusters is often restricted to particular beds and the arrangement of elements normally mirrors those of bedding-plane assemblages, we consider most, at least, to be diagenetic.

Bedding-plane assemblages and clusters are uncommon, and most species are known only from their disassociated skeletal elements. However, it is possible to use morphologic, distributional, and

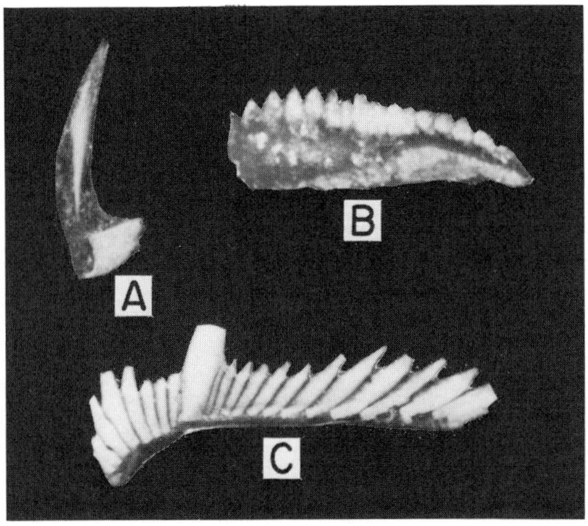

Fig. 4. Examples of the distribution of white matter. A. Coniform element. ×40. B. Pectiniform element. ×80. C. Ramiform element. ×25

statistical evidence to reconstruct the apparatuses of these species. Serious attempts at this began in the mid-1960s (Walliser 1964, Bergström and Sweet 1966, Webers 1966), and it is now standard practice in conodont taxonomy to relate together those elements that are considered to come from members of a single biologic species. Apparatuses reconstructed in this way are generally illustrated by a set of representative elements, which normally have not been derived from one animal but

which display the variety of forms possessed by members of the species. There are many traps for the unwary in the reconstruction procedure, and assessments are best carried out on the basis of repeated associations of elements in large collections. As well as taking account of the variation in preservation potential of elements and the effects of postmortem transport, it is important to be aware that different elements may change at different rates in evolving lineages. Our limited knowledge of the biology of conodonts also limits our interpretations. It has, for instance, been suggested that some conodonts may have been dimorphic (Jeppsson 1972; Merrill and Merrill 1974), and it is conceivable that dimorphs of the same species differed in all, some, or none of their elements. Additionally, it is possible that elements were added at different stages of ontogeny, or that some were lost and replaced repeatedly through life (Carls 1977), factors that would markedly affect relative frequencies in fossil collections. However, despite the difficulties, the apparatuses of several species have now been confidently reconstructed and progress is continually being made in determining or refining the multielement constructions of additional taxa.

It is apparent that most conodont apparatuses conform in their structure to a limited number of basic plans (Jeppsson 1971; Klapper and Philip 1971; Barnes et al. 1979). Hence, element associations exhibited by bedding-plane assemblages and clusters, together with those in well-established apparatus reconstructions, can serve as

Fig. 5. Bedding-plane assemblage (Nottingham University no. 5545/015) from the Pennsylvanian shales of Bailey Falls, Illinois. Back-scattered electron image at 10° tilt, ×25. The assemblage comprises a set of ramiform elements (in the upper part of the picture) and two pairs of pectiniform elements.

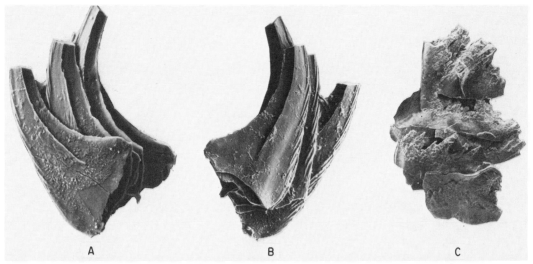

Fig. 6. Scanning electron micrographs of conodont clusters. A. and B. Lateral views of a cluster of coniform elements from the Ordovician of North Greenland. Geologisk Museum, Copenhagen no. MGUH 15071. ×250. C. A cluster of pectiniform and ramiform elements from Silurian of Indiana. Indiana Geological Survey no. 15169. ×60. Illustrated with the permission of Dr. C. B. Rexroad.

models in the analysis of new collections. Additional finds of clusters and assemblages give direct evidence of the relative numbers and dispositions of the component elements of apparatuses, and allow testing and evaluation of existing hypotheses.

Several attempts have been made to categorize apparatuses (e.g., Klapper and Philip 1971; Lindström 1973; Barnes et al. 1979). A complete review was provided by Sweet (1981, p. 16), who advocated a descriptive approach emphasizing the number of different kinds of element present. Several authors have also proposed notational schemes for the positions of elements within an apparatus (e.g., Sweet 1970; Jeppsson 1971; Klapper and Philip 1971; Barnes et al. 1979), but the scheme first published by Cooper (1975) is applied in the *Treatise on Invertebrate Paleontology* and has gained some currency. This system is based principally on seximembrate (six-component) apparatuses (Fig. 7). In septimembrate apparatuses an additional Sd position may be recognized, occupied, for example, by a quadriramate ramiform element. Although the scheme has wide applicability, there are several apparatus types for which it is inappropriate. This is the case, for instance, with the Icriodontidae (Nicoll 1982) and for some apparatuses of coniform elements (Barnes et al. 1979). Separate notational schemes are necessary for these, and it seems unlikely that a single system could be devised that would serve for all kinds of apparatus.

THE CONODONT ANIMAL

The soft parts of conodonts are known only from specimens from the Lower Carboniferous of the Edinburgh District, Scotland. Only the first specimen discovered has been described (Briggs et al. 1983). Three additional specimens from the original locality await detailed investigation at the time of writing. The conodont animal had an elongate flattened body with posterior ray-supported fins (Fig. 8). A distinct axial line runs along the trunk, which is divided into serially repeated structures. The preserved configuration of these structures just anterior to the fins (Briggs et al. 1983, fig. 2) suggests that they might be V-shaped, the apex of the V anterior.

The conodont apparatus lies in the anterior of the animal, posterior of a pair of terminal lobes flanking a median indentation that may mark the site of the mouth. The ramiform (M and S) elements lie anteriormost, followed by the paired arched blade (Pb) and platform (Pa) elements. This order is the reverse of that assumed or implied by several authors (e.g., Du Bois 1943; Rhodes 1952, 1962; Rhodes and Austin 1981) before the soft-bodied fossils were known. The apparatus is compacted in lateral aspect, with the ramiform elements oriented at a high angle and the pectiniform pairs near normal to the axis of the body.

FUNCTION OF CONODONT APPARATUSES

Recent discussion of the function of elements has centered on two hypotheses: They were either teeth or supports for the ciliated tentacles of a filter-feeding system. Fundamental to this discussion is the requirement that there be tissue external to the elements to secrete the growth lamellae. This requirement appeared to some authors to ne-

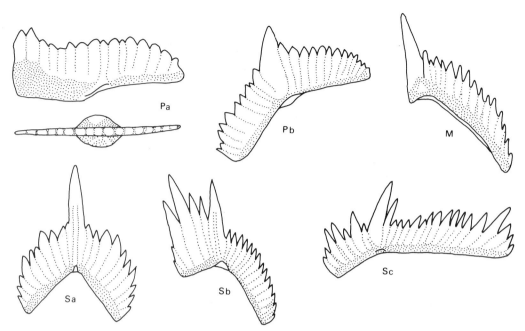

Fig. 7. The element types in the Silurian genus *Ozarkodina,* illustrating the notation of elements in a seximembrate apparatus. The Pa and Pb elements are characteristically pectiniform, the M a pick-shaped ramiform. The Sa to Sc positions are occupied by a symmetry transition series of ramiform elements. In a complete apparatus, all elements except the Sa are represented by mirror-image pairs; there may be more than one pair of Sc elements.

gate, or at least render unlikely, the possibility that the elements could function as exposed teeth (e.g., Lindström 1964; Priddle 1974). The alternative tentacle hypothesis also gained currency with Lindström's reconstruction (1974) of a hypothetical external lophophore and Conway Morris's "conodont"-bearing lophophorate, *Odontogriphus* (1976), from the Middle Cambrian Burgess Shale (see Chapter 13 by Briggs and Conway Morris, this volume). The discovery of the Edinburgh conodont animals has shown, however, that Lindström's interpretation is in error and that *Odontogriphus* is clearly not related to the euconodonts. In addition, Jeppsson (1979, 1980) and Bengtson (1976, 1980, 1983b) in particular have argued that the possibility that elements could be exposed to function as teeth cannot be dismissed on the basis of "the mistaken notion that the model is incompatible with euconodont mode of growth" (Bengtson 1980).

Considerations of function have addressed both individual elements (e.g., Bengtson 1976; Jeppsson 1979) and whole apparatuses (e.g., Schmidt 1934; Nicoll 1977, 1984; Hitchings and Ramsay 1978). Schmidt (1934) suggested that the elements represented the mandibles, hyal teeth, and gill-basket of a placoderm fish. Hitchings and Ramsay (1978) produced a filter-feeding model based on the element arrangement proposed by Rhodes (1952, 1962), with the pectiniform and M elements set anterior to the ramiform S elements.

This arrangement is not borne out by the Edinburgh conodont animals, and a filtering system of this type cannot be sustained. Briggs et al. (1983) did not attempt a detailed interpretation of the function of the apparatus but considered that a tooth hypothesis resulted in a "more credible integrated functional system with the anterior ramiform elements for grasping prey which is processed by the posterior elements."

Before the discovery of soft-bodied conodont animals, Jeppsson (1971) and Nicoll (1977), among others, had correctly anticipated the order of the elements. Nicoll (1977) proposed a type of filter-feeding function for the apparatus and, contrary to Briggs et al. (1983), subsequently (Nicoll 1985) presented a reconstruction of an animal with the apparatus functioning in this way. Nicoll (1985) assumed that the "working surface" of the elements was tissue covered and "probably ciliated" to "move water or particulate matter." He interpreted the M, Sa, Sb, Sc, and Sd elements as a "sieve for initial sorting of food particles at the anterior end of the food groove." In Nicoll's reconstruction the elements are aligned parallel to the axis of the animal, the cusp and denticles of the S elements pointing ventrally. Thus the anteriorly positioned cusps would have deflected away large particles. The alignment of the ramiform elements is such that the water would not have passed through a sieve, as in a true filter, but would have been carried along a ciliated surface

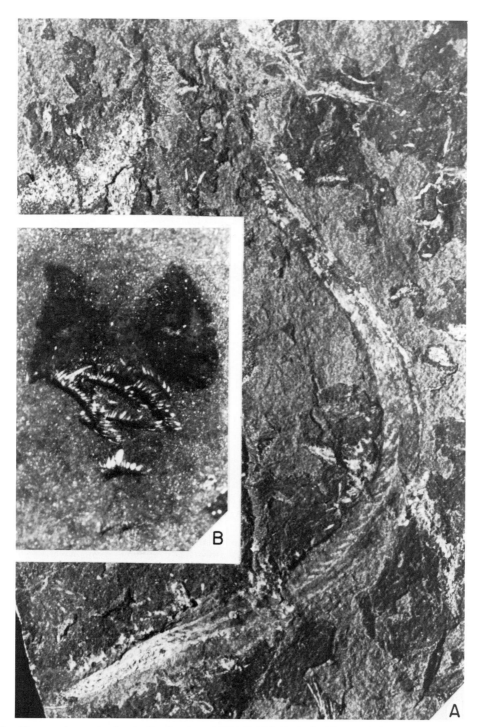

Fig. 8. The conodont animal. A. Specimen no. 13821, British Geological Survey, Edinburgh. The first complete conodont discovered. ×7. B. The head of the counterpart, no. 13822, showing the postion and orientation of the apparatus. ×25.

233

capable of trapping particles [see discussion of suspension feeding in Jørgensen (1966)]. The maximum size of particles entering the food groove would be determined by the lateral spacing of the cusps. Nicoll pointed out that the morphology of the P elements indicates a different function. Their larger cusps and denticles render a filtering function complementary to that postulated for the ramiforms unlikely. The Pa elements are constructed so as to allow the opposing surfaces of the pair to come into contact. Nicoll suggested that "they served to gently squash soft food particles being eaten by the animal"—in effect, a tooth function. He did not specifically discuss the Pb elements but considered that "the same function could have been performed by a shearing action between the opposed lateral surfaces." However, the "teeth" would presumably be rendered less effective for comminuting food by the postulated covering of soft tissue. Furthermore, the Pb and Pa elements are rather large in relation to the probable size of "particulate matter" (Nicoll 1985) captured by the ramiform elements.

It is not, of course, impossible for a filter-feeding organism to have teeth. The mandibles of simple filter-feeding Crustacea have grinding molar processes. More advanced forms acquire an incisor process as well, as in the syncarid *Anaspides,* but this is mainly a modification for eating large soft food as well as filtering (Manton 1977, p. 77). Thus, where well-developed incisors occur, they are a modification for dealing with larger food than that acquired by filtering. The large size of the Pb elements and of the blades of the Pa elements were likewise equipped to deal with bigger particles than would be collected by filtering.

Nicoll's postulated function for his Upper Devonian conodont apparatus (1985) is difficult to reconcile with the morphology and arrangement of the elements. One important reason for his conclusion, however, is the configuration of fused clusters of ramiform elements. They appear to be aligned parallel, with the denticles pointing in the same direction rather than opposed. The Pa and Pb elements, on the other hand, are commonly fused with their denticles ("upper surfaces") opposed. This difference in orientation between ramiforms and posterior elements is largely borne out by the configuration in the Edinburgh conodont animals and in natural assemblages. If the ramiform elements were not opposed, how could they have functioned as teeth? Bengtson (1983a) provided a possible solution to this paradox. He suggested that the apparatus is preserved in a resting position; the denticles were opposed and aligned symmetrically when the "pharynx" was opened out anteriorly to catch prey. Bengtson's model is based to some extent on prey capture in chaetognaths to which, in his view, conodonts may have been related. Regardless of the affinities of conodonts, the suggestion that the active and

resting positions of the apparatus may have differed is perfectly compatible with the requirement for a grasping organ to open and close.

While a filter-feeding model is difficult to sustain, considerations of element morphology lend further support to the tooth hypothesis. Jeppsson (1979) pointed out the similarity between the arrangement of alternating-sized denticles in ramiform conodonts, the teeth in some fishes and polychaetes (scolecodonts), and the raptorial appendage of stomatopods (Crustacea). One phrase in Schäfer's discussion (1972, pp. 67, 68) of the functional morphology of shark and ray teeth almost perfectly fits the morphology and orientation of an Sc element: "It is possible to set up a morphologically ordered sequence for grasping teeth, beginning with a high, upright incisor on a narrow base, and ending with a small hook-like tooth which is directed backward and rests on a wide root." Jeppsson (1979) also observed the similarity between angulate (Pb) elements and some shark incisor teeth (Schäfer 1972, pp. 65–68). He noted that the paired Pa elements occlude, the high asymmetrically attached blade acting as a guide. The blade could also have continued the shearing or cutting process initiated by the Pb element while the platform functioned in grinding. An analogy is provided by the incisor and molar processes of the mandible of some crustaceans, for example those of archaeostracans (Phyllocarida), which are also asymmetric as pointed out by Dzik (1980). Thus the "standard" seximembrate apparatus can be interpreted as an integrated functional system. The morphology of the animal itself also appears to support the tooth model. Most suspension-feeding organisms are sessile or stationary while feeding. Those pelagic crustaceans that filter-feed are slow swimmers. The conodont animal, however, was apparently adapted for relatively fast swimming and as such is more likely to have been an active predator (Jeppsson 1979; Bengtson 1983b; Briggs et al. 1983).

To what extent do conclusions based largely on Upper Paleozoic apparatuses apply to earlier forms? Evidence for the structure of Lower Paleozoic euconodont apparatuses depends largely on multielement reconstructions established by analyses of collections of disjunct elements, supplemented by a few fused clusters. Sufficient progress has been made for major patterns to be recognized, and it is apparent that apparatuses similar to those discussed above had evolved by Middle Ordovician times. Barnes et al. (1979) reviewed Ordovician apparatuses and recognized seximembrate and septimembrate associations (types IVA and IVB) essentially the same as those of Carboniferous natural assemblages. The earliest broadly comparable types were the prioniodontid apparatuses of the Early Ordovician, which comprised ramiform and pectiniform elements that may have performed the same grasping and grinding functions as their counterparts in later taxa. Similar seximembrate and septimembrate apparatuses occur throughout the Ordovician and are common in the Silurian (e.g., Jeppsson 1971) and Devonian (Klapper and Philip 1971, types 1 and 2). As in the Upper Paleozoic, there are also seximembrate or modified seximembrate apparatuses from the Early Ordovician onward in which all the elements are ramiform in style. These may have served for grasping prey that was consumed without further processing.

The Early Ordovician saw the greatest variability in apparatus types in the whole of conodont history. There are several variants on the seximembrate–septimembrate theme (types IVC–E of Barnes et al. 1979). These may represent grasping sets, as may other ramiform elements with large cusps and short processes that formed trimembrate or quadrimembrate apparatuses (types IIA and IIB). The type V apparatuses of Barnes and colleagues comprise only pectiniform elements, normally with conspicuous platform development of the processes. In may be argued that ramiform elements were, in fact, associated with the pectiniforms in life but have not been recognized as being from the same apparatuses, or perhaps have not been preserved. However, rich collections of these pectiniforms have been examined without yielding any evidence of associated ramiforms (Bergström 1983), and the existence of genuine platform-only apparatuses in the Ordovician seems probable. These apparatuses must represent a different feeding strategy, such as deposit-feeding or scavenging.

Considerable diversity is also exhibited by Ordovician coniform euconodont apparatuses (Barnes et al. 1979, types 1A–C, IIB, IIA–C). Some have elements of simple morphology resembling protoconodont elements, which in turn are strikingly similar to chaetognath grasping spines (Müller and Andres 1976; Repetski and Szaniawski 1981; Szaniawski 1982, 1983). While the relevance of these similarities to a discussion of conodont affinities is debatable, they do suggest a functional analogy. Thus, many coniform apparatuses may have been adapted for grasping. Others, however, consist of elements with more complicated morphology, often with sharp edges, keels, or prominent costae. These may have performed a cutting or shearing function in addition to grasping.

Several Ordovician apparatus types did not persist into the Silurian, which was characterized by various types of coniform apparatus and by seximembrate or modified seximembrate apparatuses derived from Ordovician "type IV" arrangements. The existence of post-Ordovician platform-only apparatuses is debatable, but icriodontid species with platformlike elements accompanied only by small conical elements were common in the Late Silurian and Devonian. Nicoll (1982) has shown from fused clusters that the apparatus of *Icriodus expansus* comprised one pair of icriodontan elements and > 140 cones. He explained this apparently excessive number by suggesting that the cone elements of icriodontids could be compared with the denticles on the processes of ramiform elements in other apparatuses (Nicoll 1982, p. 208). The possibility of a functional analogy cannot be ruled out.

PALEOECOLOGY

It is evident that conodont element morphology and apparatus structure are related to some degree to environment. A review has been presented by Clark (1981), and several generalizations have been recorded by other authors. In particular, nearshore high-energy environments are widely characterized by apparatuses of robust elements with large basal cavities (e.g., Austin 1976; Le Fèvre et al. 1976; Moskalenko 1976; Aldridge and Jeppsson 1984). In deep-water, slope, or basinal environments conodonts may be absent or, in Silurian strata at least, represented by delicate elements, with coniforms particularly abundant (Aldridge and Jeppsson 1984). Conodonts as a group had a considerable ecologic range within the marine realm, but different biozonal schemes are commonly required for different biofacies. There is also some evidence that environmental changes may have resulted in phenotypic variation of elements. Barnes et al. (1979, p. 143) noted that in Ordovician species such variation may affect size, shape, amount of white matter, and the size of the basal cavity, but pointed out that "the exact environmental factors influencing such changes are uncertain."

Clark (1981, 1983) has also reviewed the diversification and extinction of conodonts. He identified two "major evolutionary bursts": Early–Middle Ordovician and Late Devonian. The early radiation coincides with the second diversification of metazoan taxa in the Early Phanerozoic, which affected taxa that "appear as a whole to have been more specialized with narrower ecologic requirements" (Sepkoski 1979, p. 246). There is no obvious morphologic change coincident with the Late Devonian radiation, and it is not reflected in the pattern of diversity of metazoan taxa as a whole, at least at the family level (Sepkoski 1979). Clark's study of conodont extinction (1983) demonstrates that their disappearance by the end of the Triassic is not predicted from survivorship plots (although the data base may be inadequate). It is not clear to what extent environmental changes from low-energy basinal to higher-energy shelf conditions at the time of their final disappearance reflect the cause(s) of their decline. The final extinction may have been the result of chance (*sensu* Raup 1978, 1981).

THE AFFINITIES OF CONODONTS

Briggs et al. (1983) discussed the affinities of conodonts in the light of the discovery of a soft-bodied fossil of the conodont animal and pointed out similarities to two major groups, the chordates and chaetognaths. They concluded, however, that the conodonts should be assigned to the phylum Conodonta, mainly on the basis of the unique nature of the elements, as evidenced by the lack of success in identifying homologous structures in any other known group during 125 years of research (see Robison 1981). The lack of knowledge of the three-dimensional appearance of the animal prevented an assessment of where its closest affinities lay. Primitive chordates are laterally flattened in life, chaetognaths dorsoventrally, and therefore an interpretation of the orientation of the fossil is all-important in distinguishing between features similar to both.

Thus the first Edinburgh specimen did not resolve the question of conodont affinities, and the debate continued. Gould (1983) agreed that the conodonts warrant separate phylum status. Bengtson (1983a), however, impressed by the need to postulate a different feeding position to that in which the apparatus is preserved, argued that the Edinburgh specimen provides evidence of chaetognath affinity:

It may be more than a coincidence that the new conodont animal conforms in so many aspects to chaetognath anatomy. . . . it is exactly what should be expected from the near identity in morphological and histological detail between Cambrian protoconodonts and the grasping spines of Recent chaetognaths (Szaniawski 1982) and the postulated evolutionary transition from protoconodonts through paraconodonts to euconodonts.

The structural similarity between conodont elements and vertebrate teeth, on the other hand, has received considerable attention (e.g., Jeppsson 1979; Chapter 16 by Dzik, this volume). Janvier (1983) considered that "c'est avec les vertébrés que cet animal présente le plus de ressemblances" with its chevron-shaped muscle blocks and tail fins. Rigby (1983) also postulated a vertebrate relationship and even considered the conodonts closest to gnathostomes. His conclusions were based on an interpretation of the elements as teeth and on identifying parallels between structures apparent in the head region of the Edinburgh conodont animal and cartilages in the vertebrate skull. If the affinities of the conodont animal lie with the vertebrates, however, it is more likely to be an agnathan than a gnathostome; the symmetry of the conodont apparatus does not correspond to that of the gnathostome jaw.

Preliminary observations on the three additional specimens of the conodont animal so far discovered suggest that the apparent segments are indeed V-shaped, opening posteriorly; they are thus similar to the myotomes in amphioxus and fish. As Briggs et al. (1983, p. 11) pointed out, "the apparent segmentation is difficult to reconcile with a chaetognath affinity." This point has also been conceded by those who advocate a chaetognath relationship: "the oblique (possibly V-shaped) regularly repeated structures in the posterior part of the trunk . . . cannot easily be matched with any known chaetognath character" (Bengtson 1983b, p. 15). The questions of orientation and the distribution of the fins have yet to be resolved but this may be possible with the aid of the new material. The conodonts may turn out to be a separate group of jawless craniates.

Acknowledgments

Financial support for our work on conodont paleobiology is provided by Research Grant GR/3/5105 awarded by the Natural Environment Research Council. We are grateful to Dr. R. S. Nicoll for access to his 1985 manuscript before publication and for useful discussions in correspondence. We also thank Dr. Nicoll, Dr. M. P. Smith, and Dr. E. N. K. Clarkson for constructive comments on our typescript. Photographic assistance was provided by Mr. D. Jones, cartography by Mrs. J. Wilkinson, and typing by Mrs. J. Angell.

REFERENCES

Aldridge, R. J. and Jeppsson, L. 1984. Ecological specialists among Silurian conodonts. *Spec. Pap. Palaeont.* **32**, 141–149.

Austin, R. L. 1976. Evidence from Great Britain and Ireland concerning West European Dinantian conodont paleoecology. *In* Barnes, C. R. (ed.). Conodont paleoecology. *Spec. Pap. Geol. Assoc. Canada* **15**, 201–224.

———— and Rhodes, F. H. T. 1969. A conodont assemblage from the Carboniferous of the Avon Gorge, Bristol. *Palaeontology* **12**, 400–405.

Avcin, M. J. 1974. Des Moinesian conodont assemblages from the Illinois Basin. Unpubl. Ph.D. Dissert., Univ. of Illinois at Urbana-Champaign, 152 pp.

Barnes, C. R. and Fåhraeus, L. E. 1975. Provinces, communities, and the proposed nekto-benthic habit of Ordovician conodontophorids. *Lethaia* **8**, 133–149.

————, Kennedy, D. J., McCracken, A. D., Nowlan, G. S., and Tarrant, G. A. 1979. The structure and evolution of Ordovician conodont apparatuses. *Lethaia* **12**, 125–151.

————, Sass, D. B., and Poplawski, M. L. S. 1973. Conodont ultrastructure: the family Panderodontidae. *R. Ont. Mus. Life Sci. Contrib.* **90**, 36 pp.

Behnken, F. H. 1975. Leonardian and Guadalupian (Permian) conodont biostratigraphy in

Western and Southwestern United States. *J. Paleont.* **49**, 284-315.

Bengtson, S. 1976. The structure of some Middle Cambrian conodonts, and the early evolution of conodont structure and function. *Lethaia* **9**, 185-206.

———. 1980. Conodonts: the need for a functional model. *Lethaia* **13**, 320.

———. 1983a. A functional model for the conodont apparatus. *Lethaia* **16**, 38.

———. 1983b. The early history of the Conodonta. *Fossils and Strata* **15**, 5-19.

Bergström, S. M. 1983. Biogeography, evolutionary relationships, and biostratigraphic significance of Ordovician platform conodonts. *Fossils and Strata* **15**, 35-58.

——— and Sweet, W. C. 1966. Conodonts from the Lexington Limestone (Middle Ordovician) of Kentucky and its lateral equivalents in Ohio and Indiana. *Bull. Am. Paleont.* **50**, 269-441.

Briggs, D. E. G., Clarkson, E. N. K., and Aldridge, R. J. 1983. The conodont animal. *Lethaia* **16**, 1-14.

Carls, P. 1977. Could conodonts be lost and replaced? *N. Jb. Geol. Paläont. Abh.* **155**, 18-64.

Clark, D. L. 1981. Biological considerations and extinction: paleoecology. *In* Robison, R. A. (ed.). *Treatise on Invertebrate Paleontology.* Part W, suppl. 2, *Conodonta.* Geological Society of America and Univ. of Kansas Press, Lawrence, pp. W83-W91.

———. 1983. Extinction of conodonts. *J. Paleont.* **57**, 652-661.

Conway Morris, S. 1976. A new Cambrian lophophorate from the Burgess Shale of British Columbia. *Palaeontology* **19**, 199-222.

Cooper, B. J. 1975. Multielement conodonts from the Brassfield Limestone (Silurian) of Southern Ohio. *J. Paleont.* **49**, 984-1008.

Du Bois, E. P. 1943. Evidence on the nature of conodonts. *J. Paleont.* **17**, 155-159.

Dzik, J. 1980. Isolated mandibles of Early Palaeozoic phyllocarid Crustacea. *N. Jb. Geol. Paläont. Mh.* **1980**, 87-106.

Fåhraeus, L. E. 1983. Phylum Conodonta Pander, 1856 and nomenclatural priority. *Syst. Zool.* **32**, 455-459.

Gould, S. J. 1983. Nature's great era of experiments. *Nat. Hist.* **7/83**, 12-21.

Hass, W. H. 1962. Conodonts. *In* Moore, R. C. (ed.). *Treatise on Invertebrate Paleontology.* Part W, *Miscellanea.* Geological Society of America and Univ. of Kansas Press, Lawrence, pp. W3-W69.

Hitchings, V. H. and Ramsay, A. T. S. 1978. Conodont assemblages: a new functional model. *Palaeogeogr., Palaeoclimatol., Palaeoecol.* **24**, 137-149.

Janvier, P. 1983. L'"animal-conodonte" enfin demasqué? *La recherche* **14**(145), 832-833.

Jeppsson, L. 1971. Element arrangement in conodont apparatuses of *Hindeodella* type and in similar forms. *Lethaia* **4**, 101-123.

———. 1972. Some Silurian conodont apparatuses and possible conodont dimorphism. *Geol. et Palaeont* **6**, 51-69.

———. 1979. Conodont element function. *Lethaia* **12**, 153-171.

———. 1980. Function of the conodont elements. *Lethaia* **13**, 228.

———. 1982. Are conodonts animals, or . . . ? *In* Jeppsson, L. and Löfgren, A. (eds.). Third European conodont symposium (ECOS III) Abstracts. *Publ. Inst. Mineral. Paleont. Quat. Geol. Lund Univ.* **238**, 24-25.

Jørgensen, C. B. 1966. *Biology of Suspension Feeding.* Pergamon, London. 357 pp.

Klapper, G. and Philip, G. M. 1971. Devonian conodont apparatuses and their vicarious skeletal elements. *Lethaia* **4**, 429-452.

Landing, E. 1977. *"Prooneotodus" tenuis* (Müller, 1959) apparatuses from the Taconic allochthon, eastern New York; construction, taphonomy and the protoconodont "supertooth" model. *J. Paleont.* **51**, 1072-1084.

Lange, F. G. 1968. Conodonten-Gruppenfunde aus Kalken des tieferen Oberdevon. *Geol. et Palaeont.* **2**, 37-57.

Le Fèvre, J., Barnes, C. R., and Tixier, M. 1976. Paleoecology of late Ordovician and early Silurian conodontophorids, Hudson Bay Basin. *In* Barnes, C. R. (ed.). Conodont paleoecology. *Spec. Pap. Geol. Assoc. Canada* **15**, 69-89.

Lindström, M. 1964. *Conodonts.* Elsevier, Amsterdam. 196 pp.

———. 1973. On the affinities of conodonts. *In* Rhodes, F. H. T. (ed.). Conodont paleozoology. *Geol. Soc. Am. Spec. Pap.* **141**, 85-102.

———. 1974. The conodont apparatus as a food-gathering mechanism. *Palaeontology* **17**, 729-744.

——— and Ziegler, W. 1971. Feinstrukturelle Untersuchungen an Conodonten. 1. Die Überfamilie Panderodontacea. *Geol. et Palaeont.* **5**, 9-33.

——— and ———. 1981. Surface micro-ornamentation and observations on internal composition. *In* Robison, R. A. (ed.) *Treatise on Invertebrate Paleontology.* Part W, suppl. 2, *Conodonta.* Geological Society of America and Univ. of Kansas Press, Lawrence, pp. W41-W51.

Manton, S. M. 1977. *The Arthropoda, Habits, Functional Morphology and Evolution.* Clarendon Press, Oxford, 527 pp.

Mashkova, T. V. 1972. *Ozarkodina steinhornensis* (Ziegler) apparatus, its conodonts and biozone. *Geol. et Palaeont.* SB **1**, 81-90.

Merrill, G. K. and Merrill, S. M. 1974. Pennsylvanian non-platform conodonts, IIa: the dimorphic apparatus of *Idioprioniodus*. *Geol. et Palaeont.* **8**, 119–130.

Moskalenko, T. A. 1976. Environmental effects on the distribution of Ordovician conodonts of the western Siberian Platform. *In* Barnes, C. R. (ed.). Conodont paleoecology. *Spec. Pap. Geol. Assoc. Canada* **15**, 59–67.

Müller, K. J. 1959. Kambrische Conodonten. *Z. Dtsch. Geol. Ges.* **111**, 434–485.

———— and Andres, D. 1976. Eine Conodontengruppe von *Prooneotodus tenuis* (Müller, 1959) in natürlichem Zusammenhang aus dem Oberen Kambrium vom Schweden. *Paläont. Z.* **50**, 193–200.

———— and Nogami, Y. 1971. Über den Feinbau der Conodonten. *Mem. Fac. Sci. Kyoto Univ. Ser: Geol. Mineral.* **38**, 1–88.

Nicoll, R. S. 1977. Conodont apparatuses in an Upper Devonian palaeoniscid fish from the Canning Basin, Western Australia. *BMR J. Aust. Geol. Geophys.* **2**, 217–228.

————. 1982. Multielement composition of the conodont *Icriodus expansus* Branson & Mehl from the Upper Devonian of the Canning Basin, Western Australia. *BMR J. Aust. Geol. Geophys.* **7**, 197–213.

————. 1985. Multielement composition of the conodont species *Polygnathus xylus xylus* Stauffer, 1940 and *Ozarkodina brevis* (Bischoff & Ziegler, 1957) from the Upper Devonian of the Canning Basin, Western Australia. *BMR J. Aust. Geol. Geophys.* **9**, 133–147.

Norby, R. D. 1976. Conodont apparatuses from Chesterian (Mississippian) strata of Montana and Illinois. Unpubl. Ph.D. Dissert. Univ. of Illinois at Urbana-Champaign, 203 pp.

Nowlan, G. S. 1979. Fused clusters of the conodont genus *Belodina* Ethington from the Thumb Mountain Formation (Ordovician), Ellesmere Island, District of Franklin. *Curr. Res. Part A, Geol. Surv. Can. Pap.* **79-1A** 213–218.

Pander, C. H. 1856. *Monographie der fossilen Fische des Silurischen Systems der Russisch-Baltischen Gouvernements.* St. Petersburg, 91 pp.

Pietzner, H., Vahl, J., Werner, H., and Ziegler, W. 1968. Zur chemischen Zusammensetzung und Mikromorphologie der Conodonten. *Palaeontographica* **128**, 115–152.

Pollock, C. A. 1969. Fused Silurian conodont clusters from Indiana. *J. Paleont.* **43**, 929–935.

Priddle, J. 1974. The function of conodonts. *Geol. Mag.* **111**, 255–257.

Ramovš, A. 1978. Mitteltriassische Conodontenclusters in Slowenien, NW Jugoslawien. *Paläont. Z.* **52**, 129–137.

Raup, D. M. 1978. Approaches to the extinction problem. *J. Paleont.* **52**, 517–523.

————. 1981. Extinction: bad genes or bad luck? *Acta Geol. Hispanica* **16**, 25–33.

Repetski, J. E. 1980. Early Ordovician fused conodont clusters from the western United States. *In* Schönlaub, H. P. (ed.). Second European conodont symposium (ECOS II), Guidebook, Abstracts. *Abh. Geol. Bundesanst.* **35**, 207–209.

———— and Szaniawski, H. 1981. Paleobiologic interpretation of Cambrian and earliest Ordovician conodont natural assemblages. *U.S. Geol. Surv. Open-File Rep.* **81-743**, 169–172.

Rexroad, C. B. and Nicoll, R. S. 1964. A Silurian conodont with tetanus? *J. Paleont.* **38**, 771–773.

Rhodes, F. H. T. 1952. A classification of Pennsylvanian conodont assemblages. *J. Paleont.* **26**, 886–901.

————. 1962. Recognition, interpretation, and taxonomic position of conodont assemblages. *In* Moore, R. C. (ed.). *Treatise on Invertebrate Paleontology.* Part W, *Miscellanea.* Geological Society of America and Univ. of Kansas Press, Lawrence, pp. W70–W83.

———— and Austin, R. L. 1981. Natural assemblages of elements: interpretation and taxonomy. *In* Robison, R. A. (ed.) *Treatise on Invertebrate Paleontology.* Part W, suppl. 2, *Conodonta.* Geological Society of America and Univ. of Kansas Press, Lawrence, pp. W68–W78.

Rigby, J. K., Jr. 1983. Conodonts and the early evolution of vertebrates. *Geol. Soc. Am. Abstr. Progr.* **15**, 671.

Robison, R. A. (ed.) 1981. *Treatise on Invertebrate Paleontology.* Part W, suppl. 2, *Conodonta.* Geological Society of America and Univ. of Kansas Press, Lawrence, 202 pp.

Schäfer, W. 1972. *Ecology and Paleoecology of Marine Environments.* Univ. of Chicago Press, Chicago, 568 pp.

Schmidt, H. 1934. Conodonten-Funde in ursprunglichen Zusammenhang. *Paläont. Z.* **16**, 76–85.

———— and Müller, K. J. 1964. Weitere Funde von Conodonten-Gruppen aus dem oberen Karbon des Sauerlandes. *Paläont. Z.* **38**, 105–135.

Scott, H. W. 1934. The zoological relationships of the conodonts. *J. Paleont.* **8**, 448–455.

————. 1942. Conodont assemblages from the Heath formation, Montana. *J. Paleont.* **16**, 293–300.

Sepkoski, J. J., Jr. 1979. A kinetic model of Phanerozoic taxonomic diversity II. Early Phanerozoic families and multiple equilibria. *Paleobiology* **5**, 222–251.

Smith, M. P. 1984. Fused euconodont clusters from the Ordovician of East Greenland. *Geol. Soc. Am. Abstr. Prog.* **16**, 197.

Sweet, W. C. 1970. Uppermost Permian and Lower Triassic conodonts of the Salt Range and Trans-Indus Ranges, West Pakistan. *In* Kummel, B. and Teichert, C. (eds.). *Stratigraphic*

Boundary Problems, Permian and Triassic of West Pakistan. Univ. of Kansas Dept. of Geol. spec. publ. 4, pp. 207–275.

————. 1981. Macromorphology of elements and apparatuses. *In* Robison, R. A. (ed.). *Treatise on Invertebrate Paleontology.* Part W, suppl. 2, *Conodonta.* Geological Society of America and Univ. of Kansas Press, Lawrence, pp. W5–W20.

Szaniawski, H. 1982. Chaetognath grasping spines recognized among Cambrian protoconodonts. *J. Paleont.* **56,** 806–810.

————. 1983. Structure of protoconodont elements. *Fossils and Strata* **15,** 21–27.

Walliser, O. H. 1964. Conodonten des Silurs. *Abh. Hess. Landesamt Bodenforsch.* **41,** 1–106.

Webers, G. F. 1966. The Middle and Upper Ordovician conodont faunas of Minnesota. *Minnesota Geological Survey spec. publ.* **4,** 123 pp.

16. CHORDATE AFFINITIES OF THE CONODONTS

JERZY DZIK

Until recently virtually nothing was known about the morphology of conodont animals, and zoologic affinities of the group are still regarded as obscure. The present contribution evaluates the hypothesis that agnathan vertebrates are the closest relatives of conodonts. The microstructure of element of conodont apparatuses, their organization into skeletal structures, and the inferred morphology of the soft body parts are compared to the anatomy of early vertebrates. Other problematic phosphatic dermal sclerites are also reviewed.

A complete phosphatic element of the conodont apparatus consists of two distinct parts. The oral part (crown of Nicoll 1977) is developed by centrifugal external accretion. It is rather compact in structure, and its surface is smooth or regularly ornamented. The basal part (basal filling), which is developed by inner accretion, varies greatly in microstructure and morphology. The mode of secretion of the crown tissue is here considered diagnostic for the group; fossil sclerites secreted differently are not considered to be conodont elements (see Bengtson 1976, 1983a, for opposing view).

CROWN TISSUE

The crown of conodont elements is composed of relatively large, elongated crystallites of apatite. They are parallel to each other, and normal to the element surface (*Fryxellodontus, Cordylodus*) or, most commonly, parallel to the cusp or denticle axis. Frequent disconformities in distribution of lamellae marking subsequent stages in secretion of the crown clearly indicate that the surface of at least platform elements was periodically worn out (Müller and Nogami 1971; Müller 1981). Whether this was caused by mechanical abrasion or chemical resorption, cannot be determined. The shape of apatite crystallites and the pattern of their distribution persisted from the Late Cambrian *Teridontus nakamurai* (Nogami 1967) (see Landing et al. 1980) to the extinction of conodonts (Pietzner et al. 1968; Lindström and Ziegler 1971; Barnes et al. 1973; Müller 1981; Bitter and Merrill 1983). The internal surface of the crown, visible in the basal cavity, always shows distinct growth lines (Pietzner et al. 1968; Lindström and Ziegler 1981).

A basic problem is to explain the function of the conodont elements by allowing for the presence of soft tissue above the crown tissue. Several solutions have been proposed. It is suggested that conodont elements (1) formed a skeleton of the ciliary apparatus (Lindström 1973, 1974; Conway Morris 1976, 1980; Nicoll 1977), (2) grew within epithelial pockets and were protruded while functional (Bengtson 1976; Carls 1977), (3) were covered by a hard horny tissue secreted above the epithelium (Priddle 1974), or (4) originated in a manner analogous to vertebrate teeth and were replaced in ontogeny (Carls 1977; Jeppsson 1980). None of these interpretations is convincing. Bengtson's pocket hypothesis can hardly explain healed fractures of conodont cusps (see Müller and Nogami 1971) and brushlike appearance of platform elements not uncommon in the Devonian (Müller 1981). Carls' concept of ontogenetic replacement implies the occurrence of smaller, regenerating elements in at least some natural assemblages. However, nothing like this has ever been reported. The high rate of evolution and repeated development of platform elements, with their molarlike appearance, are evidence against the filtratory apparatus concept (Dzik 1976; Jeppsson 1979a). Neither were scleritized organic caps covering conodont elements recognized in articulated apparatuses, nor were phosphatic nuclei identified in teeth of Paleozoic hagfishes, to support Priddle's model of the conodont element.

BASAL FILLING

Unlike the crown, apatite crystallites in the basal filling are roughly isometric and usually do not show any preferred orientation (Pietzner et al. 1968; Lindström and Ziegler 1981). The boundary between the crown and basal filling tissues is always sharp and distinct (Fig. 1).

Most conodont elements lack any remnants of the basal filling. Sometimes, however, a black, carbonized spongy tissue penetrated by meandering canals can be found inside the basal cavity. It may represent shrunken remnants of unmineralized organic matrix of the basal filling tissue (Fig. 1B). In some other cases the basal filling tissue is well mineralized, but its morphology suggests that partial shrinkage occurred during the early diagenesis (Lindström and Ziegler 1971, 1981). Irregular canals and tubuli penetrate this kind of tissue (Barnes et al. 1973). When the tissue underwent more advanced mineralization while alive, the

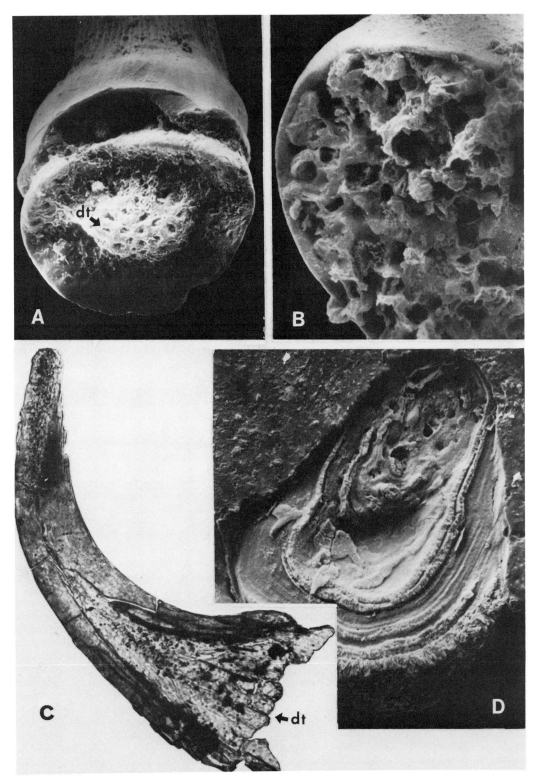

Fig. 1. Microstructure of the basal filling in elements of the conodont *Semiacontiodus cornuformis* (Sergeeva 1963) from the Llanvirnian of the Baltic area (glacial erratic boulders). A. Basal view, openings of dentine tubuli (dt) visible, sample E-079. ×300. B. Basal view of a black, unmineralized basal fillilng, sample E-113. ×500. C. Medial thin section, dentine tubuli (dt) in longitudinal section, sample E-079. ×200. D. Etched oblique section through conspicuous callus of the basal filling, sample E-079. ×300.

tubuli appear to be confined to the axial parts of the element and they open basally (Fig. 1A, C). Most commonly, the basal filling tissue shows lamellar organization. Except for growth lines, no other internal structures can be recognized (Müller and Nogami 1971; Lindström and Ziegler 1981). In the histogeny of some primitive Early Ordovician conodonts, such a laminated basal filling tissue followed the spongy tissue (Fig. 1D). Most post-Ordovician conodonts had compact basal filling tissue only. This variation in structure of the basal filling tissue seems to represent consecutive stages in its mineralization.

Generally, the basal filling tissue contributes only a small portion of the total mass. There is only a single group of Ordovician conodonts with the basal filling tissue dominant. *Archaeognathus primus* Cullison 1938 from the Dutchtown Formation (Late Llanvirnian?) of Missouri is the best known representative of this group (Cullison 1938), which is widespread and abundant in North America (Mosher and Bodenstein 1969) and Siberia (Moskalenko 1972, 1976; Barskov et al. 1982). The crown tissue is restricted to the working edge of the element, and it forms only a thin crown, frequently split into coniform denticles. The basal filling reveals a complex internal structure with wide, branching canals (Barskov et al. 1982). Although these so-called coleodontids still await detailed taxonomic description and reconstruction, their conodont nature seems to be well established.

CONODONT APPARATUSES

The original organization of apparatuses preserved on rock bedding plane (natural assemblages) generally is obscured by sediment compression during lithification. This has prompted many contradictory reconstructions (Jeppsson 1971; Collinson et al. 1972; Lindström 1973, 1974; Nicoll 1977; Hitchings and Ramsay 1978). A fortunate exception in the mode of preservation is represented by the natural assemblage of *Ozarkodina steinhornensis* (Ziegler 1956) described by Mashkova (1972) from Early Devonian limestones. Its elements collapsed obliquely from the sides and are preserved with some relief. As judged after Mashkova's and other natural assemblages (see also Pollock 1969), conodont elements were arranged along both sides of the medial plane, with their processes oriented dorsally and ventrally and the cusps directed toward each other in every pair (Dzik 1976). Two pairs of the elements (*sp* and *oz*) were somehow separated from the remainder of the apparatus and possibly obliquely oriented (Fig. 2).

This general pattern has been confirmed by subsequent discovery of a cluster of seven coniform elements representing one-half of the apparatus of a Late Ordovician species (Aldridge 1982). The clusters illustrated by Briggs et al. (1983) indicate that, despite different arrangement of processes, the cusps of all the elements point in the same direction. Because the *sp* element, at one end of the apparatus, usually is the most robust one (frequently with a platform), while the *ne* element, at the opposite end, usually bears a sharp, swordlike cusp, they were inferred to represent the posterior and anterior ends of the apparatus, respectively (Dzik 1976). A functional gradient along the apparatus, from the grasping function of the *ne* element to the masticatory function of the *sp* element, was proposed. This has been corroborated by the discovery of soft body parts in the conodont animal (Briggs et al. 1983). The pattern of conodont apparatus thus is well established (Fig. 2), although there still are some doubts regarding the position of symmetric (*tr*) elements inside or outside the apparatus. In Mashkova's (1972) apparatus, the *tr* element (identified by Jeppsson 1979b) has the same orientation as the others, and even though its counterpart is not preserved, there is no evidence for its medial positon as proposed by Jeppsson (1971, 1979b). Jeppsson cited the commonness of numerical underrepresentation of *tr* elements to support his reconstruction. Such underrepresentation, however, may be caused by hydrodynamic factors alone, for *tr* elements usually are the most ramified and gracile in the apparatus.

Apparatuses of ramiform conodont elements can be easily homologized in the Early Ordovician (Dzik 1983) through Late Triassic (Sweet 1981). They have 7 pairs of elements each, except for some Late Triassic apparatuses with possibly 10 pairs of elements each [5 pairs of hindeodelliform elements; Ramovš (1978), Dzik and Trammer (1980)] and Devonian *Icriodus* apparatuses, which bear hundreds of coniform elements (Nicoll 1982) and which perhaps represent disintegrated denticles of originally ramiform elements.

The closest functional analog to apparatuses of coniform conodont elements is the grasping apparatus of the Chaetognatha (Fig. 3A; Szaniawski 1982), while apparatuses of ramiform elements have their closest analog in the buccal apparatus of hagfish (Fig. 3B; Priddle 1974). The buccal apparatus of *Myxine* resembles conodont apparatuses (especially *Archaeognathus*) in being bilaterally symmetric and composed of transverse rows of denticles (horny instead of phosphatic) and also in working by scissorlike closing of its lateral parts (Dawson 1963).

CONODONT SOFT BODY

The first fossils with preserved soft-part anatomy that were claimed to represent the conodont animal include *Lochriea wellsi* Melton and Scott 1973 and other "conodontochordates" from the Carboniferous Bear Gulch Limestone of Montana

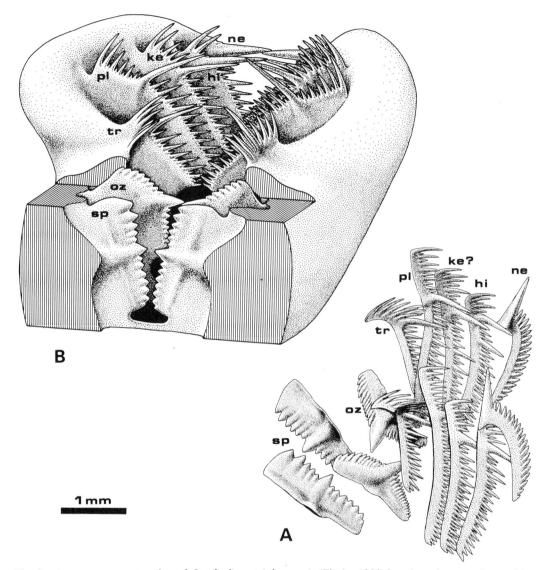

Fig. 2. Apparatus reconstruction of *Ozarkodina steinhornensis* (Ziegler 1956) based on the natural assemblage described by Mashkova (1972) from the Early Devonian of Tadjikistan.   A. Reconstructed complete set of elements as found on the bedding plane (cf. Mashkova 1972, pl. 1). Identification of the left *tr* element according to Jeppsson (1979b).   B. Proposed spatial arrangement of elements in the oral part of the conodont animal (modified from Dzik 1976, fig. 10a), venter up. Note that frequency distribution data suggest presence of unpaired *tr* element.

(Melton and Scott 1973). In having a hypocercal (?) caudal fin with delicate rays, they resemble the Late Devonian, possibly planktic anaspide *Endeiolepsis;* on the other hand, the general shape of the body and the concentration of visceral organs into a globular dense structure (ferrodiscus) indicate the Recent salpae as at least the ecologic analog of these animals. "Conodontochordates" are now generally believed to be conodont-eaters rather than true conodont animals (Lindström 1973, 1974; Conway Morris 1976).

The second fossil with preserved soft-body

parts proposed to represent the conodont animal was the Middle Cambrian *Odontogriphus omalus* Conway Morris 1976 from the Burgess Shale of British Columbia (Conway Morris 1976; Chapter 13 by Briggs and Conway Morris, this volume). It is poorly preserved but it shows a looplike organ with coniform teeth in the oral region, two lateral organs with complex internal organization on both sides of the head, a longitudinal digestive (?) tract, and metameric bands in the trunk region. Contrary to the reconstruction by Conway Morris (1976), metamery is invisible at the margin of the

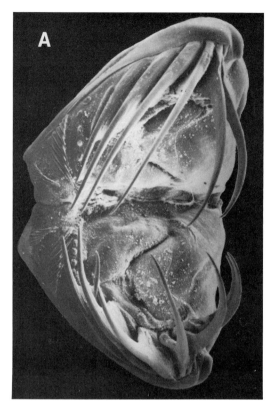

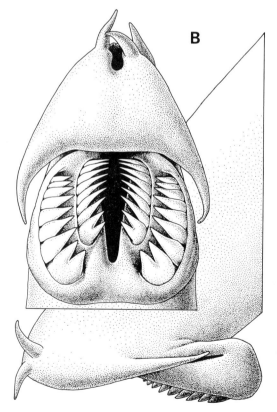

trunk, while transverse bands end laterally in dark round spots arranged in rows along both sides of the central, metameric area of the trunk. The caudal area is flexed and partially hidden under the trunk, with no evidence of metamerization.

Leaving aside the problem of the unknown original mineral composition of the denticles of *Odontogriphus,* chordate affinities of this animal seem more plausible than its lophophorate nature proposed originally by Conway Morris (1976, 1980). All Recent and fossil lophophorates are sedentary organisms with a relatively large tentacular apparatus that never has an internal skeleton comparable to the conodont apparatus. By contrast, *Odontogriphus* was free-living and not a suspension feeder. Its body shape seems to have resembled early agnathans, and all its preserved internal structures can be at least tentatively homologized with heterostracans. Metameric bands ending in black spots occur in *Lanarkia* and *Turinia* (Turner 1982), where they are interpreted as remnants of branchial sacs and their openings. The number of proposed branchial sacs in *Odontogriphus* (at least 25) exceeds that in the Heterostraci but still is significantly smaller than in *Amphioxus.* The lateral organs of *Odontogriphus* may represent something related to the osteostracan lateral line organs. Even if its mouth really bore a tentacular organ as proposed by Conway Morris (1976), similar tentacles occur in *Amphioxus. Odontogriphus* thus may be a Cambrian naked relative of heterostracans. Its relationship to conodonts cannot be seriously considered unless the microstructure of the oral denticles is known.

The most convincing discovery of a fossil conodont animal with preserved soft-part remnants is *Clydagnathus?* cf. *cavusformis* Rhodes et al. 1969 from the Early Carboniferous of Scotland (Briggs et al. 1983; Chapter 15 by Aldridge and Briggs, this volume). This fossil indicates that the conodont animal was lamprey shaped, probably with a hypocercal caudal fin supported by delicate rays. Like cyclostome agnathans, known also from slightly younger Carboniferous strata (Bardack and Richardson 1977), the conodont animal had V-shaped myomeres. *Clydagnathus?* cf. *cavusformis* has an almost complete apparatus preserved in the oral part of the body. Elements of the apparatus lie transversely to the longitudinal axis of the animal's body as in Mashkova's natural assemblage (1972). The pairs of *sp* and *oz* elements were possibly hidden in the throat, while other elements supported two oval lobes (Fig. 4; Briggs et al. 1983).

←

Fig. 3. Closest Recent analogs to conodont apparatuses. A. Grasping apparatus of the chaetognath *Eukrohnia hamata* (Möbius 1875). ×25. B. Jaw apparatus of the hagfish *Myxine* sp. ×5.

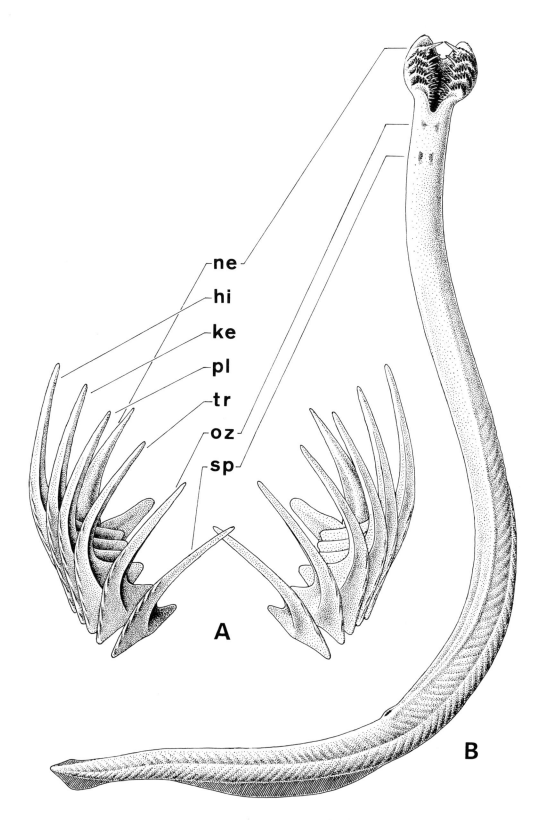

ne

hi

ke

pl

tr

oz

sp

A

B

Fig. 4. A. Apparatus of the conodont *Besselodus arcticus* Aldridge 1982 from the Late Ordovician of Greenland, reconstructed in life position. Tentative homology of the elements in Jeppsson's notation (1971). Based on data from Aldridge (1982). B. Reconstruction of the body of *Clydagnathus?* cf. *cavusformis* Rhodes et al. 1969 from the Early Carboniferous of Scotland. Based on Briggs et al. (1983).

HYPOTHESIS OF CHAETOGNATH
AFFINITIES OF THE CONODONTS

Morphologic (Figs. 3 and 4A) and proposed functional (Jeppsson 1979a) similarities between conodont apparatuses and grasping apparatuses of the Chaetognatha are striking (Szaniawski 1982). The two animal groups show much similarity in body shape and size (Briggs et al. 1983), and probably occupied similar niches in Paleozoic ecosystems (Seddon and Sweet 1971). As shown by Szaniawski (1982), phosphatized chaetognath grasping spines frequently occur in Cambrian rocks. Thus, chaetognaths stratigraphically precede the oldest true conodonts.

Despite these similarities, however, conodont elements and chaetognath grasping spines developed in basically different ways. The crown tissue of conodont elements was secreted from the outside of the basal cavity, while the tissue of the grasping spine is secreted from the inside. Bengtson (1976) proposed a homology between the conodont basal filling and the tissue of the grasping spine. According to this concept, the grasping spine has become embedded into an epithelial pocket in the course of evolution from chaetognaths to conodonts. The epithelium of the pocket subsequently secreted the crown tissue. The Late Cambrian to Middle Ordovician westergaardodinids ("paraconodonts") have been proposed to represent an intermediate stage of the evolution (see also Andres 1981).

The hypothesis deriving conodont elements from chaetognath grasping spines encounters serious problems, however. First of all, the accidental occurrence and originally late mineralization of the basal filling tissue contradict its proposed primitivism relative to the crown tissue. Among primitive panderodontid and protopanderodontid conodonts, the basal filling tissue was not laminar at the early stages of either phylogeny or histogeny. Moreover, secretion did not take place within the basal cavity in the early westergaardodinid *Furnishina.* Nothing could be homologized with the basal filling. Elements of *Furnishina* developed centrifugally from both the outside and the inside at the very beginning of their histogeny, but in contrast to true conodonts, there is no sharp boundary between the crown and the basal filling tissues. In spite of a similarity in shape, there is a major difference in histogeny between the conodonts, westergaardodinids, and chaetognaths, and there is no evidence for transitions between these groups.

The crucial counterargument to the hypothesis of chaetognath affinities of conodonts is the presence of V-shaped muscular somites in the Carboniferous conodont animal (Briggs et al. 1983; Chapter 15 by Aldridge and Briggs, this volume). This is absolutely incompatible with the mode of locomotion and the body plan of the Chaetognatha.

Moreover, the asymmetric fin of the conodont animal suggests that it worked laterally instead of vertically. These two features ultimately reject the hypothesis of conodont relationships to the Chaetognatha.

HYPOTHESIS OF CHORDATE AFFINITIES
OF THE CONODONTS

Neither the presence of an internal skeleton nor its phosphatic composition defines a chordate. The peculiarity of primitive chordate dermal sclerites consists in their secretion within epithelial sacs from the outside by epithelial ameloblasts and from the inside by mesodermal odontoblasts. Calcium and phosphate ions in chordate tissues are persistently in supersaturated solution, but they can precipitate as hydroxyapatite only on matrices that provide nucleation centers. β-proteins secreted by ameloblasts serve the function of a matrix for precipitation of hydroxyapatite in the enamel (Little 1973). In therian mammals, hydroxyapatite crystalizes into relatively large crystallites organized more-or-less regularly into prisms (Poole 1971). Teeth of lower tetrapods and crossopterygian fish usually have an enamel cover rather homogeneous in structure but with distinct increment lines (Poole 1971; Meinke and Thomson 1983). Other fish teeth are capped with enameloid (or durodentine) tissue that originates by mineralization of mixed collagen (secreted by odontoblasts) and proteins. The enameloid differentiates and begins to calcify before the dentine formation starts; at that time, collagen fibers revert to a labile form and hence are never incorporated into the mineralized tissue (Poole 1971; Shellis and Miles 1976).

The mineralization of dentine occurs on a collagen matrix secreted by odontoblasts. Odontoblasts are not incorporated into mineralized dentine, but their processes (Tomes' fibers) penetrate dentine layers and frequently also the enamel (Schmidt and Keil 1971). The presence of Tomes' fibers in the enamel (and also in the enameloid) seems to be related to the late calcification of the matrix and usually is confined to its deeper, earliest part (Kerr 1955; Schmidt and Keil 1971; Shellis and Miles, 1976). The boundary between the parts of the tooth secreted by ameloblasts and odontoblasts is usually distinct. Sometimes, however—perhaps when the mineralization is very delayed—there is a complete transition between the two tissues. This is why most authors consider enameloid caps to be built with a special kind of dentine (Ørvig 1967; Schmidt and Keil 1971; Poole 1971; Taylor and Adamec 1977). Shellis and Miles (1976), however, demonstrated that at least the external part of the enameloid cap is homologous to the enamel of mammals.

Fossil agnathans with calcified dermal sclerites display diverse histogenetic methods of forming of sclerite caps. In *Astraspis desiderata* Walcott 1892, one of the oldest histologically studied agnathans, the crowns of the dermal tubercles are covered with a tissue that is almost homogeneous and does not contain any fibers or canals (Ørvig 1958; Denison 1967; Halstead 1969). It consists of small, elongated, randomly oriented apatite crystallites (Fig. 5B). The cap is distinctly separated from the underlying dentine (Fig. 5A), which is perforated by numerous tubuli (Halstead 1969) and consists of granular apatite (Fig. 5C). Growth lines visible in the dentine frequently are not parallel to the boundary between the tissues (Denison 1967), suggesting that the dentine grew out from the center of the cavity of the cap after its completion. *Astraspis* ranges from the Early (Lehtola 1973) to Late Caradocian (Denison 1967).

The tubercles of *Pycnaspis splendens* Ørvig 1958, which temporally but not spatially co-occurred with *Astraspis,* differ from those of *Astraspis* in having their caps penetrated by dentine tubuli (Ørvig 1958), but the boundary between the cap and the underlying dentine still is sharp. Even

more dentinelike are the caps of tubercles in another Late Ordovician agnathan, *Eriptychius americanus* Walcott 1892. The boundary separating the cap disappeared almost completely, although a thin enamel-like cover still can be recognized in some tubercles (Denison 1967; Halstead 1969). This seems to be typical of post-Ordovician agnathans. An external nondentine cap of the tubercle can be recognized only in polarized light (Schmidt and Keil 1971). In supposedly primitive Silurian thelodonts, scales are almost entirely penetrated by dentine tubuli (Fig. 6; Gross 1967), and the boundary between the parts that originated ectodermally and mesodermally is visible only at its external surface (Fig. 6A). Mineralization probably took place only after the formation of matrix of the whole scale, and a subsequent epithelial increment was insignificant.

Stratigraphic data thus suggest that, contrary to Ørvig (1967), the distinction between enamel and dentine was primitive in vertebrates, while aspidin and true bone were later derivatives of the dentine (Halstead 1969; Meinke and Thomson

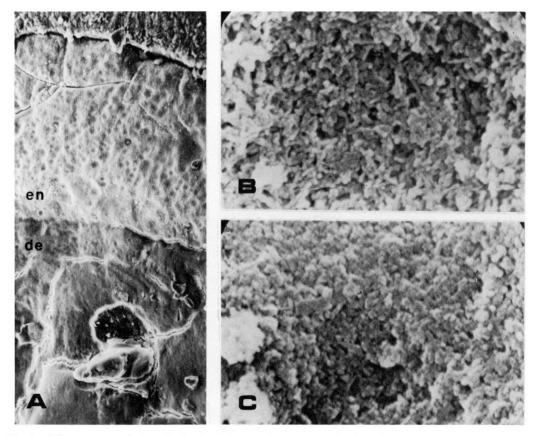

Fig. 5. Microstructure of a dental tubercle of the agnathan *Astraspis desiderata* Walcott 1892 from the Late Caradocian Harding Sandstone, Colorado. A. Etched vertical (not exactly medial) section through the cap. Note distinct boundary between the enamel(oid) (en) and the dentine (de). ×500. B. Enamel. ×24,000. C. Dentine. × 24,000.

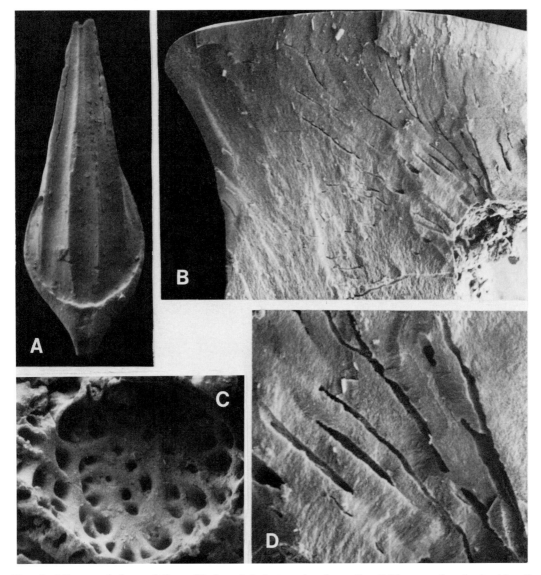

Fig. 6. Micromorphology of Late Silurian thelodont scales from the Baltic area (erratic boulder E-279). A. *Logania cuneata* (Gross 1947), scale in upside view. Note distinct boundary between smooth area formed by the epithelium and rough basal part of mesodermal origin (cf. Fig. 1A). ×100. B. Longitudinally broken scale of *Thelodus parvidens* Sgassiz 1844, with well-preserved branching dentine tubuli (cf. Fig. 1C). × 300. C. Pulp cavity of *Thelodus trilobatus* Hoppe 1931, with openings of bunches of dentine tubuli (cf. Fig. 1A). ×750. D. Same as in B. ×1000.

1983). Even more primitive sclerites might then be expected to have a homolog of the enamel as the dominant tissue.

A homology between the conodont crown tissue and the enamel and between the basal filling and the dentine was proposed by Schmidt and Müller (1964; see also Scott 1969; Dzik 1976). A sharp structural distinction between the crown tissue and the basal filling seems to be primitive in the conodonts (but see Bengtson 1976, 1983b). In most conodonts, sporadic mineralization of the basal filling tissue occurs secondarily, usually in large elements probably belonging to older adults. The main argument against agnathan affinities of the conodonts used to refer to the lack of dentine tubuli, spaces after osteocytes, or collagen fibers in tissues of the conodont element (Gross 1954; Denison 1967). As shown above, this argument is no longer valid. Incorporation of collagen fibers, cell processes, or whole cells is not a necessary feature of these mineralization mechanisms in hard tissues of the vertebrates. Furthermore, the basal filling tissue of early conodonts seems to have been penetrated by processes of mesodermal cells.

As shown by the presence of distinct growth

lines in the conodont crown tissue, mineralization of organic matrix occurred immediately after or even during its secretion. This makes the crown tissue different from durodentine or enameloid caps of fish and agnathans. However, the enamel cap in *Astraspis,* one of the oldest histologically studied vertebrates, differs from the conodont crown tissue only in smaller size and random orientation of its constituent crystallites (Fig. 5B). It is generally accepted that hydroxyapatite crystallites follow in their orientation the collagen fibers of the matrix (Schmidt and Keil 1971; Shellis and Miles 1976). The crown tissue thus might originate on a collagen matrix, similar to the enameloid of Recent fish. By analogy to the evolution of the enamel in mammals (Koenigswald 1981), one may suggest that the crystallite size and orientation in the conodont cusp were related to its supposed mechanical function (Jeppsson 1979a).

Crystallites in the basal filling tissue differ from those of the crown tissue. At least in the primitive coniform elements, they were arranged in spherulitic aggregates and randomly oriented at the beginning of mineralization. After calcification, numerous tubuli remained in the basal filling (Fig. 1A, C). They do not differ in diameter, mode of branching, and distribution from the tubuli in the dentine of thelodont scales (compare Figs. 1 and 6). The main difference between the basal filling tissue of conodont elements and the agnathan dentine consists in complete disappearance of the tubuli at later histogenetic stages in the former. In this respect, the basal filling tissue can be regarded as an analog of the aspidin.

The hypothesis of chordate affinities of the conodonts has now received much support through the recognition of V-shaped myomeres in the conodont animal body (Briggs et al. 1983).

CONNECTING LINKS

Agnathans with body armor have been recorded in the Llanvirnian (Ritchie and Gilbert-Tomlinson 1977) and even earlier (Nitecki et al. 1975; Bockelie and Fortey 1976; Repetski 1978), although conclusive histologic evidence still is lacking. The oldest unquestionable conodonts occur in the Franconian (Miller 1980). Except for the doubtful Middle Cambrian *Odontogriphus,* the only older fossil that could be related to true conodonts is *Fomitchella infundibuliformis* Missarzhevsky 1969 from the Early Tommotian of Siberia. *Fomitchella* was interpreted, on purely morphologic grounds, as ancestral to the Ordovician *Pseudoonetodus* (Dzik 1976). Bengtson (1983b) demonstrated that the sclerites of *Fomitchella* had grown centrifugally. He also noted that "the lamellae consist of finely granular apatite without preferred crystallographic orientation" (p. 12). Being similar in shape to early conodont elements and lacking any mineralized

mesodermal tissue, the sclerites of *Fomitchella* have almost the same microstructure as the enamel cap of *Astraspis*. It is plausible to derive regularly oriented crystallites of both *Fryxellodontus* type (with crystallite axes normal to the surface) and *Panderodus* type (with crystallite axes parallel to the surface) from the pattern of *Fomitchella*.

The morphologic gap between *Fomitchella* and *Astraspis* can be filled up with a group of Cambrian to Early Ordovician phosphatic sclerites with more balanced contributions of ectodermal and mesodermal mineral tissues. This group includes *Hadimopanella apicata* Wrona 1982 from the Early Cambrian *Bonnia-Olenellus* Zone of Spitsbergen. These sclerites are much smaller in size than *Fomitchella*, but their hyaline, centrifugally growing caps are easily recognizable. Their wide basal cavity is filled by a callus of the basal filling, which is penetrated by an irregular network of very thin horizontal and vertical canals (Wrona 1982). These canals, usually filled by acid-resistant minerals, may represent spaces left by the collagen fibers or organic matrix. *Hadimopanella apicata*, with its single-tipped sclerites, is an end member of a morphocline including the Early Cambrian *Lenargyrion knappologicum* Bengtson 1977 (with several irregularly distributed apices; Bengtson 1977), the Middle Cambrian *Hadimopanella oezgueli* Gedik 1977, the Late Cambrian *Utahphospha sequina* Müller and Miller 1976, and the Early Ordovician *U. cassiniana* Repetski 1981 (with very regular crowns of tubercles: Müller and Miller 1976; Repetski 1981; Boogaard 1983). There is evidence that the caps of *Utahphospha* grew from the outside, centrifugally (Müller and Miller 1976), while the tissue that fills up their basal cavities and links individual sclerites into a continuous body cover has a rather spongy appearance. In *Lenargyrion* the cap is composed of very small, randomly arranged crystallites (Bengtson 1977), as in *Fomitchella* and *Astraspis*.

In *Lenargyrion,* very small sclerites, like those of *Hadimopanella apicata,* have only a single apex, but the number of apices increases proportionately with increase in sclerite size (Bengtson 1977). This ontogenetic pattern resembles the histogeny of some conodont elements (Dzik and Trammer 1980). The correlation of the number of apices to the sclerite size seems to hold also among different species. Sclerites of much larger size than the *Hadimopanella–Utahphospha* group may thus be expected to have more numerous apices. Such a pattern occurs indeed in the Late Cambrian to Late Ordovician *Milaculum* (Fig. 7A).

The external surface of *Milaculum* sclerites is smooth and the basal surface is rough as in *Utahphospha* (Fig. 7C), but a distinct cap tissue has not been recognized. The sclerites are penetrated by vertical canals (Müller 1973) resembling dentine

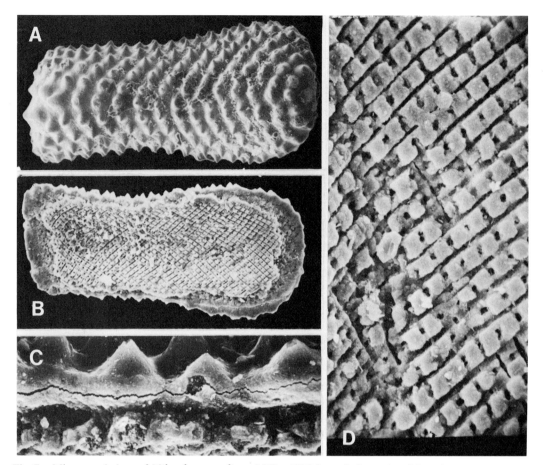

Fig. 7. Micromorphology of *Milaculum scandicum* Müller 1973 from the latest Arenigian of Sweden. A. Sclerite in upside view. Note regular distribution of tubercles (cf. Fig. 8B–F). ×200. B. Basal view of the same specimen. Note mineralized basal part. ×200. C. Boundary between the smooth upper surface and the spongy basal tissue. ×1000. D. Basal tissue with layers of perpendicularly oriented horizontal canals (probably free spaces after collagen cords) and vertical dentine (?) tubuli. ×1000.

tubuli. The basal part, which probably was weakly mineralized and thus is rarely preserved, has a rather complex internal structure (Fig. 7D). It is penetrated by horizontal, parallel canals. In each subsequent layer, the canals are oriented perpendicularly to those of the underlying layer. This pattern resembles scolecodonts, that is, jaws of the Eunicida. This may be merely analogy, as is the case, for instance, with the well-known identity in distibution of the collagen fibrils in the cornea of the mammal eye and in the graptolite periderm (Towe and Urbanek 1972). The peculiar distribution of collagen cords in *Milaculum* might be derived from the pattern of collagen cords in *Hadimopanella.* More importantly, the basal part of *Milaculum* sclerites, with its structure obviously related to the mechanical functions of body cover, was incorporated into the skin. This demonstrates that the sclerites were dermal.

CONCLUSIONS

This overview of phosphatic dermal sclerites suggests that it is the skeletal tissue of ectodermal origin that developed first in phylogeny. There is little basis for speculation concerning its original function. With the initially disorderly arrangement of crystallites and the rather low-conical shape of sclerites taken into account, one may suppose that even if they armored the oral part of the body, they hardly could function as a grasping apparatus. The inferred functional analogy between panderodontid conodonts and chaetonaths may rather reflect subsequent convergent evolution.

The dermal tubercles of *Astraspis* have an enamel microstructure that is similar to that of *Fomitchella* and *Hadimopanella* (Fig. 8). Subsequent histogenetic delay in the mineralization al-

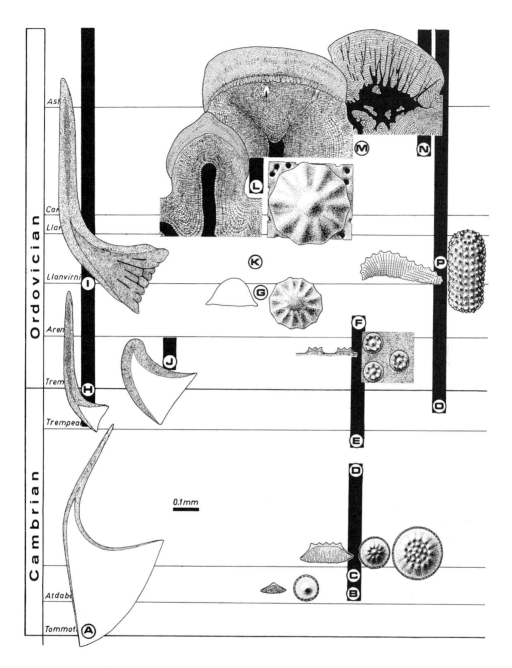

Fig. 8. Stratigraphic distribution of centrifugally growing phosphatic dermal sclerites in the Cambrina and Ordovician. Semidiagrammatic drawings of medial sections and oral views of the best known forms. A. *Fomitchella infundibuliformis* Missarzhevsky 1969. Early Tommotian, Siberia (Bengtson 1983b). B. *Hadimopanella apicata* Wrona 1982. *Bonnia-Olenellus* Zone, Spitsbergen (Wrona 1982). C. *Lenargyrion knappologicum* Bengtson 1977. Atdabanian, Siberia (Bengtson 1977). D. *Hadimopanella oezgueli* Gedik 1977. Middle Cambrian, Turkey and Spain (Boogaard 1983). E. *Utahphospha sequina* Müller and Miller 1976. *Elvinia* Zone, Utah (Müller and Miller 1976). F. *Utahphospha cassiniana* Repetski 1981. El Paso Group, Texas (Repetski 1981). G. *?Astraspis* sp. Oil Creek Formation, Oklahoma (Nitecki et al. 1975; Ethington and Clark 1981). H. *Teridontus nakamurai* (Nogami 1967). Wiberns Formation, Texas (Miller 1980; Landing et al. 1980). I. *Semiacontiodus cornuformis* (Sergeeva 1963). Kundan, Baltic area. J. *Fryxellodontus inornatus* Miller 1969. Rabbitkettle Formation, District of Mackenzie (Landing et al. 1980). K. *Arandaspis prionotolepsis* Ritchie and Gilbert-Tomlinson 1977. Stairway Sandstone, Australia (Ritchie and Gilbert-Tomlinson 1977). L. *Astraspis desiderata* Walcott 1892. Harding Sandstone, Colorado (Halstead 1969; Lehtola 1973). M. *Pycnaspis splendens* Ørvig 1958. Harding Sandstone, Wyoming (Ørvig 1958; Denison 1967). N. *Eriptychius americanus* Walcott 1892. Harding Sandstone, Wyoming (Denison 1967; Halstead 1969) O. *Milaculum perforatum* Müller 1973. Mila Group, Iran (Müller 1973) P. *Milaculum scandicum* Müller 1973. Kundan, Sweden.

lowed for the penetration of the enamel by Tomes' fibers and, finally, for the complete structural unification of the enamel with the dentine. This process probably was repeated and reversed several times in chordate phylogeny. One of those lineages where the boundary between ectodermal and mesodermal tissues disappeared was the *Milaculum* lineage. The Agnatha thus developed body armor composed of phosphatic sclerites.

Quite opposite in direction was the evolution of the conodonts. Their sclerites were confined to the oral region of the body. Perhaps the Middle Cambrian *Odontogriphus* was a primitive conodont animal with conical elements in the oral apparatus and with a naked but heterostracan-shaped body. When the oral sclerites began to function in catching prey, as in Recent myxinoids or chaetognaths, a strong selection pressure appeared for elongation and strengthening of the sclerites. It lead to the parallel orientation of crystallites in the cusp. Elongation of the cusp increased in phylogeny, while the basal cavity became progressively shallow. The crown tissue began to dominate. In the primitive coniform conodont elements, the basal filling frequently was penetrated by dentine tubuli, but it soon became compact and rigid.

Evolution of the conodont grasping apparatus probably corresponded to a shift from necto-benthic scavenger to pelagic carnivore niche. This is consistent with the differences in body shape between the Cambrian *Odontogriphus* and the Carboniferous *Clydagnathus*.

There is virtually no feature of the conodonts that would contradict their classification as vertebrates. The structure of skeletal tissues, the mode of their origin, the pattern of medially symmetric oral apparatus, the body shape, and its internal organization with distinct V-shaped somites—all are known in chordates.

Acknowledgments

SEM photographs were taken at the Ohio State University at Columbus, with technical assistance of Mr. Maureen L. Lorenz (Figs. 3A and 7), Geological Institute in Warszawa, Poland (Fig. 5), and Nencki's Institute of Experimental Biology in Warszawa, Poland (Figs. 1 and 6). Specimens of *Astraspis* have been generously provided by Dr. Stig M. Bergström. Dr. Hubert Szaniawski critically read the manuscript and made many valuable and helpful comments.

REFERENCES

Aldridge, R. J. 1982. A fused cluster of coniform conodont elements from the Late Ordovician of Washington Land, western North Greenland. *Palaeontology* **25**, 425–430.

Andres, D. 1981. Beziehungen zwischen kam-

brischen Conodonten und Euconodonten (vorläufige Mitteilung). *Berliner Geowiss. Abh.* **32A,** 19–31.

Bardack, D. and Richardson, E. S., Jr. 1977. New agnathan fishes from the Pennsylvanian of Illinois. *Fieldiana: Geol.* **33**, 489–510.

Barnes, C. R., Sass, D. B., and Poplawski, M. L. S. 1973. Conodont ultrastructure: the family Panderodontidae. *R. Ont. Mus. Life Sci. Contrib.* **90,** 1–36 pp.

Barskov, I. S., Moskalenko, T. A., and Starostina, L. P. 1982. Novye dokazatelstva prinadlezhnosti konodontoforid k pozvonochnym. *Paleont. Zh.* **1982** (2), 80–86.

Bengtson, S. 1976. The structure of some Middle Cambrian conodonts, and the early evolution of conodont structure and function. *Lethaia* **9**, 185–206.

————. 1977. Early Cambrian button-shaped phosphatic microfossils from the Siberian platform. *Palaeontology* **20**, 751–762.

————. 1983a. A functional model for the conodont apparatus. *Lethaia* **16**, 38.

————. 1983b. The early history of the Conodonta. *Fossils and Strata* **15**, 5–19.

Bitter, P. H. von, and Merrill, G. K. 1983. Late Palaeozoic species of *Ellisonia* (Conodontophorida): evolution and palaeozoological significance. *R. Ont. Mus. Life Sci. Contrib.* **136**, 1–57.

Bockelie, T. and Fortey, R. A. 1976. An Early Ordovician vertebrate. *Nature* 260, 36–38.

Boogaard, M. van den. 1983. The occurrence of *Hadimopanella oezgueli* Gedik in the Lancara Formation in NW Spain. *Proc. K. Nederl. Akad. Wetensch., Ser. B* **86**, 331–341.

Briggs, D. E. G., Clarkson, E. N. K., and Aldridge, R. J. 1983. The conodont animal. *Lethaia* **16**, 1–14.

Carls, P. 1977. Could conodonts be lost and replaced? *N. Jb. Geol. Paläont. Abh.* **155**, 18–64.

Collinson, C., Avcin, M. J., Norby, R. D., and Merrill, G. K. 1972. Pennsylvanian conodont assemblages from La Salle county, northern Illinois. *Ill. State Geol. Surv. Guidebook Ser.* **10**, 1–37.

Conway Morris, S. 1976. A new Cambrian lophophorate from the Burgess Shale of British Columbia. *Palaeontology* **19**, 199–222.

————. 1980. Conodont function: fallacies of the tooth model. *Lethaia* **13**, 107–108.

Cullison, J. S. 1938. Dutchtown fauna of southeastern Missouri. *J. Paleont.* **12**, 219–228.

Dawson, J. A. 1963. The oral cavity, the "jaws," and horny teeth of *Myxine glutinosa*. In Brodal, A. and Fänge, R. (eds.). *The Biology of Myxine*. Universitetsforlaget, Oslo, pp. 231–255.

Denison, R. H. 1967. Ordovician vertebrates from western United States. *Fieldiana: Geol.* **16**, 131–192.

Dzik, J. 1976. Remarks on the evolution of Ordovician conodonts. *Acta Palaeont. Polon.* **21**, 395–455.

————. 1983. Relationships between Ordovician Baltic and North American Midcontinent conodont faunas. *Fossils and Strata* **15**, 59–85.

———— and Trammer, J. 1980. Gradual evolution of conodontophorids in the Polish Triassic. *Acta Palaeont. Polon.* **25**, 55–89.

Ethington, R. L. and Clark, D. L. 1981. Lower and Middle Ordovician conodonts from the Ibex area, western Millard County, Utah. *Brigham Young Univ. Geol. Stud.* **28**, 1–155.

Gross, W. 1954. Zur Conodonten-Frage. *Senckenb. Leth.* **35**, 73–85.

————. 1967. Über Thelodontier-Schuppen. *Palaeontographica A* **127**, 1–67.

Halstead, L. B. 1969. Calcified tissues in the earliest vertebrates. *Calcif. Tissue Res.* **3**, 107–124.

Hitchings, V. H. and Ramsay, A. T. S. 1978. Conodont assemblages: a new functional model. *Palaeogeogr., Palaeoclimatol., Palaeoecol.* **24**, 137–150.

Jeppsson, L. 1971. Element arrangement in conodont apparatuses of *Hindeodella* type and in similar forms. *Lethaia* **4**, 101–123.

————. 1979a. Conodont element function. *Lethaia* **12**, 153–171.

————. 1979b. Growth, element arrangement, taxonomy, and ecology of selected conodonts. *Publ. Inst. Mineral. Palaeont. Quat. Geol. Lund Univ.* **218**, 1–42.

————. 1980. Function of the conodont elements. *Lethaia* **13**, 228.

Kerr, T. 1955. Development and structure of the teeth in the dogfish *Squalus acanthias* L. and *Scylliorhinus caniculus* (L.). *Proc. Zool. Soc. Lond.* **125**, 93–113.

Koenigswald, W. von. 1981. Zur Konstruktion des Schmelzes in Säugetierzähnen. *In* Reif, W. E. (ed.). *Funktionsmorphologie.* Paläontologische Gesellschaft, München, pp. 85–97.

Landing, E., Ludvigsen, R., and Bitter, P. H. von. 1980. Upper Cambrian to Lower Ordovician conodont biostratigraphy and biofacies, Rabbitkettle Formation, District of Mackenzie. *R. Ont. Mus. Life Sci. Contrib.* **126**, 1–42.

Lehtola, K. A. 1973. Ordovician vertebrates from Ontario. *Contrib. Mus. Paleont. Univ. Mich.* **24**, 23–30.

Lindström, M. 1973. On the affinities of conodonts. *In* Rhodes, F. H. T. (ed.). Conodont paleozoology. *Geol. Soc. Am. Spec. Pap.* **141**, 85–102.

————. 1974. The conodont apparatus as a food-gathering mechanism. *Palaeontology* **17**, 729–744.

———— and Ziegler, W. 1971. Feinstrukturelle Untersuchungen an Conodonten. 1. Die Überfamilie Panderodontacea. *Geol. et Palaeont.* **5**, 9–33.

———— and ————. 1981. Surface micro-or-namentation and observations on internal composition. *In* Robison, R. A. (ed.). *Treatise on Invertebrate Paleontology.* Part W, suppl. 2, *Conodonta.* Geological Society of America and Univ. of Kansas Press, Lawrence, pp. W41–W52.

Little, K. 1973. *Bone Behaviour.* Academic Press, London.

Mashkova, T. V. 1972. *Ozarkodina steinhornensis* (Ziegler) apparatus, its conodonts and biozone. *Geol. et Palaeont.* **1**, 81–90.

Meinke, D. K. and Thomson, K. S. 1983. The distribution and significance of enamel and enameloid in the dermal skeleton of osteolepiform rhipidistian fishes. *Paleobiology* **9**, 138–149.

Melton, W. G. and Scott, H. W. 1973. Conodont-bearing animals from the Bear Gulch Limestone, Montana. *Geol. Soc. Am. Spec. Pap.* **141**, 31–65.

Miller, J. F. 1980. Taxonomic revisions of some Upper Cambrian and Lower Orodvician conodonts, with comments on their evolution. *Univ. Kans. Paleont. Contrib.* **99**, 1–10.

Mosher, L. G. and Bodenstein, F. 1969. A unique conodont basal structure from the Ordovician of Alabama. *Geol. Soc. Am. Bull.* **80**, 1401–1402.

Moskalenko, T. A. 1972. O polozhenii Neurodontiformes sredi konodontov. *Trudy Inst. Geol. Geofiz. Sib. Otd. AN SSSR* **112**, 72–74.

————. 1976. Unikalnye nakhodki ostatkov konodontoforid v ordovike Irkutskovo Amfiteatra. *Doklady AN SSSR* **229**(1), 193–194.

Müller, K. J. 1973. *Milaculum* n.g., ein phosphatisches Mikrofossil aus dem Altpaläozoikum. *Paläont. Z.* **47**, 217–228.

————. 1981. Internal structure. *In* Robison, R. A. (ed.). *Treatise on Invertebrate Paleontology.* Part W, suppl. 2, *Conodonta.* Geological Society of America and Univ. of Kansas Press, Lawrence, pp. W20–W41.

———— and Miller, J. F. 1976. The problematic microfossil *Utahphospha* from the Upper Cambrian of the western United States. *Lethaia* **9**, 391–395.

———— and Nogami, Y. 1971. Über den Feinbau der Conodonten. *Mem. Fac. Sci. Kyoto Univ. Ser. Geol. Mineral.* **38**, 1–88.

Nicoll, R. S. 1977. Conodont apparatuses in an Upper Devonian paleoniscid fish from the Canning Basin, Western Australia. *BMR J. Aust. Geol. Geophys.* **2**, 217–218.

————. 1982. Multielement composition of the conodont *Icriodus expansus* Branson & Mehl from the Upper Devonian of the Canning Basin, Western Australia. *BMR J. Aust. Geol. Geophys.* **7**, 197–213.

Nitecki, M., Gutschick, R. C., and Repetski, J. E. 1975. Phosphatic microfossils from the Ordovician of the United States. *Fieldiana: Geol.* **35**, 1–9.

Ørvig, T. 1958. *Pycnaspis splendens,* new genus, new species, a new ostracoderm from the Upper Ordovician of North America. *Proc. U.S. Natl. Mus. Nat. Hist.* **108,** 1–23.

————. 1967. Phylogeny of tooth tissues: evolution of some calcified tissues in early vertebrates. *In* Miles, A. E. W. (ed.). *Structural and Chemical Organization of Teeth,* vol. 1. Academic Press, London, pp. 45–110.

Pietzner, H., Vahl, J., Werner, H., and Ziegler, W. 1968. Zur chemischen Zusammensetzung und Mikromorphologie der Conodonten. *Palaeontographica A,* **128,** 115–152.

Pollock, C. A. 1969. Fused Silurian conodont clusters from Indiana. *J. Paleont.* **43,** 929–935.

Poole, D. F. G. 1971. An introduction to the phylogeny of calcified tissues. *In* Dahlberg, A. A. (ed.). *Dental Morphology and Evolution,* Univ. of Chicago Press, Chicago, pp. 65–79.

Priddle, J. 1974. The function of conodonts. *Geol. Mag.* **111,** 255–257.

Ramovš, A. 1978. Mitteltriassische Conodontenclusters in Slowenien, NW Jugoslawien. *Paläont. Z.* **52,** 129–137.

Repetski, J. E. 1978. A fish from the Upper Cambrian of North America. *Science* **200,** 529–530.

————. 1981. An Ordovician occurrence of *Utahphospha* Müller & Miller. *J. Paleont.* **55,** 395–400.

Ritchie, A. and Gilbert-Tomlinson, J. 1977. First Ordovician vertebrates from the Southern Hemisphere. *Alcheringa* **1,** 351–368.

Schmidt, H. and Müller, K. J. 1964. Weitere Funde von Conodonten-Gruppen aus dem oberen Karbon des Sauerlandes. *Paläont. Z.* **38,** 105–135.

Schmidt, W. J. and Keil, A. 1971. *Polarizing Microscopy of Dental Tissues.* Pergamon, Braunschweig.

Scott, H. W. 1969. Discoveries bearing on the nature of the conodont animal. *Micropaleontology* **15,** 420–426.

Seddon, G. and Sweet, W. C. 1971. An ecological model for conodonts. *J. Paleont.* **45,** 869–880.

Shellis, R. P. and Miles, A. E. W. 1976. Observations with the electron microscope on enameloid formation in the common eel (*Anguilla anguilla:* Teleostei). *Proc. R. Soc. Lond. (B.)* **194,** 253–269.

Sweet, W. C. 1981. Families Ellisoniidae and Xaniognathidae. *In* Robison, R. A. (ed.). *Treatise on Invertebrate Paleontology.* Part W, suppl. 2, *Conodonta.* Geological Society of America and Univ. of Kansas Press, Lawrence, pp. W152–W157.

Szaniawski, H. 1982. Chaetognath grasping spines recognized among Cambrian protoconodonts. *J. Paleont.* **56,** 806–810.

Taylor, K. and Adamec, T. 1977. Tooth histology and ultrastructure of a Paleozoic shark *Edestus heinrichi. Fieldiana: Geol.* **33,** 441–470.

Towe, K. M. and Urbanek, A. 1972. Collagen-like structures in Ordovician graptolite periderm. *Nature* **237,** 443–445.

Turner, S. 1982. A new articulated thelodont (Agnatha) from the Early Devonian of Britain. *Palaeontology* **25,** 879–889.

Wrona, R. 1982. Early Cambrian phosphatic microfossils from southern Spitsbergen (Hornsund region). *Palaeont. Polon.* **43,** 9–16.

AUTHOR INDEX

GENERA/SPECIES INDEX

SUBJECT INDEX